Lecture Notes in Physics

The Lecture Notes in Physics

The series Lecture Notes in Physics (LNP), founded in 1969, reports new developments in physics research and teaching – quickly and informally, but with a high quality and the explicit aim to summarize and communicate current knowledge in an accessible way. Books published in this series are conceived as bridging material between advanced graduate textbooks and the forefront of research to serve the following purposes:

• to be a compact and modern up-to-date source of reference on a well-defined topic;

• to serve as an accessible introduction to the field to postgraduate students and nonspecialist researchers from related areas;

• to be a source of advanced teaching material for specialized seminars, courses and schools.

Both monographs and multi-author volumes will be considered for publication. Edited volumes should, however, consist of a very limited number of contributions only. Proceedings will not be considered for LNP.

Volumes published in LNP are disseminated both in print and in electronic formats, the electronic archive is available at springerlink.com. The series content is indexed, abstracted and referenced by many abstracting and information services, bibliographic networks, subscription agencies, library networks, and consortia.

Proposals should be sent to a member of the Editorial Board, or directly to the managing editor at Springer:

Dr. Christian Caron
Springer Heidelberg
Physics Editorial Department I
Tiergartenstrasse 17
69121 Heidelberg/Germany
christian.caron@springer.com

Malte Henkel   Michel Pleimling
Roland Sanctuary   (Eds.)

Ageing and the
Glass Transition

Springer

Editors

Malte Henkel
Laboratoire Physique des Matériaux
Université Henri Poincaré Nancy I
B.P. 239
F-54506 Vandœuvre lès Nancy Cedex
France
E-mail: henkel@lpm.u-nancy.fr

Roland Sanctuary
Laboratoire de Physique des Matériaux
Université de Luxembourg
162a, avenue de la Faïencerie
L-1511 Luxembourg
Luxembourg
E-mail: roland.sanctuary@uni.lu

Michel Pleimling
Department of Physics
Robertson Hall
Virginia Polytechnic Institute
and State University
Blacksburg VA 24061-0435
USA
E-mail: michel.pleimling@vt.edu

M. Henkel et al., *Ageing and the Glass Transition*, Lect. Notes Phys. 716 (Springer, Berlin Heidelberg 2007), DOI 10.1007/b11815419

Library of Congress Control Number: 2006940658

ISSN 0075-8450
ISBN-10 3-540-69683-0 Springer Berlin Heidelberg New York
ISBN-13 978-3-540-69683-4 Springer Berlin Heidelberg New York

Springer is a part of Springer Science+Business Media
springer.com
© Springer-Verlag Berlin Heidelberg 2007

Typesetting: by the authors and techbooks using a Springer LATEX macro package
Cover design: WMXDesign GmbH, Heidelberg

Printed on acid-free paper   SPIN: 11815419   54/techbooks   5 4 3 2 1 0

Preface

This volume has grown from the invited lectures given at the summer school "Ageing and the glass transition" held at the university of Luxemburg in September 2005. After a short introduction to the notion of ageing, this volume begins with several experimental lectures describing the current state of experimental investigations into several distinct types of glassy behaviour – spin glasses, structural glasses and granular systems are studied. Then follow introductions to a broad variety of theoretical methods applicable to ageing phenomena and glassy systems, including simulational techniques, exactly solvable models, field-theoretical methods. We thank our lecturers for their considerable effort to produce clear and understandable lectures and pedagogical and readable lecture notes. Participants of the school contributed with shorter talks and posters, to be found in a freely accessible proceedings volume.[1]

In the Grande Région Saarland-Luxembourg-Lorraine, there is a long-standing tradition of frontier-transgressing inter-university cooperation. Since a few years, the universities of Saarbrücken, Luxembourg and Nancy have been offering a fully integrated curriculum in physics (SLLS), leading to a tri-national final degree recognized in all partner countries as a national degree, several years before the beginning of the Bologna processes and also today going much further in integration.[2] Our summer school was also for our students an opportunity to come into direct contact with leading experts in a specific field. Nothing is more informative than direct experience, be it in learning foreign languages, coming to know another country or a new academic field.

---

[1] M. Henkel, M. Pleimling and R. Sanctuary (eds.), *Statistical mechanics of ageing phenomena and the glass transition*, J. Phys.: Conf. Series, Vol. 40, Institute of Physics, Bristol (2006). See http://www.iop.org/EJ/toc/1742-6596/40/1.

[2] See http://ci.physik.uni-saarland.de for further information. The SLLS is supported by the Deutsch-französische Hochschule/Université Franco-allemande and by the European Interreg IIIC/eBird programme.

Organizing a summer school is only possible with the help of many people. It is therefore a pleasure to thank J. Baller, E. Apel, M. Heinen-Krumreich and R. Wagener for their contributions. Financially, this summer school was made possible through grants of the Deutsch-französische Hochschule/Université Franco-allemande, of the Université de Luxembourg and by the Institute of Physics, whom we thank sincerely for their support. We are very grateful to C. Caron and Springer Verlag for the willingness to accept these lectures in the Springer Lecture Notes and for generous support. MH thanks the Isaac Newton Institute Cambridge (England) and the INFN and the Dipartimento di Fisica of the Università di Firenze (Italy) for warm hospitality, where the last stages of editing this volume were done.

Nancy, Blacksburg and Luxembourg   *Malte Henkel*
September 2006   *Michel Pleimling*
                                                    Roland Sanctuary

Contents

List of Contributors

P. Alnot
Laboratoire Européen de Recherche
Universitaire Sarre-Lorraine-
(Luxembourg)

R. Bactavatchalou
Université du Luxembourg
Laboratoire de Physique des
Matériaux
162a, avenue de la Faïencerie
L-1511 Luxembourg, Luxembourg

J. Baller
Université du Luxembourg
Laboratoire de Physique des
Matériaux
162a, avenue de la Faïencerie
L-1511 Luxembourg, Luxembourg

Olivier Dauchot
Service de Physique de l'État
Condensé
CEA-Saclay
L'Orme des merisiers
F-91191 Gif-sur-Yvette cedex
France
olivier.dauchot@cea.fr

S. Dorosz
Université Henri Poincaré
Nancy 1 Boulevard des Aiguillettes
Nancy, France

Claude Godrèche
Service de Physique Théorique
CEA Saclay
F-91191 Gif-sur-Yvette cedex
France
godreche@dsn-mail.saclay.cea.fr

Wolfhard Janke
Institut für Theoretische Physik
Universität Leipzig
Augustusplatz 10/11
D-04109 Leipzig
Germany
wolfhard.janke@itp.uni-
leipzig.de

M. Kolle
Universität des Saarlandes
Experimentalphysik
POB 151150
D-66041 Saarbrücken, Germany

Jan Kristian Krüger
FR 7.2 Experimentalphysik
Universität des Saarlandes
Geb. 38
POB 151150
D-66041 Saarbrücken
Germany
jan.krueger@uni.lu

S. P. Krüger
Université Henri Poincaré
Nancy 1 Boulevard des Aiguillettes
Nancy, France

U. Müller
Université du Luxembourg
Laboratoire de Physique des
Matériaux
162a, avenue de la Faïencerie
L-1511 Luxembourg, Luxembourg

M. Philipp
Université du Luxembourg
Laboratoire de Physique des
Matériaux
162a, avenue de la Faïencerie
L-1511 Luxembourg, Luxembourg

W. Possart
Universität des Saarlandes
Werkstoffwissenschaften
POB 151150
D-66041 Saarbrücken
Germany

Uwe C. Täuber
Department of Physics
Centre for Stochastic Processes
in Science and Engineering
Virginia Polytechnic Institute
and State University
Blacksburg, Virginia 24061-0435
USA
tauber@vt.edu

Ch. Vergnat
Universität des Saarlandes
Experimentalphysik
POB 151150
D-66041 Saarbrücken, Germany

Eric Vincent
Service de Physique de l'État
Condensé (CNRS URA 2464)
DSM/DRECAM/SPEC
CEA Saclay
F-91191 Gif-sur-Yvette cedex
France
eric.vincent@cea.fr

1

Introduction

M. Henkel[1], M. Pleimling[2], and R. Sanctuary[3]

[1] Laboratoire de Physique des Matériaux, Université Henri Poincaré Nancy I,
  B.P. 239, F-54506 Vandœuvre-lès-Nancy Cedex, France
[2] Department of Physics, Virginia Polytechnic Institute and State University,
  Blacksburg, Virginia 24061-0435, USA
[3] Laboratoire de Physique des Matériaux, Université du Luxembourg,
  162a, avenue de la Faïencerie, L-1511 Luxembourg

Understanding cooperative phenomena far from equilibrium poses one of the most challenging research problems of present-day many-body physics. At the same time, the practical handling of many of these materials has been pushed to great sophistication, and a lot of practical knowledge about them exists since prehistoric times. Glasses are one example of such systems. In many cases, they are made by rapidly cooling ("quenching") a molten liquid to below some characteristic temperature-threshold. If this cooling happens rapidly enough, normal crystallization no longer takes place and the material remains in some non-equilibrium state. These non-equilibrium states may at first and even second sight look very stationary – everyone has probably seen in archaeological museums intact specimens of Roman glass or even older tools from the Paleolithic or old-stone-age – after all, obsidian or fire-stone is a quenched volcanic melt. But since the material is not at equilibrium, at least in principle it is possible (and it does happen very often in practice) that over time the properties of the material change – in other words, the material ages.[1] Finally, one may wonder what happens from a thermodynamic point of view to a glass-forming material quenched to below its characteristic threshold for glass-formation. Is the hardening of the quenched glass-former merely a kinetic effect such that a glass-transition can be arbitrarily said to occur when motion has slowed down over so many orders of magnitude that any ongoing motion simply escapes the attention or patience of the experimentalist? Or does the system pass through a true thermodynamic phase-transition and enters a new physical regime with a qualitatively different behaviour?

Since glasses constitute a sort of paradigmatic example of systems undergoing physical ageing, it is perhaps helpful to spell out a little more what is meant precisely by this concept. Although ageing was first observed in glassy

---

[1] To avoid misunderstandings: *physical ageing* as it is understood here is caused by reversible microscopic processes, whereas chemical or biological ageing comes from the action of essentially irreversible (bio-)chemical processes.

M. Henkel et al.: *Introduction*, Lect. Notes Phys. **716**, 1–6 (2007)
DOI 10.1007/3-540-69684-9_1   © Springer-Verlag Berlin Heidelberg 2007

systems [1], it has been realized in recent years that very similar phenomena already occur in non-disordered and non-frustrated systems which hence are also commonly referred to as ageing. For the purposes of illustration, we consider a simple Ising model, made from spin variables $\sigma_i = \pm 1$ attached to each site i of a hypercubic lattice and which are meant to describe magnetic moments aligned with respect to some axis. The classical hamiltonian describes the usual nearest-neighbour interactions $\mathcal{H} = -J \sum_{(i,j)} \sigma_i \sigma_j$ where $J > 0$ is the exchange integral and the sum extends over pairs of nearest neighbours. The motion of the spins is generated by coupling the model to a thermal bath of temperature T. A possible way of realizing this is through the so-called heat-bath dynamics which is defined by the stochastic rule

$$\sigma_i(t + \Delta t) = \pm 1 \quad , \quad \text{with probability } \left[1 \pm \tanh(h_i(t)/T)\right]/2 \qquad (1.1)$$

where Δt is the time increment, the local time-dependent field is $h_i(t) = \sum_{y(i)} \sigma_{y(i)}(t)$ and $y(i)$ runs over the nearest neighbours of the site i. It is well-known that this rule satisfies detailed balance and hence the system evolves towards the equilibrium probability distribution $P_{eq} = Z^{-1} \exp(-\mathcal{H}/T)$, where Z is the canonical partition function [2,3]. The system is prepared at some initial temperature T_{ini} well above the critical temperature $T_c > 0$. The initial time $t = 0$ is defined by coupling the system to the thermal bath at some low temperature $T < T_c$ and starting the dynamics. During the simulation, the temperature T is kept fixed and one observes the time-dependence of observables such as correlation functions or susceptibilities.[2] Qualitatively, the behaviour of the system can be illustrated through the equilibrium free energies at the temperatures T_{ini} and T, see Fig. 1.1. Before the quench, the system is at equilibrium with respect to the initial temperature $T_{ini} > T_c$ and sits at the minimum of the free energy, as indicated by the black ball in Fig. 1.1a. After perturbing the state of the system, rapid relaxation with a finite relaxation time $0 < \tau < \infty$ occurs. On the other hand, immediately after the quench the system did not yet have had the time to evolve but, with respect to the new equilibrium, its free energy is no longer minimal, see Fig. 1.1b. Rather, two new local minima of the free energy appeared, which correspond to the two equivalent ordered states of the system. Because of the competition between these two equivalent equilibrium states, the system as a whole cannot relax rapidly to one of them but rather undergoes a slow dynamics, with formally infinite relaxation times. Locally, each spin will be subject to the time-dependent field $h_i(t)$ coming from its neighbours and this field will tilt the balance between the two equivalent equilibrium states of Fig. 1.1b in favour of one or the other. Physically, this means that the system will rapidly decompose into ordered domains and the slow long-time dynamics of this domain growth will be determined by the motion of the domain walls between these ordered domains (see Janke's lecture in this volume for

---

[2] The chosen dynamics is such that the total average magnetization $M(t) = \langle \sigma(t) \rangle = \sum_i \langle \sigma_i(t) \rangle$ remains at its initial value $M(0) = 0$.

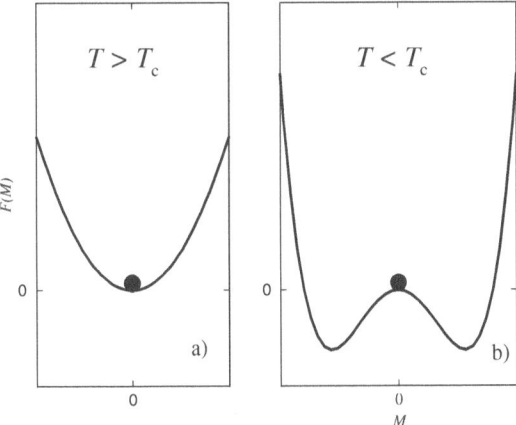

Fig. 1.1. Free energy of a simple ferromagnet at (**a**) an initial high temperature $T > T_c$ before the quench and (**b**) after the quench to a low temperature $T < T_c$

illustrative figures). This slow (non-exponential) dynamics is the first defining property of ageing systems.

Another aspect of this non-equilibrium dynamics (since in a spatially infinite system none of the equilibrium states will be reached in a finite time) becomes apparent if one considers a quantity like the two-time autocorrelation function of spins at site i at times t and s

$$C(t, s) = \langle \sigma_i(t)\sigma_i(s) \rangle \tag{1.2}$$

which by spatial translation-invariance is independent of the chosen site i. In Fig. 1.2a data for $C(t, s)$ plotted over against the time difference $t - s$ are displayed for the three-dimensional Ising model. For a fixed value of the smaller time s, the autocorrelation relaxes rapidly to a plateau $C_{eq} \simeq M_{eq}^2$ and only for large values of $t - s$ falls off to zero. Furthermore, for different values of s the data clearly fall on different curves which means that time-translation invariance is broken. Together with the slow dynamics mentioned above, this breaking of time-translation invariance is the second defining property of ageing systems. While in principle this could mean that the details of the dynamics of ageing systems might depend on the entire prehistory of the sample under study, a great simplification, due to dynamical scaling, is apparent in Fig. 1.2b where the *same* data for $C(t, s)$, when plotted over against t/s, neatly collapse onto a single curve, if only the time s is large enough. Since in domain coarsening one expects that the linear size of the ordered domains is $L = L(t) \sim t^{1/z}$ when t is large enough and z is the dynamical exponent [4], the collapse in Fig. 1.2b means that $C(t, s) = f(L(t)/L(s))$, or in other words $L(t)$ is the only relevant length-scale at time t. Dynamic scaling is the third essential property of ageing systems.

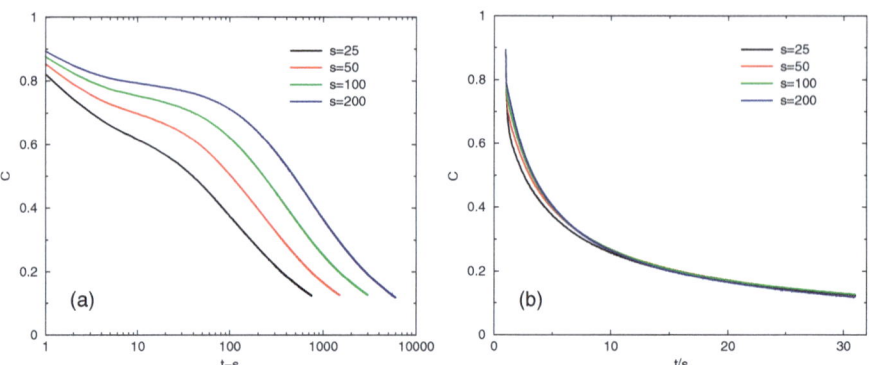

Fig. 1.2. (a) Ageing and (b) dynamical scaling of the two-time autocorrelation function $C(t, s)$ in the three-dimensional Glauber-Ising model quenched to $T = 3 < T_c$, for several values of the waiting time s

These three basic properties of ageing systems are also found in glassy systems. An important property of glasses is the presence of frustrations which prevent the relaxation of all local degrees of freedom. In consequence, the free-energy landscape of glasses can be very complex, with many local minima. The classic example for ageing behaviour was observed by Struik [1] in studying the mechanical properties of polymeric glasses which after a quench from the molten phase to low temperatures (i) relax very slowly (typical time-scale of years), (ii) show clear evidence of the breaking of time-translation invariance and furthermore, (iii) the experimental data for the time-dependent creep curves of the mechanical response can all be mapped onto a single master curve. Remarkably, that master curve turned out to be *the same* for materials as different as polymers such as PVC or PMMA, sugar or even metals like lead! Evidently, there are universal scaling functions in ageing which exactly because of their universality one may hope to be able to understand theoretically. The similarities and differences of the dynamical scaling observed experimentally in real glasses will be one of the topics of the lectures, see especially E. Vincent's lecture in this volume.

In order to understand better the profound interrelation between ageing and the glass transition, the experimental background is crucial – after all physics is based on experimental information on how Nature behaves. We have therefore arranged the experimental lectures as being the first ones in this volume. In Chap. 2, E. Vincent gives an introduction and overview over the time-dependent behaviour of magnetic spin glasses. After introducing the concept of a spin glass and having illustrated the basic aspects of glassiness in these systems, the time- and frequency-dependent scaling behaviour of magnetic correlations and susceptibilities of disordered and frustrated magnets with a glassy behaviour is described. Ageing, rejuvenation and memory effects are discussed in great detail and the lectures culminate with the description of recent efforts to characterize the ageing behaviour in terms of a time-dependent, growing correlation length. Chapter 3 is devoted to an in-depth discussion on the nature of the thermal glass-transition in structural glasses. To some extent, theoretical ideas are informed by the windows through which we have to look at a certain phenomenon. The lectures by J.K. Krüger describe what can be seen about structural glasses through a so far less-considered window, namely by using data on the elasticity properties, especially at high frequencies as obtained from Brillouin spectroscopy. Do these data provide objective criteria by which one could distinguish between purely kinetic effects and a genuine phase-transition? After having recalled the traditional view on the thermal glass transition (being inferred from the regime of low relaxation frequencies), an introduction to the methods of Brillouin spectroscopy is given. These methods are then applied to discuss in detail the glass transition, going from purely thermal to dynamical aspects. The observation of generalized Cauchy relations of the elastic constants and the study of the opto-acoustic dispersion function are used to argue that close to the glass transition there is more than mere kinetic effects. The nature of a thermodynamical glass-transition is studied through an analysis of the Grüneisen parameters. A theoretical assessment of these new experimental results does not yet exist. In Chap. 4 O. Dauchot discusses a rather different type of system with glassy behaviour: granular systems which are usually considered as a-thermal. Having reviewed the similarities with thermal glassy systems which arise in the context of gentle compaction of grains, the main aspects of phenomenology and current attempts at a thermodynamical formulation are discussed, notably Edward's proposal for the definition of a configurational entropy which may in turn be used to define the analogue of a temperature. These ideas are systematically compared with the available experimental evidence. We hope that in future works parallels between the more conventional glassy systems, as described in the first two lectures, and granular systems may become fruitful.

One of the main tools of the theorist dealing with complex systems is nowadays through numerical simulation. W. Janke gives in Chap. 5 a thorough introduction to the techniques of Monte Carlo simulation and the analysis of the resulting data through scaling methods. Temperature-dependent and finite-size scaling methods are presented and a detailed presentation of the tricks of the trade of Monte Carlo simulation, the choice of the algorithm and the related statistical analysis is given, resulting in a manual on modern Monte Carlo methods. These methods are illustrated through a recent study on simple spin systems undergoing ageing in the context of phase-ordering kinetics and the calculation of two-time correlation and response functions. The result provides new evidence in support of recent ideas trying to generalize dynamical scaling to a larger group of *local* scale-transformations. Although generally applicable, purely numerical methods always suffer from some systematic uncertainty. That is only one reason why exactly solvable models are important. More generally, exactly solvable systems allow for an in-depth analysis in a well-defined context which is extremely important in developing general ideas about the underlying physical processes. In equilibrium critical phenomena, the Ising model has in this way been raised from an exotic curiosity to a textbook standard. In Chap. 6 C. Godrèche describes the zero-range process as an extremely simple-looking, but quite subtle model on which many static and dynamic properties of glassy systems can be analyzed in fine detail. It is an attractive feature of this model that the technical complexity of the tools needed is less formidable than might have been hoped or feared. The model shows a condensation transition in its non-equilibrium stationary state and furthermore its dynamics shares many aspects with the one of truly glassy systems. At present, this kind of model is very actively studied and might well become a paradigmatic reference for future investigations of glassy dynamics. More general than specific exactly solvable models, the field-theoretic renormalization group is another member of the essential set of theoretical tools. In Chap. 7 U.C. Täuber shows how the familiar methods of the field-theoretical renormalization-group of equilibrium systems can be extended to systems far from equilibrium. After reviewing both equilibrium and non-equilibrium critical dynamics, the passage from the Langevin equation to field-theory through the dynamical functional is presented. It is then shown how, starting from perturbation theory, a renormalization-group calculation for a non-equilibrium system can be set up. This is illustrated in applications to reaction-diffusion systems, population dynamics and branching and segregation phenomena. This chapter serves as a manual to non-equilibrium field-theoretic renormalization-group methods. One of the strengths of the renormalization-group is that it tells what quantities are universal, that is independent of almost all "details" of a specific system (e.g. critical exponents), and which are not (e.g. location of critical points).

References

1. L.C.E. Struik, *Physical ageing in amourphous polymers and other materials*, Elsevier (Amsterdam 1978)
2. N.G. van Kampen, *Stochastic Processes in physics and chemistry*, 2nd edition, North Holland (Amsterdam 1992)
3. O. Narayan and A.P. Young, Phys. Rev. **E64**, 021104 (2001)
4. A.J. Bray, Adv. Phys. **43**, 357 (1994)

2

Ageing, Rejuvenation and Memory:
The Example of Spin-Glasses

E. Vincent

Service de Physique de l'Etat Condensé (CNRS URA 2464),
DSM/DRECAM/SPEC, CEA Saclay, F-91191 Gif sur Yvette Cedex, France
eric.vincent@cea.fr

Abstract. In this paper[1], we review the general features of the out-of-equilibrium dynamics of spin-glasses. We use this example as a guideline for a brief description of glassy dynamics in other disordered systems like structural and polymer glasses, colloids, gels etc. Starting with the simplest experiments, we discuss the scaling laws used to describe the isothermal ageing observed in spin-glasses after a quench down to the low temperature phase (these scaling laws are the same as established for polymer glasses). We then discuss the rejuvenation and memory effects observed when a spin-glass is submitted to temperature variations during ageing, and show some examples of similar phenomena in other glassy systems. The rejuvenation and memory effects and their implications are analyzed from the point of view of both energy landscape pictures and of real space pictures. We highlight the fact that both approaches point out the necessity of hierarchical processes involved in ageing. We introduce the concept of a slowly growing and strongly temperature-dependent dynamical correlation length, which is discussed at the light of a large panel of experiments.

2.1 What is a Spin-Glass?

A spin-glass is a disordered and frustrated system. From the theorist's point of view, the definition of the spin-glass is very simple: it is a set of randomly interacting magnetic moments on a lattice. The total energy is simply the sum over interacting neighbours (S_i, S_j) of all coupling energies $J_{ij}S_iS_j$, where the $\{J_{i,j}\}$ are random variables, gaussian or $\pm J$ distributed:

$$H = -\sum_{i,j} J_{i,j}S_iS_j .$$

(2.1)

---

[1] This paper is written on the basis of a course given at the summer school "Ageing and the glass transition" in the University of Luxembourg. The results presented here are the joint work of the Saclay group: F. Bert, J.-P. Bouchaud, V. Dupuis, J. Hammann, D. Hérisson, F. Ladieu, M. Ocio, D. Parker, E.V., and others.

E. Vincent: *Ageing, Rejuvenation and Memory: The Example of Spin Glasses*, Lect. Notes Phys. **716**, 7–60 (2007)
DOI 10.1007/3-540-69684-9_2

The impressive number of publications devoted to the spin-glass problem these last decades (see references in e.g. [1–4]) is in sharp contrast to the rather simple formulation as described by Eq. (2.1).

From the experimentalist's point of view, the way to obtain a set of randomly interacting magnetic moment is usually to dilute magnetic ions. The canonical example is that of intermetallic alloys, like for instance $Cu:Mn_{3\%}$, in which 3% of (magnetic) Mn atoms are thrown by random in a (non-magnetic) Cu matrix. The Mn magnetic atoms sit at random positions, therefore are separated by random distances, and the oscillating character of the RKKY interaction with respect to distance makes their coupling energy take a random sign. This class of systems corresponds to the historical discovery of spin-glasses, which traces back to the studies of strongly diluted magnetic alloys and the Kondo effect [3].

Later on, spin-glasses have been identified within insulating compounds. An example that we have studied in details at our laboratory is the the Indium diluted Chromium thiospinel $CdCr_{2x}In_{2(1-x)}S_4$, with superexchange magnetic interactions between the Cr ions [5]. For $x = 1$, this compound is a ferromagnet with $T_c = 80$ K. The nearest neighbour interactions are ferromagnetic and dominant for $x = 1$, but the next-nearest ones are antiferromagnetic. Hence, when some (magnetic) Cr ions are substituted by (non-magnetic) In ions, some ferromagnetic bindings are suppressed, and the effect of other antiferromagnetic interactions is enhanced. The balance that globally favours ferromagnetism for zero or small In-dilution is disturbed, and the ferromagnetic phase is replaced by a spin-glass phase for $x \leq 0.85$.

The phase diagram of the CrIn thiospinel is shown in Fig. 2.1a, together with the magnetic behaviour corresponding to various values of x in Fig. 2.1b [5, 6]. As usual, the "FC" curves correspond to a measurement procedure in which the sample is cooled in presence of the measuring field, and the "ZFC" curves are obtained after cooling in zero field, applying the field at the lowest temperature and measuring the magnetization while increasing the temperature step by step. In Fig. 2.1b, the $x = 1$ curve shows a very abrupt increase of the magnetization when approaching $T_c = 80$ K from above, that is characteristic of the ferromagnetic transition. At lower temperatures, magnetic irreversibilities are observed (splitting of the ZFC and FC curves), which are probably due to defects. In the $x = 0.95$ and $x = 0.90$ curves, the ferromagnetic transition is progressively rounded as the level of dilution increases, and the splitting of the FC and ZFC curves at low temperature indicates the reentrance of a spin-glass phase, that has been characterized in other studies [5, 6]. For $x = 0.85$, the ferromagnetic phase has disappeared, and at $T_g = 16.7$ K the system undergoes a transition from a paramagnetic to a spin-glass phase that presents the same features as observed in intermetallic spin-glasses.

Figure 2.2 shows in more details the typical results of a ZFC/FC measurement on a spin-glass. It is important to emphasize that a low-temperature splitting of the ZFC/FC curves is not by itself characteristic of a spin-glass.

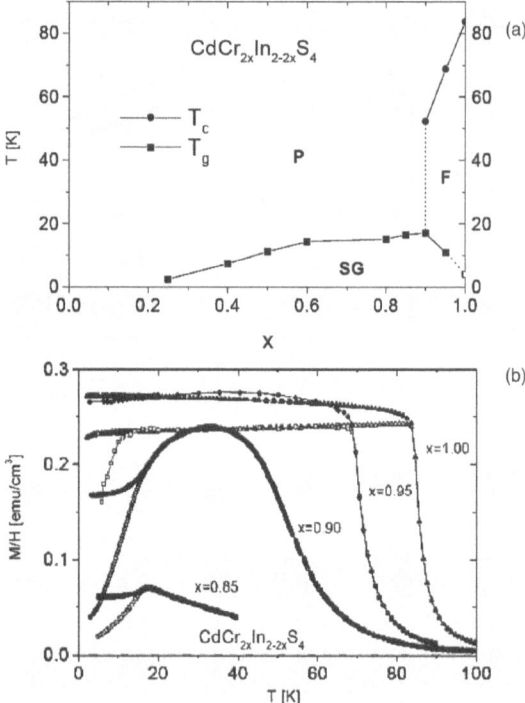

Fig. 2.1. CdCr$_{2x}$In$_{2-2x}$S$_4$ thiospinel compound: **a)** Phase diagram [5, 6], showing the paramagnetic (P), ferromagnetic (F) and spin-glass (SG) phases; **b)** zero-field cooled (ZFC, *open symbols*) and field-cooled (FC, *filled symbols*) magnetizations

This is only the signature of the onset of magnetic irreversibilities, which are not necessarily related to a collective behaviour, as for instance happens with superparamagnetic nanoparticles whose magnetization fluctuations are blocked by the effect of individual anisotropy barriers [7]. However, the (approximate) flatness of the FC curve that is observed here below T_g shows that, when going from the paramagnetic region to low temperatures, the susceptibility increase is rather sharply stopped. This is suggestive of a collective behaviour, and is indeed observed in *concentrated* systems of nanoparticles, where the dipole-dipole interactions are at the origin of a (super-)spin-glass-like transition [8–10].

While the FC curve can be measured upon decreasing or as well increasing the temperature in presence of the field, because the magnetization value can be considered at equilibrium (in a first approximation, usually within 1%), the ZFC one is fully out of equilibrium. After cooling in zero field and applying the field at some $T < T_g$, the magnetization ZFC(t) relaxes upwards as a function of time. In a symmetric way, starting from a FC state at T, if the field is turned to zero, the "thermo-remanent" magnetization (TRM) relaxes downwards. It has been observed in the early studies of slow dynamics in

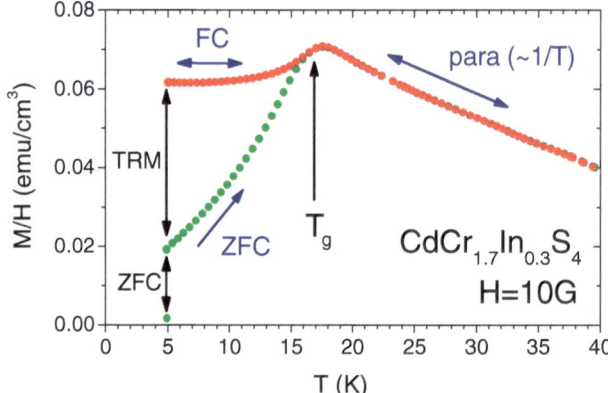

Fig. 2.2. Zero-field cooled (ZFC) and field-cooled (FC) magnetization curves of the CdCr$_{1.7}$In$_{0.3}$S$_4$ thiospinel spin-glass

spin-glasses that, for sufficiently low-fields, these two "mirror experiments" do yield mirror results: $ZFC(t) + TRM(t) = FC$ (this relation even holds if a slight relaxation of the FC magnetization occurs, $FC \equiv FC(t)$) [11].

Another well-studied example of insulating spin-glass is the Sr-diluted Eu sulfur Eu$_x$Sr$_{1-x}$ (e.g. $x = 0.3$) [3], in which the alteration from the EuS ferromagnet to a spin-glass phase occurs in the same way as in the thiospinel.

These various examples of spin-glasses are helpful for understanding how the situation of randomly interacting moments is realized in "real" spin-glass samples. However, what we want to stress out is that there is a generic spin-glass behaviour which is common to all these systems and independent of the details of the sample chemistry, which the reader will be allowed to forget at least in a first approximation. Metallic as well as insulating spin-glasses show in 3d a well defined phase transition at T_g (attested by the critical behaviour of some quantities), and slow dynamics is observed in the spin-glass phase with the occurrence of such interesting phenomena as ageing, rejuvenation and memory effects. Regarding these different aspects, no difference can be traced out between metallic and insulating spin glasses, although the latter are magnetically more concentrated and have shorter range interactions. Certain systematic differences as a function of spin anisotropy have indeed been observed and are explained later in this paper, but, to the best of our present understanding, they are not directly related to their metal/insulator character or to any obvious chemical feature.

Finally, let us note that there is indeed a basic difference between the theoretical spin-glass, in which there is a spin at each lattice node, and the experimental spin-glass, which is site-diluted. It is not yet clear how far this type of difference may be relevant (for a recent review on the question of universality, see for instance [12]). As will be occasionally evoked along this paper (which is devoted to "experimental" spin-glasses), the comparison of "real" (experimental) with "theoretical" spin-glasses is not yet totally understood, but significant progresses have been made these last years, as well analytically as numerically [1, 13–16].

2.2 Slow Dynamics and Ageing

A crucial feature of the spin-glass behaviour (and of glassy dynamics in general) is the existence of relaxation processes at all time scales, from the microscopic times ($\sim 10^{-12}$ s in spin-glasses) to, at least, as long as the experimentalist can wait. The slow relaxation processes are particularly spectacular: in a spin-glass, any field change causes a very long-lasting relaxation of the magnetization, and the response to an *ac* field is noticeably delayed. The basic experiments in which glassy dynamics is commonly investigated can be presented in 3 general classes: *dc* response, *ac* response, and spontaneous fluctuations (noise).

2.2.1 DC Experiments

The study of the relaxation of the magnetization after a small field change has brought a lot of informations about the glassy features of the spin-glass dynamics. For now we only consider the case of "small fields", that are excitation fields which remain in the limit of linear response, or in other words fields that act as a non-perturbative probe. Usually, this field range (depending on the sample) is limited up to 1 or 10 Oe, a few percents of the field needed to surmount the interactions and recover a paramagnetic state (usually 100–1000 Oe).

Let us first consider the case of the relaxation of the thermo-remanent magnetization (TRM). Magnetization relaxations reveal a "waiting time" dependence of the dynamics that is singled out as "ageing" [17–19]. In the experimental procedures, the ageing time becomes another degree of freedom. As sketched in Fig. 2.3, the sample is rapidly cooled in a small field H from above T_g to $T < T_g$, and the sample is kept under field at temperature T during a waiting time t_w, after which the field is cut (at $t = 0$). Then the relaxation is measured as a function of the observation time t. Figure 2.4 shows the results, which demonstrate the 2 basic features of spin-glass dynamics:

(i) the magnetization relaxation is slow, roughly logarithmic in time (glassy state)
(ii) it strongly depends on the waiting time: the longer t_w, the slower the relaxation (ageing).

Hence, time-translation invariance is lost in the slow dynamics of the spin glass: the relaxation depends on both t_w and t, not only on t (non-stationary dynamics). For increasing t_w, the response to cutting off the magnetic field

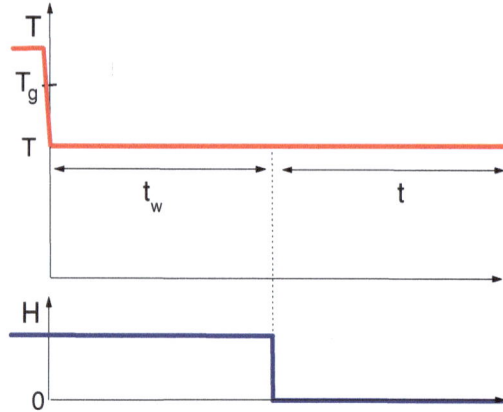

Fig. 2.3. Sketch of the TRM measurement procedure

becomes slower and slower on two respects: the initial fall-off of the magnetization becomes smaller, and the position of the inflection point of the curves shifts towards longer times. This inflection point approximately occurs at times t of the order of t_w itself. When plotted as a function of t/t_w (inset of Fig. 2.4), the curves are gathered together (but they do not superimpose exactly onto each other, with a systematic t_w-dependent departure). In a first approximation, we may consider that the curves obey a t/t_w scaling.

The same phenomenon of "ageing" has been known for a long time for the mechanical properties of a wide class of materials called "glassy polymers" [20]. When a piece of e.g. PVC is submitted to a mechanical stress, its response (elongation, ...) is logarithmically slow. And the response depends on the time elapsed since the polymer has been quenched below its freezing temperature. Like in spin-glasses, for increasing ageing time the response becomes slower and slower, which was called "physical ageing" (as opposed to "chemical ageing"). The t_w-dependence of the dynamics of glassy polymers has been expressed as a scaling law that can be precisely applied to the case of spin-glasses, as is explained below (see also [21]).

Figure 2.5 presents the mirror experiment, in which the sample is cooled in zero field, the field being applied after waiting t_w (ZFC relaxation). The same t_w-dependence is observed as in TRM relaxations.

Following the suggestion of L. Lundgren et al. [22], we also plot (bottom part of Fig. 2.6) the logarithmic derivative $dM/d \log t$ of the magnetization M. The curves are bell shaped, with a broad maximum in the region $t \sim t_w$. These curves have an interesting physical interpretation which has been proposed by L. Lundgren and the Uppsala group [22]. The magnetization relaxations are slower than exponential, they can be modelled by a sum of exponential decays $\exp(-t/\tau)$, the decay times τ being distributed as a certain function $g(\tau)$ which is defined in this way as an effective density of relaxation times. Taking the derivative $dM/d \log t$ introduces a $t/\tau \exp(-t/\tau)$ term in the integrand,

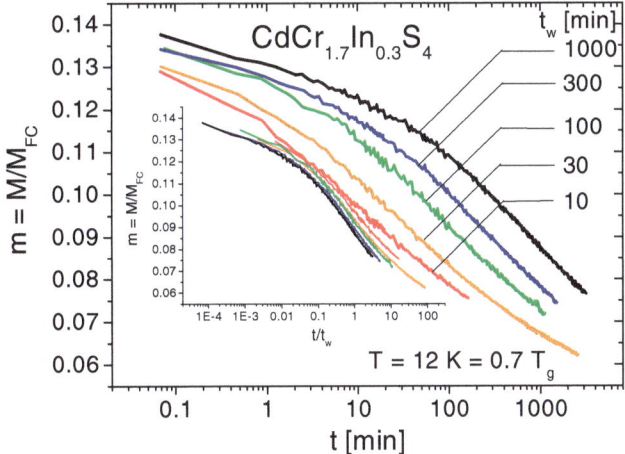

Fig. 2.4. Example of TRM relaxations measured for various values of the waiting time t_w (thiospinel spin-glass). The inset shows the same curves, plotted as a function of t/t_w

which is sharply peaked around $t = \tau$. Approximating this peaked function by a δ-function allows bringing out $g(\tau)$ from the integral over τ and yields

$$dM_{tw}/d\log t \propto g_{tw}(\tau = t) . \qquad (2.2)$$

We have now labelled M_{tw} and g_{tw} by t_w to emphasize that each relaxation curve, taken for a given t_w, gives access through its logarithmic time derivative to the density of relaxation times that represents the dynamics of the spin-glass at a time of the order of t_w after the quench. Thus, each derivative $dM_{tw}/d\log t$ gives an estimate of the density $g_{tw}(\tau = t)$, and as t_w increases $g_{tw}(\tau)$ shifts towards longer times. This gives a physical picture of the 2 important features listed above:

(i)   the effective relaxation times are widely distributed (glassy state)
(ii) this distribution peaks around $\tau = t_w$, which implies that for increasing t_w's the relaxation times become longer (ageing, the spin-glass becomes "stiffer").

We mentioned above that the departures from a perfect t/t_w scaling are systematic as a function of t_w. In addition to the thiospinel example in the inset of Fig. 2.4, we show in Fig. 2.6 the example of a Ag:Mn$_{2.7\%}$ spin-glass. The same trend is observed as in Fig. 2.4: as a function of t/t_w, the large t_w relaxations decrease faster than the short t_w relaxations. That is, the TRM dependence on t_w is slightly slower than the variation of t_w itself. We call this situation "sub-ageing", as opposed to the case of "full ageing" that would correspond to full t/t_w scaling.

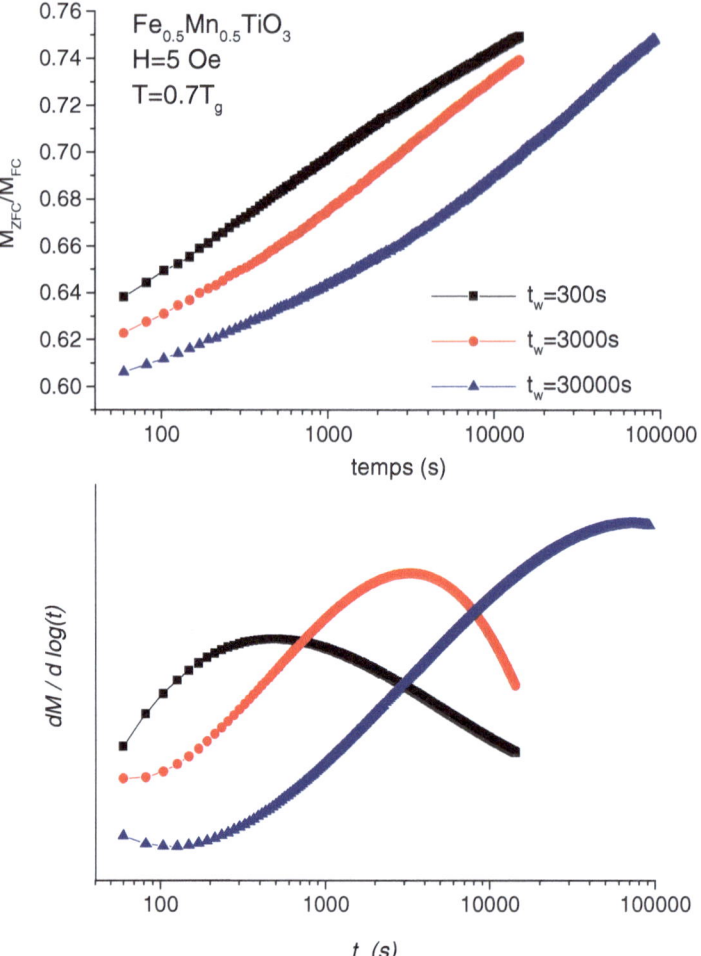

Fig. 2.5. ZFC magnetization relaxations of the $Fe_{0.5}Mn_{0.5}TiO_3$ Ising spin-glass [68], for 3 values of t_w. *Top part*: magnetization relaxations. *Bottom part*: logarithmic derivatives $dM/d\log t$ of the curves from the *top part*, displaying within a good approximation [22] the distribution of effective response times corresponding to the dynamics of the spin-glass after a time of order t_w

On a log-scale, the various t_w-relaxations are spaced by less than $\log t_w$, say by a quantity $\mu \log t_w$ (with $\mu < 1$). For increasing t_w, the shift of $g_{tw}(\tau)$ towards longer times can therefore be expressed as a shift of the relaxation times that is not exactly $\tau \sim t_w$ but rather $\tau \sim t_w^{\mu}$. But t/t_w^{μ} itself does not give a full quality scaling of the t_w-relaxations. At this point, we have to go one step further than the approximation which consists in defining a density of relaxation times $g_{tw}(\tau)$ at fixed t_w from a given t_w-relaxation. Since $g_{tw}(\tau)$ is found to vary with t_w, it varies as well during the relaxation itself as a

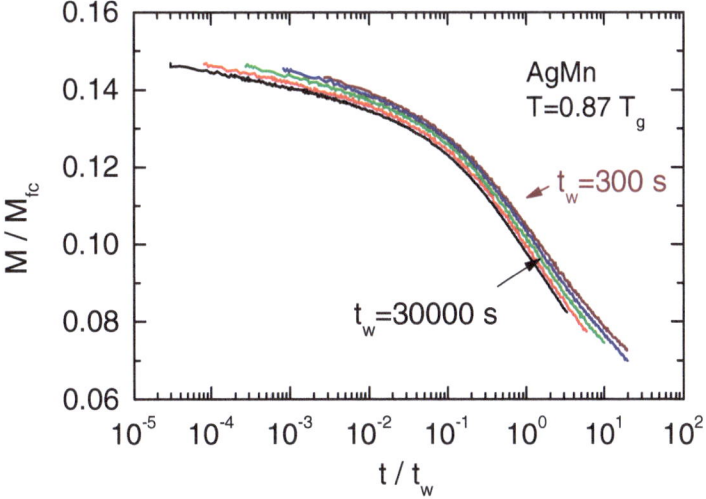

Fig. 2.6. TRM relaxation curves of the Ag:Mn$_{2.7\%}$ spin-glass, plotted as a function of t/t_w (data from [19]). A systematic departure from a t/t_w scaling ("subageing") is observed as a function of t_w

function of time t, and the shift of the relaxation times $\tau \sim t_w^\mu$ should rather be re-written $\tau \sim (t_w + t)^\mu$. This allows the definition of an effective time λ [18–20], obeying for each individual relaxation process dm/m (of relaxation time τ) to:

$$dm/m = dt/\tau = dt/(t_w + t)^\mu = d\lambda/t_w^\mu . \qquad (2.3)$$

λ defines an artificial time frame in which the spin-glass would keep a constant age t_w, whereas its age $t_w + t$ constantly increases in the laboratory time frame. Integrating Eq. (2.3) (setting $\lambda = 0$ for $t = 0$), λ reads

$$\lambda/t_w^\mu = \{1/(1 - \mu)\}\{(t_w + t)^{1-\mu} - t_w^{1-\mu}\} \qquad (2.4)$$

which reduces to $\lambda \sim t$ for $t \ll t_w$.

Then, plotting the relaxation curves of different t_w's as a function of λ/t_w^μ allows a very precise rescaling onto one unique master curve. This procedure has indeed been first suggested to account for ageing in the mechanical properties of polymers. For spin-glasses, in more details, the λ/t_w^μ scaling should be applied to the only ageing part of the relaxation, which must be separated from a stationary contribution $\chi_{eq}(\chi_{eq} \sim t^{-\alpha}, \alpha$ being a very small exponent, of the order of $0.03 - 0.1$), best evidenced in ac experiments (see below) but also present here: $\chi = \chi_{eq} + f(\lambda/t_w^\mu)$ [19]. An example of such a precise rescaling is presented in Fig. 2.7.

This rescaling procedure works very well for all known examples of spin glasses. Like in polymers, the exponent μ is always found lower than one ($\mu \sim 0.8-0.9$, subageing), even if it may sometimes get surprisingly close to 1 (see the example of AgMn in [19], in which $\mu \sim 0.97$ is found, $\mu = 1$

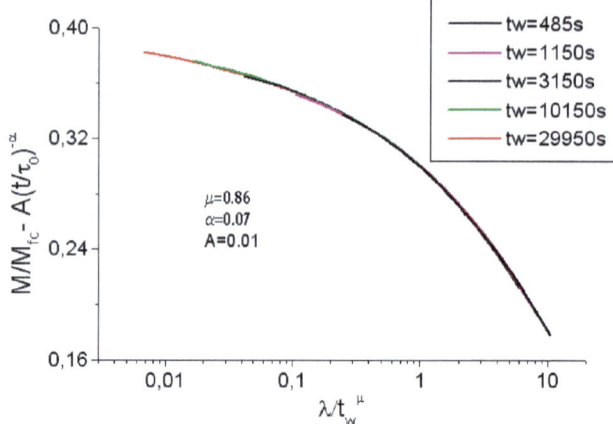

Fig. 2.7. Scaling of a set of TRM relaxation curves (thiospinel sample). The ageing part of the 5 magnetization curves (obtained by subtracting the stationary part $A(t/\tau_0)^{-\alpha}$ to the total magnetization) shows a fairly good scaling as a function of the reduced variable λ/t_w^μ

remaining excluded by the data with a large range of t_w's explored, from 300 to 30000 s). The (simpler) t/t_w scaling with $\mu = 1$ can be expected on some rather general grounds [1,23], and the question of the origin of subageing is yet unsolved [24]. It has been proposed that $\mu < 1$ arises as an effect of an initial age acquired during the necessarily finite cooling time [25–27]. If it is clear that a slower cooling yields a smaller μ, there is no sign in most results (except in the experiment of [25], and for zero cooling time in the numerics of [26]) that μ could go to 1 for very short cooling times, which always remain long in experiments when compared with microscopic paramagnetic times ($\sim 10^{-12}$ s).

The dependence of μ on the amplitude of the magnetic field H has also been carefully checked [28,29]. As shown in Fig. 2.8, $\mu(H)$ decreases for increasing field, but for vanishing field it seems very unlikely that μ goes to 1. This region could be precisely explored in experiments by Ocio and Hérisson who took data for fields as low as 0.001 Oe [29]. μ is found at a plateau value of ~ 0.85 in the range 10–0.001 Oe (five decades). On the other hand, for increasing fields, μ eventually goes to zero [28], which means that the field change is enough to erase the effect of previous ageing (H > 300 Oe in Fig. 2.8). Let us note that above this value there may still be some slow relaxations (although with no t_w-dependence), and that the instantaneous, paramagnetic-like, response to the field is only obtained for still higher fields (600 Oe in the case of Fig. 2.8).

Finally, it might well be that $\mu < 1$ be related to some finite size effects, as proposed in [30,31], as the result of a saturation of ageing in some small parts of the sample (grains?), while larger parts would obey $\mu = 1$ for astronomical times. This possible explanation could however not be confirmed experimentally.

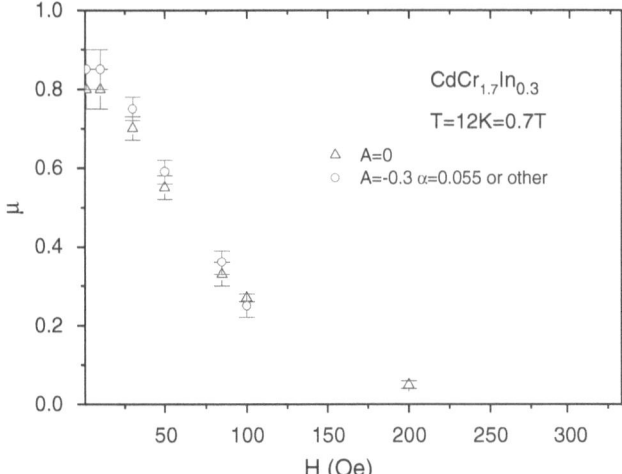

Fig. 2.8. Field dependence of the subageing exponent $\mu(H)$ in the thiospinel sample (circles). The triangles show the $\mu(H)$ values obtained when the stationary part of the magnetization is *not* subtracted

An amusing example of subageing has been studied in the rheology of a microgel paste [32], which is of the type used as toothpaste. Here the notion of a freezing temperature is not relevant, but instead the initial state of ageing is obtained by applying a strong shear stress, which turns the paste into a fluid, whose viscosity then progressively increases with time (so toothpaste does flow out of the tube when it is pressed, but does not flow from the toothbrush to the ground).

At low stresses, the response to a shear excitation is a long-time creep curve, which is slower when the experiment is performed after a longer waiting time. The resulting curves (Fig. 2.9) have been scaled together as a function of t/t_w^μ (not far from λ/t_w^μ), and μ is found to decrease as a function of increasing stress, like in glassy polymers, and like in spin-glasses as a function of the amplitude of the magnetic field (Fig. 2.8). Similar results have been obtained these last years in various examples of colloidal gels [33–35].

2.2.2 AC Susceptibility

Slow dynamics and ageing in the spin-glass phase can also be observed by *ac* susceptibility measurements, in which a small *ac* field (~ 1 Oe) is applied all along the measurement. Again, the starting point of ageing experiments consists in cooling the spin-glass from above T_g, down to some $T < T_g$ at which the *ac* response is measured as a function of the time elapsing, which is the "age" of the system (equivalent to $t_w + t$ in the *dc* procedures). We find here the same 2 characteristics as observed in *dc* experiments:

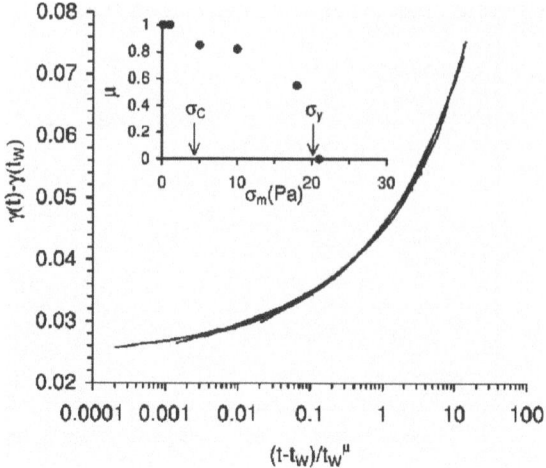

Fig. 2.9. From [32], creep curves of a soft deformable microgel (similar to tooth-paste), measured at different waiting times ranging from $t_w = 15$ s to 10000 s, for a probe stress $\sigma_m = 10$ Pa greater than the yield stress σ_c, above which the suspension begins to flow. The inset shows the evolution of the subageing exponent μ as the probe stress increases up to σ_y, above which ageing disappears ($\mu = 0$)

(i) the *ac* response is delayed, i.e. the susceptibility has 2 components: an in-phase one χ', and an out-of-phase one χ''. χ'' is zero above T_g (para-magnetic phase), and rises up as the sample is cooled into the spin-glass phase.

(ii) the susceptibility relaxes down, signing up the occurrence of ageing. This relaxation is visible on both χ' and χ'', but is more important in relative value $\Delta\chi/\chi$ in the out-of-phase component χ''.

Figure 2.10 shows the χ''-relaxation as a function of time for different frequencies ω. A very clear frequency dependence is seen in Fig. 2.10: the amplitude (in the fixed experimental time window) of the observed relaxation increases as the frequency ω decreases. On the other hand, the infinite time limit of χ'' seems very convincingly to be non-zero, pointing out to a finite χ''_{eq} stationary limit. Once shifted vertically by an arbitrary amount (that should correspond to χ''_{eq}) and plotted as a function of the reduced variable $\omega.t$, the curves can be superposed. Actually, in this *ac* experiment, $1/\omega$ is the typical observation time and plays the same role as t in the *dc* relaxation procedures. The total age of the system is here the time t along which the *ac* relaxation is measured after cooling, equivalent to $t_w + t$ in the *dc* experiment. Hence, the present $\omega.t$ scaling is equivalent to the t/t_w scaling of the *dc* experiments [19]. Strangely enough, there is no sign of subageing (t_w^μ in place of t_w) in the scaling behaviour of the *ac* data. Indeed, the superposition of the *ac* curves is not as constraining as that of a series of TRM relaxations over a large range of t'_ws. But any attempts of an $\omega.t^\mu$ scaling of the *ac* data have favoured

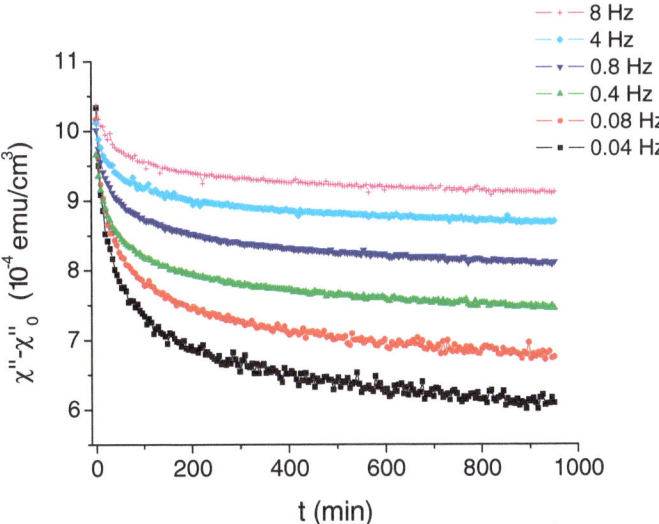

Fig. 2.10. Time decay of the out-of-phase susceptibility χ'' of the thiospinel sample after a quench (ageing), for different frequencies. The curves have been shifted vertically by an arbitrary amount χ_0'' for the sake of clarity

$\mu \sim 1$. One difference with dc experiments which may be pointed out is that ac measurements are necessarily performed in the $w.t \geq 1$ regime (sometimes called "quasi-stationary" regime), that corresponds to the limited region of $t/t_w < 1$ in dc experiments. The possibility of a link between the observation of a subageing behaviour and the time regime explored in the experiments remains open [19].

Similar ac procedures are used in the study of structural and polymer glasses. For instance, in [36], the dielectric constant ε of glycerol has been measured following the same procedures as above. The out-of-phase susceptibility ε'' shows a strong relaxation as a function of the time following the quench (Fig. 2.11).

The relaxation has at least the same qualitative frequency dependence as observed in spin-glasses: the lower the frequency ω, the larger the relaxation in a given time window. The authors state that no ωt-like scaling is obeyed [36]; however, in the case of this structural glass, one cannot exclude that the influence of the cooling time, probably stronger than in spin-glasses, may bring corrections to the effective value t_{eff} of t which could finally yield an ωt_{eff} scaling.

2.2.3 Noise Measurements

The measurement of noise in spin-glasses has been a high-level challenge for the experimentalists, because the spontaneous magnetic fluctuations are tiny

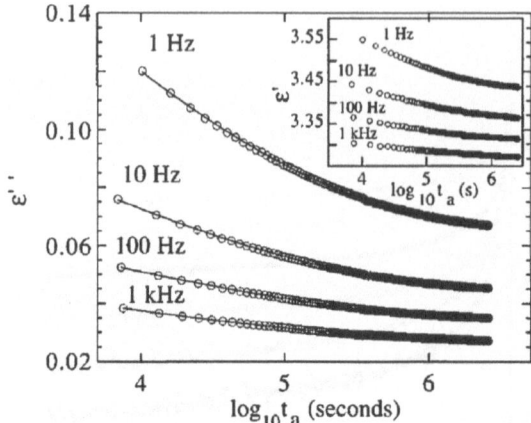

Fig. 2.11. From [36], relaxation of the dielectric constant of glycerol at 178 K, as a function of the time following the quench from above $T_g = 190$ K (ageing), for various frequencies. The main part of the figure shows the out-of-phase component ε'', the inset shows the in-phase component ε'. The amplitude of the relaxation is larger at lower frequencies (same qualitative trend as in spin-glasses)

when compared with the magnetization obtained in response to an external field (in the recent experiment described below, they are equivalent to the response to a field of $\sim 10^{-7}$ Oe). We only recall here the general lines of these remarkable experiments, developed by M. Ocio. The interested reader will refer to his corresponding papers [29, 37].

The response to a magnetic field, whose investigation was detailed above, is related to the spontaneous magnetic fluctuations via the Fluctuation-Dissipation relation (FDR), established for ergodic systems at equilibrium. In its integrated form, it relates the relaxation function $\sigma(t', t)$ ($\sigma = m/h$, response at t after cutting off a field h at t', same as the TRM) to the auto-correlation C of the fluctuations of the magnetization m, namely $C(t', t) = \langle m(t').m(t) \rangle$:

$$\sigma = C/k_B T . \qquad (2.5)$$

A lot of work has been devoted to extensions of FDR to non-equilibrium situations, for which the ageing regime of the spin-glass is archetypal [38, 39]. A prominent result by Cugliandolo and Kurchan [38] is a modified FD relation which reads

$$\sigma = C.F(C)/k_B T \qquad (2.6)$$

where $T/F(C)$ takes the meaning of an effective temperature. In this approach, for large t', the obtained correction factor $F(C)$ is a function of the autocorrelation C only, i.e. it does not explicitly depend on t and t' but has a time dependence through the value of $C(t', t)$ only.

This result was one of the strong motivations of the recent noise experiments performed by M. Ocio and D. Hérisson [29]. A decade before, the very first noise measurements were performed by M. Ocio and Ph. Refregier in collaboration with H. Bouchiat and Ph. Monod [37]. In these pioneering experiments, the Fourier transform of the noise could be measured, and compared with the *ac* susceptibility, that was measured in another setup. This early work suffered two limitations: firstly, the comparison between noise and response could only be made up to an unknown calibration factor, and secondly the time regime was limited to the quasi-stationary region $\omega t > 1$ (as opposed to the "strongly ageing" regime explored in TRM experiments). The results was that the FDR was obeyed as far as could be checked [37].

In the new set of experiments [29], a special setup which allows both types of measurements in the same geometry has been built. For noise measurements, the pickup coil (3rd order gradiometer geometry) which contains the sample is "simply" connected to a *dc* SQUID, and the full signal is recorded as a function of time (not only its Fourier transform). The response function is investigated in the strongly ageing regime by means of TRM-relaxation recordings. One bright idea was to use the pickup coil itself as an excitation coil, through which the field is applied by induction of a current in the pickup loop. Thus, the magnetic geometry (rather complex in a gradiometer) is exactly the same for the detection of the magnetization fluctuations as for applying the excitation field, allowing a direct comparison between fluctuations and response.

In order to cancel the self-inductive response to the field variation which triggers the TRM relaxation, a bridge configuration is used for response measurements, in which the main branch involving the sample is balanced by an equivalent one without sample, excited oppositely. The whole experiment is placed in a magnetic shield which lowers the residual magnetic field below 10^{-3} Oe. Important care was also taken for eliminating all electromagnetic parasite sources, as well as external low-frequency disturbances such as those accompanying the day-night cycle of the laboratory.

Finally, the measurements were made possible. An absolute calibration was realized with the help of an ultra-pure copper sample, in which the magnetic response and the fluctuations of eddy currents are related through classical (ergodic) FDR. With an ergodic sample like copper, this setup constitutes an absolute thermometer after calibration by only 1 fixed point (one of the unrealized projects of M. Ocio was to develop the use of this method for absolute thermometry).

An example of noise recordings is presented in Fig. 2.12.

Each trace shows the SQUID output (proportional to the sample magnetization, with an arbitrary offset) during one experiment, starting from above T_g and cooling. Due to the slight residual field, the trace shows the magnetization peak observed when crossing T_g. After cooling, the temperature is stabilized at say $T = 0.7T_g$, and the magnetization fluctuations are recorded as a function of time during $\sim 10^4$ s. After that the sample is re-heated above T_g. The experiment is repeated ~ 300 times. On each of the recorded traces, for any choice of times (t_w, t) the correlation $m(t_w) \cdot m(t_w + t)$ can be com-

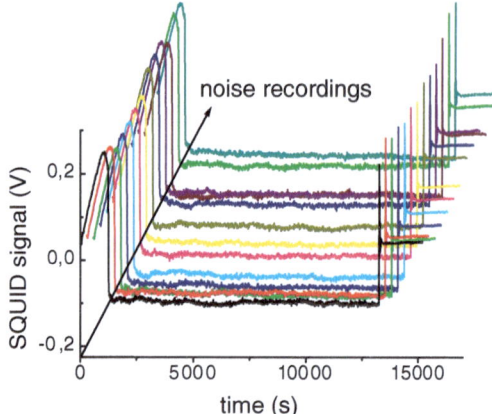

Fig. 2.12. SQUID signal (proportional to the magnetization, with an arbitrary offset voltage) in a series of successive noise recording experiments (data from [29]). Each experiment starts from above T_g; due to the slight residual field, the magnetization shows a peak when crossing T_g

puted. This value is of course strongly fluctuating from one experiment to the other, but the average over $\sim$300 measurements is taken and after properly subtracting offsets the autocorrelation $C(t_w, t) = \langle m(t_w) \cdot m(t_w + t) \rangle$ is obtained. It is represented in the top part of Fig. 2.13 in the same way as usual TRM results, that is as a function of t for various fixed values of t_w.

The bottom part of Fig. 2.13 shows in the same representation the results obtained from the TRM experiments performed in the same setup, with an excitation field of $\sim$10^{-3} Oe. The two insets show that both noise and response functions obey the same scaling law as a function of the reduced variable $\zeta = \lambda / t_w^{\mu}$ (the same fitting parameters can be used). The comparison between both sets of results is best illustrated in the plot of Fig. 2.14, in which the response function $\sigma(t_w, t)$ (or the susceptibility $\chi = 1 - \sigma$) is plotted as a function of $C(t_w, t)$ for 3 different temperatures $T = 0.6, 0.8$ and $0.9\,T_g$. See [29] for the details of normalization of $C(t_w, t)$.

For each of the 3 temperatures, the point cloud is the set of "raw" results obtained for various values of (t_w, t). The straight lines with $1/T$ slope show the expected result when the classical FDR is obeyed with no correction. There is a clear $1/T$ regime for the higher values of C, and the results show a crossover towards a weaker slope $1/T_{\text{eff}}$ with $T_{\text{eff}} > T_g$ as C decreases. These deviations show the first experimental observation in a spin-glass of deviations from the normal FDR in the ageing regime.

In order to make a more quantitative comparison with the theoretical predictions [38], it is necessary to extrapolate the results in the very long time region. An estimate of this very long time behaviour can tentatively be obtained by extrapolating the existing data to the region where the stationary part of the relaxation $t^{-\alpha}$ has relaxed to zero, i.e. by subtracting to σ the

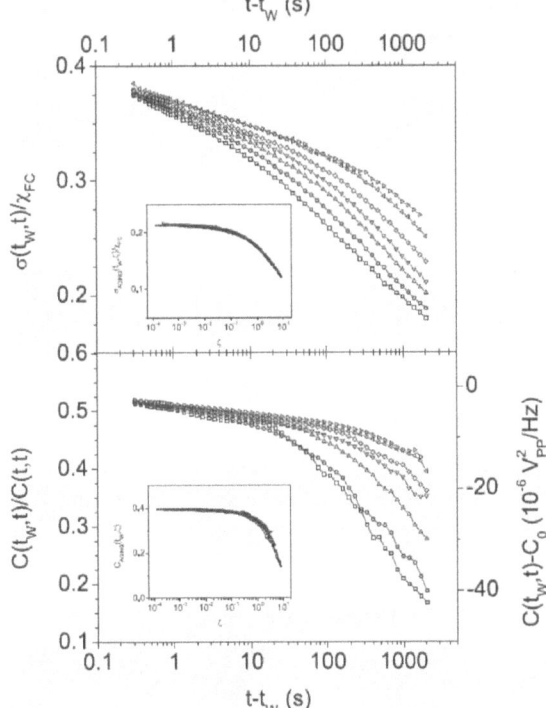

Fig. 2.13. From [29], TRM-relaxation (*top*) and autocorrelation (*bottom*) functions, recorded at 13.3 K with the thiospinel sample. The different curves correspond, from *bottom* to *top*, to $t_w = 100, 200, 500, 1000, 2000, 5000$ and 10000 s. The insets shows the respective ageing parts, deduced by the scaling analysis (see text), and plotted as a function of the reduced time variable $\zeta = \lambda/t_w^\mu$

stationary part which has been obtained on the basis of a precise rescaling of the curves (as shown in the inset of Fig. 2.13). This is shown in Fig. 2.14 in solid curves, which are indeed the superposition of the different curves obtained for various t_w's. The different curves are indistinguishable within the present accuracy, which strongly suggests (in the framework of this crude extrapolation) that the correction factor $F(C)$ to the FDR is only a function of C, as predicted in [38].

It may be risky to push much further the comparison at this stage, since the extrapolation to long times is problematic, and also there remain some difficulties with the normalization of $C(t_w, t)$ by $C(t, t)$ [39]. One point which is out of doubt is that the data in the ageing region do not tend to favour a horizontal slope, as expected in domain growth type models (infinite $T_{\rm eff}$). The observed mean slopes correspond to $T_{\rm eff}(0.6T_g) \sim 1.5T_g, T_{\rm eff}(0.8T_g) \sim 3T_g$, and $T_{\rm eff}(0.9\,T_g) \sim 4T_g$. However, the extrapolated data show some curvature, and do not look like straight lines as would be expected from 1-step RSB

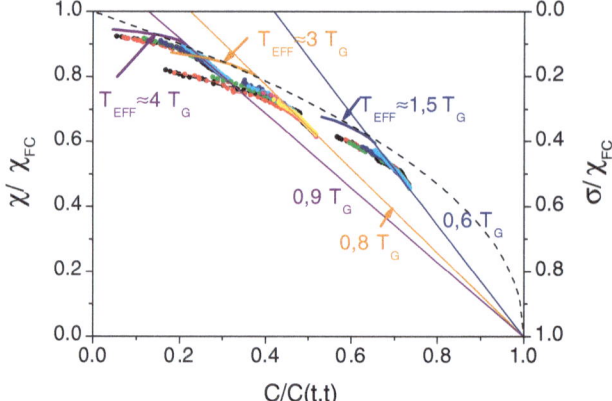

Fig. 2.14. Response function versus autocorrelation, for 3 different temperatures T (data from [29]). The *straight lines* show the $1/T$ equilibrium regime. The points are the raw results. The curves are obtained by subtracting the stationary part (equivalent to a long time extrapolation of the data). The *dashed line* is a $\chi = (1 - C)^{0.47}$ fit, in reference to the continuous RSB model [40]

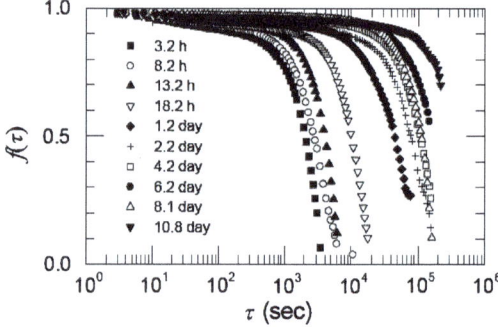

Fig. 2.15. From [34], time evolution of the dynamic structure factor of a gel, measured by multispeckle dynamic light scattering at $q = 6756\,\mathrm{cm}^{-1}$. The curves are labelled by the gel age t_w

type models of spin-glasses [1]. In continuous RSB models like the mean-field spin-glass [2], a $\chi = (1 - C)^{1/2}$ behaviour is predicted [40]. The dashed line in Fig. 2.14 shows a $\chi = (1 - C)^{0.47}$ fit which gives at least a rough account of the results. The next step in this discussion of the first directly comparable noise and response data may arise if a direct experimental determination of $C(t, t)$ is obtained, for example from neutrons scattering data [41].

The autocorrelation function may be more easily accessible in colloidal systems. In the case of colloidal gels, the technique of multispeckle dynamic light scattering allows the direct determination of the dynamical structure factor $f(q, \tau)$, which is the autocorrelation function of the density fluctuations over a

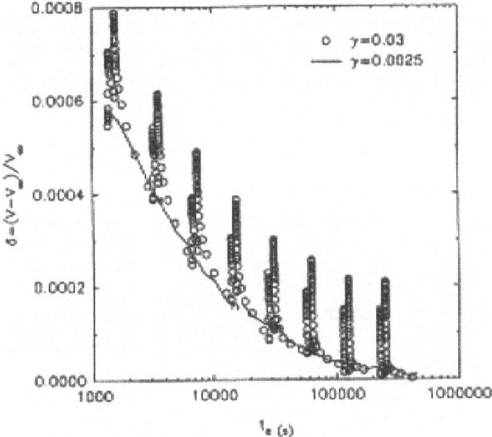

Fig. 2.16. From [42], volume relaxation (in fraction of the long time value) of an epoxy glass sample after a quench. At some times, a stress of amplitude γ is applied. For a low stress value (*solid line*), there is no visible effect. For a higher stress value (*circles*), rejuvenation can be seen as spikes

time τ and at a length scale $2\pi/q$. In [34], this autocorrelation has been found to present interesting similarities with the magnetization autocorrelation (and response function) of spin-glasses (Fig. 2.15).

At fixed q (in the above spin-glass case $q = 0$) the time decay of $f(q, \tau)$ has the unusual form $f(q, \tau) \sim \exp\{-(\tau/\tau_f)^{1.5}\}$, but like in spin-glasses the autocorrelation depends on the waiting time t_w during the gel restructuration through $\tau \sim t_w^{0.9}$ (subageing).

2.2.4 Rejuvenation by a Stress

Before turning to the rejuvenation effects which are observed in spin glasses in response to temperature changes, let us mention that a certain kind of rejuvenation effects has been known for a long time in the rheology of glassy materials in response to a mechanical stress [20], and that the equivalent of these phenomena in spin glasses can be traced out in the effect of a (sufficiently strong) variation of the magnetic field [28].

Figure 2.16 shows a typical ageing experiment in which the volume relaxation following the quench of an epoxy glass sample is measured [42].

This volume relaxation accompanies the stiffening of the mechanical properties during ageing of all structural and polymer glasses. In the experiment of Fig. 2.16, at some times a stress of amplitude γ is applied. The solid line corresponds to a low γ value, for which the stress has no visible effect. But, for a higher γ value (open circles), a phenomenon called "rejuvenation" is observed: suddenly the volume increases, and the relaxation is renewed, starting from a value corresponding to a "younger age".

A similar phenomenon can be seen in spin-glasses. Figure 2.17 shows an *ac* experiment [28] in which, after 300 min, a *dc* field H = 30 Oe is applied (in comparison, the *ac* field, which does not influence ageing here, is H_{ac} = 0.3 Oe).

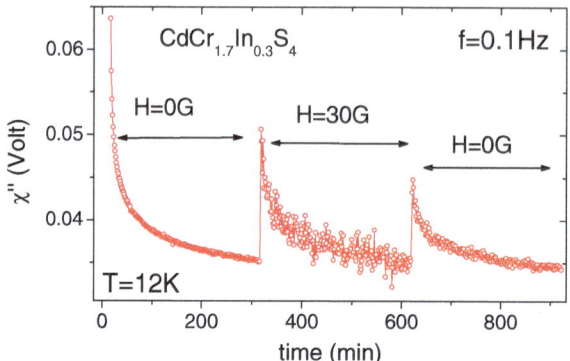

Fig. 2.17. AC out-of-phase susceptibility of the thiospinel spin-glass after a quench. An additional *dc* field is applied in the middle part of the experiment, inducing rejuvenation (data from [28])

The slow relaxation of χ'', which is characteristic of ageing, shows a sudden drop when the *dc* field is applied, and restarts from a "younger state". When the *dc* field is turned back to zero, a weaker but similar drop is observed. The Zeeman coupling of the spins to the *dc* field in this experiment is strong enough to overcome the more subtle spin rearrangements which progressively occurred during ageing as a result of the local minimization of interaction energies. Hence, part of the effect of ageing is erased by applying the *dc* field, and ageing (partly) restarts from new (rejuvenation effect). The same effect is also visible in the *dc* (TRM) experiments presented above; as shown in Fig. 2.8, the μ exponent of the t_w-scaling decreases with the amplitude of the field used for the TRM-relaxation, or in other words, the influence of t_w becomes weaker and weaker as a stronger field perturbation is applied (in the limit $\mu = 0$ there is no t_w effect). In the toothpaste experiment (Fig. 2.9 [32]), as the shear stress amplitude increases, μ also decreases.

It is likely that this effect of the magnetic field on a spin-glass is the equivalent of the effect of a mechanical stress on a glass, in which the slow rearrangements of atoms (or polymers, or micro-spheres or discs in a colloid) during ageing are partly destroyed by applying a shear or elongation stress. The rejuvenation effects as a function of temperature that we present in the next chapter pertain to a different class of phenomena, with the possibility of obtaining almost independent ageing evolutions at different temperatures, and memory effects despite rejuvenation.

2.3 Ageing, Rejuvenation and Memory

2.3.1 Cooling Rate Effects

The state of a glass is strongly influenced by the way it has been cooled. What one usually has in mind is the kind of picture that is shown in Fig. 2.18 [43], displaying the evolution of a thermodynamic quantity like the enthalpy or the specific volume as a function of temperature during the cooling process.

Above the freezing temperature T_f, the glass follows the equilibrium line in the graph of Fig. 2.18, but when crossing T_f it falls out of equilibrium, reaching a state in which the enthalpy relaxes down slowly (ageing). T_f is of course only dynamically defined: for a faster cooling, T_f is higher, and a slower cooling allows the glass to follow the equilibrium line down to lower temperatures. Following the scheme of Fig. 2.18, a state B that would be attained after rapidly cooling to A and ageing for a long time could more easily be obtained by a slower cooling.

This view of glasses was the starting point of experiments in spin-glasses in which we explored how the ageing behaviour could be influenced by the temperature history, having in mind that well-suited cooling procedures might bring the spin-glass into a strongly aged state which otherwise would require astronomical waiting times to be established [44]. These experiments have brought important surprises. The one presented in Fig. 2.19 is representative of the unexpected features which were found in the spin-glass behaviour [45].

In this experiment, we compare the relaxation of the ac susceptibility at $0.7\,T_g$ after two cooling procedures in which the region of T_g was crossed at

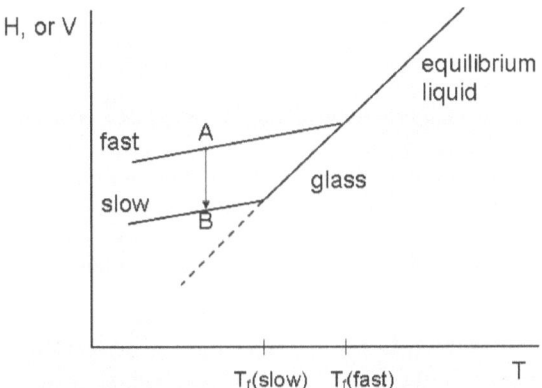

Fig. 2.18. Sketch of the typical enthalpy or volume variation with temperature in a glass (freely inspired from e.g. [43]). During cooling, the liquid falls out of equilibrium at a freezing temperature T_f which depends on the cooling rate ("fast", or "slow"), and becomes a glass. After a fast cooling to point A, ageing over very long times will eventually bring the glass to point B, which can be attained much more quickly by a slow cooling

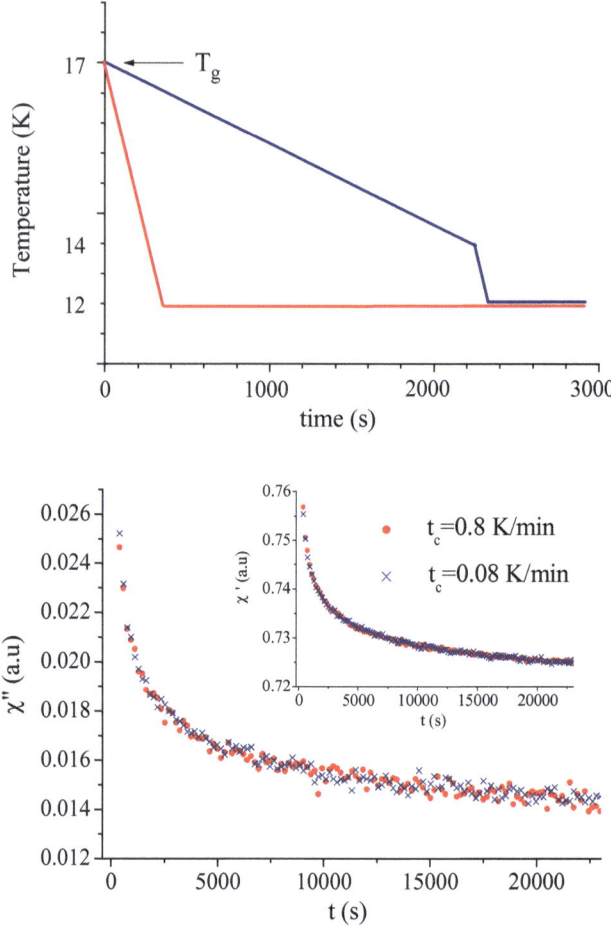

Fig. 2.19. Effect of the cooling rate on the relaxation of the ac susceptibility in the thiospinel sample (from [45]). *Top part*: sketch of the procedure, in which a fast and a slow cooling rate are used around $T_g = 16.7\,$K, before measuring at $12\,$K. *Bottom part*: relaxation of the out-of-phase (*main figure*) and in-phase (*inset*) components of the ac susceptibility, from the time $t = 0$ at which the temperature of $12\,$K has been reached. *Full circles*: fast cooling. *Crosses*: slow cooling

cooling rates differing by a factor 10. Both ageing relaxations, measured from the time at which the final temperature was reached, are exactly superimposed onto each other, as well for χ'' as for χ': a slower cooling through T_g does not help bringing the spin-glass closer to equilibrium, at least as far as can be seen in this measurement. Note that, in the slow cooling procedure, we used a fast cooling rate in the last Kelvin's; a slower final approach of the landing temperature does indeed influence further ageing at this temperature, as shown in the original publication. Our point here is that a slower cooling in the T_g region does not help ageing at a lower temperature.

We have studied this apparent "insensitivity" of the spin-glass to cooling rate effects in more systematic experiments in which the temperature is changed by steps. Figure 2.20 presents the result of a "negative temperature cycle" experiment [46].

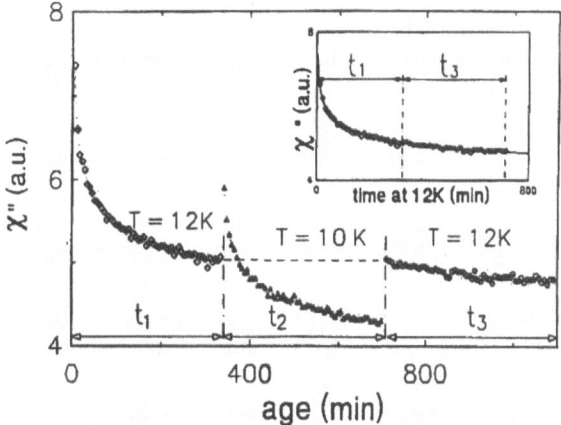

Fig. 2.20. Relaxation of the out-of-phase susceptibility χ'' during a negative temperature cycle of amplitude $\Delta T = 2\,\mathrm{K}$ (frequency 0.01 Hz), showing ageing at $12\,\mathrm{K}$, rejuvenation at $10\,\mathrm{K}$, and memory at $12\,\mathrm{K}$ (from [19, 46]). The inset shows that, despite the rejuvenation at $10\,\mathrm{K}$, both parts at $12\,\mathrm{K}$ are in continuation of each other (memory)

After a normal cooling (typically of $\sim 100\,\mathrm{s}$ from $1.3\,T_g$ to $0.7\,T_g$), the spin-glass is kept at constant temperature $T = 12\,\mathrm{K} = 0.7\,T_g$ for $t_1 = 300\,\mathrm{min.}$, during which ageing is visible in the strong relaxation of χ''. Then, the temperature is lowered one step further from $T = 12$ to $T - \Delta T = 10\,\mathrm{K}$. What is observed is not a slowing down of the relaxation, but on the contrary a jump of χ'' and a restart, which we state as a rejuvenation effect upon decreasing the temperature, as if ageing was starting anew at $T - \Delta T$. The apparent absence of influence of former ageing at T is in agreement with the previous experiment (Fig. 2.19) in which "slower cooling does not help".

One may wonder whether this renewed relaxation corresponds to a full rejuvenation of the sample: the answer is no. A first point is that the new relaxation can be identical to the one obtained after a direct quench one, but only – of course – if ΔT is sufficiently large, here $\Delta T \geq 2\text{–}3\,\mathrm{K}$. And one should not forget that this identity can only be checked in the very limited time window of the experiments, thus not proving very much concerning the *overall* state of the spin-glass. More importantly, the 3rd part of the experiment brings a definitive negative answer. When the temperature is turned back from $T - \Delta T = 10\,\mathrm{K}$ to $T = 12\,\mathrm{K}$, the χ'' relaxation restarts exactly from the point that was attained at the end of the stay at T, and goes on in precise continuity with the former one, as if nothing of relevance at T had happened at $T - \Delta T$. As shown in the inset of Fig. 2.20, this can be checked by shifting the 3rd relaxation to the end of the 1st one: they are in continuity, and can be superposed on the reference curve which is obtained in a simple ageing at T. Hence, during ageing at $T - \Delta T$ and despite the strong associated χ''-relaxation, the spin-glass has kept a "memory" of previous ageing at T, and this memory is retrieved when heating to T.

This negative temperature cycle experiment pictures in a spectacular manner the phenomenon of rejuvenation and memory in a spin-glass. When examined in more details, however, the situation is not always so simple. Figure 2.21 shows the results of negative temperature cycle experiments performed with various values of ΔT.

For $\Delta T = 1\,\mathrm{K}$, the beginning of the 3rd part relaxation shows a transient spike, which lasts for $\sim 5000\,\mathrm{s}$ before the curve merges with those, obtained for higher $\Delta T'$s, that are in continuity with the relaxation at T. Thus, for a smaller ΔT than that corresponding to full memory, there is indeed some contribution at T from ageing at $T - \Delta T$, and this contribution is "incoherent", extending over rather long but finite times (3–5000 s). Note that the data of Fig. 2.21 is taken at frequency 0.1 Hz, whereas in Fig. 2.20 it is taken at 0.01 Hz. In Fig. 2.21, the points can therefore be taken more rapidly, and a small upturn is visible for $\Delta T = 2\,\mathrm{K}$: full memory is only obtained for $\Delta T = 3$ and 4 K.

For smaller and smaller values of the temperature interval ($\Delta T < 1\,\mathrm{K}$), the observed "transient spike" decreases, changes sign (the curve merges with the reference from below), and finally vanishes [46, 47]. In this small ΔT regime, apart from the transient part, ageing at $T - \Delta T$ contributes "coherently"

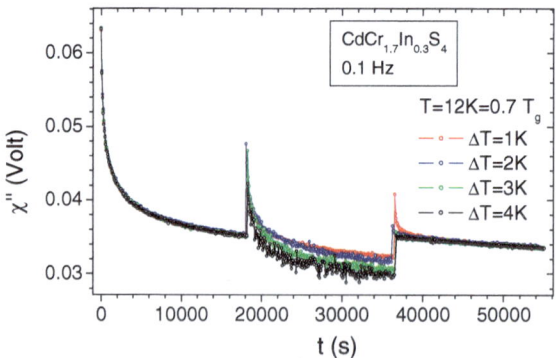

Fig. 2.21. Relaxation of the out-of-phase susceptibility χ'' during negative temperature cycles of different amplitudes (from [48], but see also [46, 47] for other examples), ranging from $\Delta T = 1\,\mathrm{K}$ (upper curve, with the prominent spike) to $\Delta T = 4\,\mathrm{K}$ (lower curve, no spike and full memory). The frequency is 0.1 Hz

to ageing at T as an additional ageing time t_{eff}, in such a way that the 3rd relaxation must be shifted by $(t_2 - t_{eff})$ to be in continuity with the 1st part. Details on the results in this regime, together with their discussion in terms of a Random Energy Model, can be found in [47].

2.3.2 Memory Dip Experiments

The ability of the spin-glass to keep a memory despite (partial) rejuvenation can be further explored in experiments with multiple temperature steps. The first (double) "memory dip experiments", suggested by P. Nordblad, have been developed in collaboration between the Uppsala and Saclay groups [45]. An example of a "multiple dip experiment" is shown in Fig. 2.22 [24, 48, 49].

This is an ac experiment in which the sample is cooled by 2 K steps of duration $\sim$1/2 hour down to 4 K, and then reheated continuously (inset of Fig. 2.22). Figure 2.22 shows χ'' as a function of temperature during this procedure, starting from $T > T_g$ where $\chi'' = 0$ (paramagnetic phase). χ'' rises up when crossing $T_g = 16.7$ K, and when the cooling is stopped, the relaxation of χ'' due to ageing is observed during 1/2 hour (successive points at the same temperature in the figure). Upon further cooling by another 2 K step, the χ'' jump of rejuvenation is seen, and the relaxation due to ageing takes place. At each new cooling step, rejuvenation and ageing can be seen, and this happens $\sim$6 times in the experiment of Fig. 2.22. In the second part of the experiment, the sample is re-heated continuously, at a slow rate ($\sim$0.001 K/s, equal to the average cooling rate) which allows the measurement of χ''. Amazingly, apart

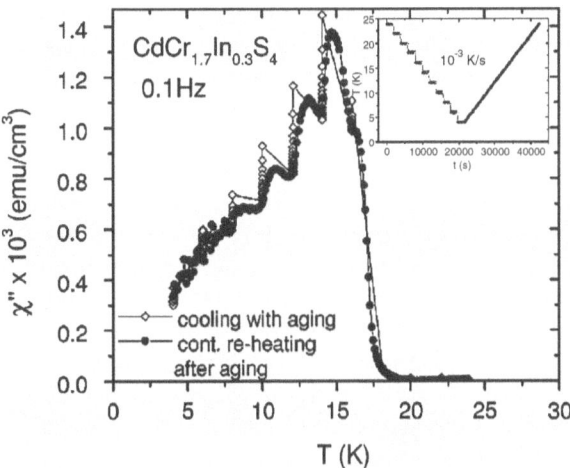

Fig. 2.22. An example of multiple rejuvenation and memory steps [24, 48, 52, 53]. The sample was cooled by 2 K steps, with an ageing of time of 2000 sec at each step (*open diamonds*). Continuous reheating at 0.001 K/s (*full circles*) shows memory dips at each temperature of ageing

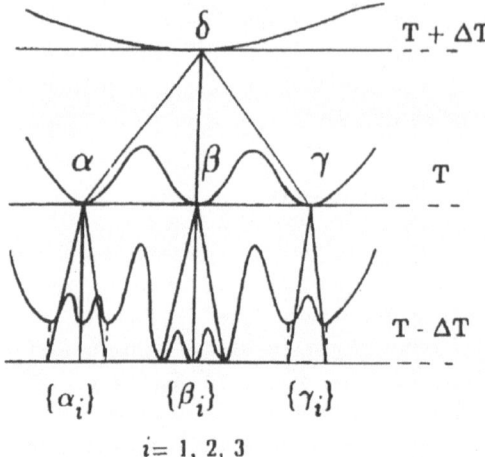

δ $T + \Delta T$

α β γ T

$T - \Delta T$

$\{\alpha_i\}$ $\{\beta_i\}$ $\{\gamma_i\}$

$i = 1, 2, 3$

Fig. 2.23. Schematic picture of the hierarchical structure of the metastable states as a function of temperature [4, 19, 44]

from the rather noisy low-T region, the memory of each of the ageing stages performed during cooling is revealed in shape of "memory dips" in $\chi''(T)$, tracing back the lower value of χ'' which was attained at each of the ageing temperatures. Thus, the spin-glass is able to keep the simultaneous memory of several (5 or 6!) successive ageings performed at lower and lower temperatures. Increasing the temperature afterwards reveals the memories, and meanwhile erases them.

This very asymmetric scheme of rejuvenation upon cooling, topped up by memory effects upon heating, has led the Saclay group to propose a description of these phenomena in terms of a hierarchical organization of the metastable states as a function of temperature, as pictured in Fig. 2.23 [19,44].

This very simple picture sketches the effect of temperature variations in terms of a modification of the free-energy landscape of the metastable states (and not only of a change in the transition rates between them). At fixed temperature T, ageing corresponds to the slow exploration by the spin glass of the numerous metastable states. When the temperature is decreased from T to $T - \Delta T$, the free-energy valleys are considered to subdivide into smaller ones, separated by new barriers. Rejuvenation arises from the transitions that are now needed to equilibrate the population rates of the new sub-valleys: this is a new ageing stage. For large enough ΔT (and on the limited experimental time scale), the transitions can only take place between the sub-valleys, in such a way that the population rates of the main valleys are untouched, keeping the memory of previous ageing at T. Hence the memory can be retrieved when re-heating and going back to the T-landscape. This tree picture, somewhat naïve, is however able to reproduce many features of the experiments when discussed in more details [47]. It has been made quantitative in developments of the Trap Model and the Random Energy Model [47, 50]. In the mean-field model of the spin-glass with full replica symmetry breaking [2], it has been shown that rejuvenation and memory effects can be expected in the dynamics [51].

Beyond this description of ageing and rejuvenation and memory effects in terms of metastable states, it is of course very intriguing to imagine what kind of spin arrangements allow such complex phenomena when the temperature is varied [52–55]. It is very natural, as proposed in the "droplet model" [56, 57], to consider that the spin-glass, initially in a random configuration after the quench, slowly builds up from neighbour to neighbour a spin-glass local order over larger and larger length scales. Frustration makes the process of minimizing the interaction energy of each spin with its neighbours very slow, making the jump from microscopic times (which are at play in the domain growth of pure ferromagnets, in which $l \sim t^{1/2}$) to macroscopic times corresponding to thermally activated crossing of free-energy barriers. In the droplet model, the spin-glass is a kind of "disguised ferromagnet", having simply two (spin reversal symmetric) ground states, which compete in the slow growth of spin-glass ordered domains during ageing. Can we see such domains in experiments? No obvious macroscopic symmetry is expected in spin-glass order, therefore no imaging of such domains could be realized until now, in contrast with the case of ferromagnetic domain growth. The only pictures that we have of the growth of a potential spin-glass order are obtained from recent numerical simulations. Figure 2.24 shows a nice example given by Berthier and Young in [15], but the reader should not be misled by the apparent simplicity of this ferromagnetic-like picture.

In Fig. 2.24 the grey scale codes the relative orientations of the spins in two copies (replicas) of the system which, starting from different random states, evolve independently by a Monte-Carlo algorithm. The snapshots taken after different waiting times t_w show the growth of uniformly coloured regions. A region with a uniform grey-level colour is a region in which the individual spins have a constant angle from one replica to the other: over this region, seen in independent Monte-Carlo evolutions, the neighbour spins build the same relative angles. This is indeed an image of regions in which the spin evolution is correlated, which are in this sense equivalent to spin glass ordered domains.

If we now come back to the multiple memory experiment in Fig. 2.22, thinking of a spin-glass order being established on longer and longer length scales during each stage of ageing, the observed rejuvenation and memory effects have some implications concerning these dynamic length scales. The restart of dissipative processes when going from T to $T - \Delta T$ indicates that the spin-spin correlations growing at $T - \Delta T$ are different from those established at T. For thermally activated processes, if correlations extend up to a given length scale L_T^* during ageing at T, the correlation length $L_{T-\Delta T}^*$ which is attained at $T - \Delta T$ during the same time should be smaller, $L_{T-\Delta T}^* < L_T^*$. The memory effect imposes here an important constraint: ageing up to $L_{T-\Delta T}^*$

Fig. 2.24. From the numerical simulations in [15]: relative orientation θ_i of the spins S_i in two copies (a, b) of a numerical Heisenberg spin-glass. The gray scale stands for $\cos\theta_i(t_w) = S_i^a(t_w) \cdot S_i^b(t_w)$. From *top* to *bottom*, three different waiting times $t_w = 52, 27$, and 57 797 s are represented, showing the slow growth of a local random ordering of the spins

should occur without changing significantly the correlations established at the scale L_T^*, that is, $L_{T-\Delta T}^* < L_T^*$. In practice, the independence of ageing at length scales $L_{T-\Delta T}^*$ and L_T^* is realized by a strong separation of the related *time* scales $\tau : \tau(L, T - \Delta T) \gg \tau(L, T)$. This necessary separation of the ageing length scales with temperature has been coined "temperature-microscope" effect by J.-P. Bouchaud [53]: in an experiment like shown in Fig. 2.22, at each stage ageing should take place at well-separated length scales $L_n^* < \ldots < L_2^* < L_1^*$, as if the magnification of the microscope was varied by orders of magnitude at each temperature step. This hierarchy of embedded length scales as a function of temperature is the "real space" equivalent of the hierarchy of metastable states in the "phase space" (Fig. 2.23).

Do we have examples of systems which present such a hierarchy of reconformation length scales? This has been proposed for the very generic case of an elastic line in presence of pinning disorder [58,59]. Here, frustration arises from the competition between elastic energy, which tends to make the line straight, and pinning energy, which tends to twist the line to go through all pinning sites. As sketched in Fig. 2.25 [59], starting from a random configuration after a quench, the line will progressively "age" by equilibrating slowly (thermally activated dynamics) over larger and larger distances.

At a given temperature T and after some ageing, the line can be pictured as a fuzzy ribbon (top of right part in Fig. 2.25) which is equilibrated over a length scale L_T^*. At smaller length scales, the line continues to fluctuate between configurations which are roughly equivalent at temperature T (thus seen as a fuzzy ribbon). However, when going from T to $T - \Delta T$, the difference between the equilibrium populations of some of these configurations may become significant, and a new equilibration at shorter length scales $L_{T-\Delta T}^* < L_T^*$ must take place. These dissipative processes will cause a rejuvenation signal. Meanwhile, processes at length scale L_T^* are frozen at $T - \Delta T$, and the memory of previous ageing remains intact despite the rejuvenation processes, which occur at smaller (and well-separated) length scales.

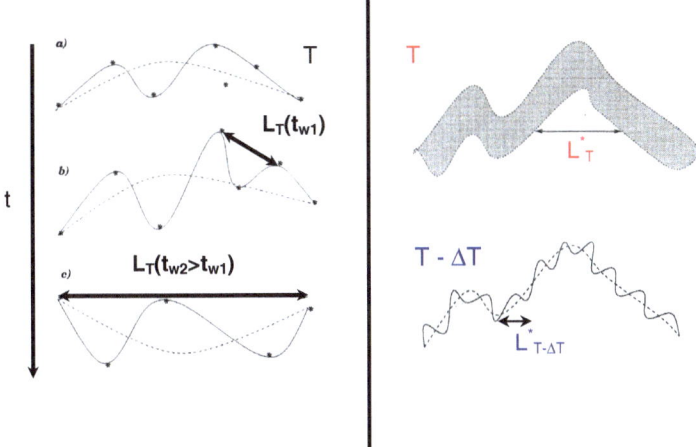

Fig. 2.25. From [59], sketch of ageing, rejuvenation and memory phenomena in terms of the dynamics of an elastic line in pinning disorder. *Left part*: at fixed temperature T, as time goes on, the line matches the pinning sites over larger and larger distances L_T. *Right part*: as the temperature is lowered from T to $T - \Delta T$, rejuvenation processes occur at a smaller length scale $L_{T-\Delta T}$, while the memory of reconformations at the larger length scale L_T is preserved

This scheme is a good candidate for the mechanism of ageing, rejuvenation and memory in spin glasses [53,59,60]. The theory of an elastic line in pinning disorder yields a hierarchy of embedded states and length scales [58]. In the spin-glass, it is not yet clear what objects could play the role of pinned elastic lines. Experiments on disordered ferromagnets show that spin-glass dynamics can indeed be observed, which is most probably due to the dynamics of the walls [6,60]. Thus, we propose that the observed slow dynamics in spin-glasses is explained in terms of wall-like dynamics, but in the present state of the art we cannot identify what are these walls, and what is the nature of the domains which are separated by these walls (see however the "sponge-like" excitations which have been characterized in numerical simulations [14]).

2.3.3 Rejuvenation and Memory Versus Cumulative Ageing

In the previous section we described a "rejuvenation and memory like" dynamics, implying a hierarchical organization of the metastable states and of the corresponding length scales. This type of dynamics is found in systems which have so many "embedded" degrees of freedom that some of them are available to excitation at any temperature, even independently from each other at sufficiently different temperatures.

In "domain growth like" dynamics, of the type occurring in a ferromagnet, the approach of equilibrium is a one way only evolution through domain growth and wall elimination, in which the size of the domains should always increase. In an ideal ferromagnet, in which no energy barriers impede the domain wall motion, the temperature does not play any role. If we think of activated processes like the pinning of walls on defects, then temperature is relevant, but domain growth should just be accelerated or slowed down by temperature changes. Ageing by domain growth processes is "temperature cumulative", in the sense that ageing continues additively ("cumulatively") from one temperature to the other. In this type of dynamics, it is not clear how rejuvenation and memory effects may arise. In the droplet theory [56] they are related to "temperature chaos" effects, a scenario introduced in [56,61] which we do not discuss here. Detailed discussions of its possible relevance can be found in [53–55].

However, this language should not be misleading, and there is indeed some part of "domain growth" in "rejuvenation and memory" dynamics [21], but in our present understanding what is growing here is an object of the nature of a pinned wall rather than a (compact) domain. For a sufficiently small temperature variation ΔT, no rejuvenation effects are seen in the spin-glass: ageing continues from T to $T - \Delta T$ (see ac experiments in Subsect. 2.3.1 and [46], or dc experiments with negative temperature cycles in [19]). In the hierarchical picture, for small ΔT's the free-energy landscape is almost identical at T and $T - \Delta T$. In more general words, for small ΔT's the length scale of the ageing processes are almost the same at T and $T - \Delta T$, and ageing is cumulative between both temperatures. As soon as ΔT is large enough, the free-energy

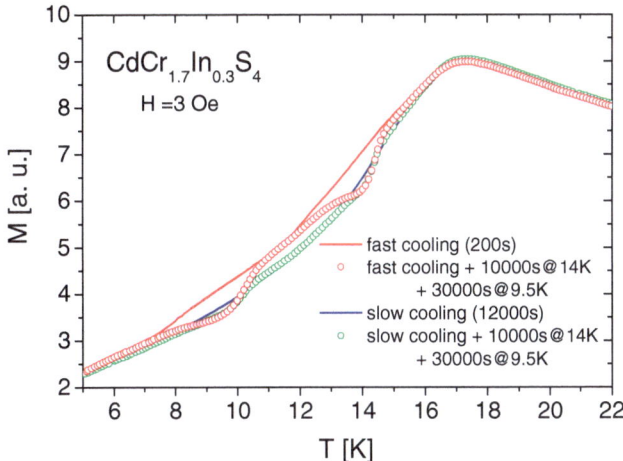

Fig. 2.26. Effect of various cooling procedures on the ZFC magnetization of the thiospinel (insulating) spin-glass [48]. Comparison of fast and slow coolings, with and without stops

landscapes become different, the ageing length scales are separated (as is clear in the example of the pinned elastic line), and rejuvenation occurs due to the existence of independent degrees of freedom.

In some spin-glass experiments like the one presented in Fig. 2.26 [48], this dual aspect of ageing dynamics shows up very clearly.

In this experiment, the sample is zero-field cooled with various thermal histories, and after applying the field at low temperature the magnetization is measured while increasing the temperature continuously at fixed speed (small steps of 0.1 K/min). On one hand, we can observe the effect of a slow cooling in comparison with that of a fast cooling: the slow-cooled curve lies below the fast one in the whole temperature range. There is indeed a cooling rate effect in spin-glasses, provided that one chooses an appropriate procedure to evidence it. On the other hand, we can evidence memory effects by stopping the cooling at two distinct temperatures and waiting during ageing of the spin-glass. The magnetization measured during re-heating after this step-cooling procedure shows clear dips at both temperatures at which the sample has been ageing (this experimental procedure, very similar to that of the ac experiment in Fig. 2.22, has been proposed by the Uppsala group [62]). In a third experiment, we can mix both effects, by slowly cooling the sample and interrupting the slow cooling by long waiting times at constant temperature. The resulting magnetization curve is lower than those obtained after faster cooling (temperature cumulative ageing), and shows memory dips on top of this lower curve.

In a similar experiment, performed with another spin-glass (Au:Fe$_{8\%}$ from [63], in Fig. 2.27, metallic sample instead of the insulator of Fig. 2.26), we

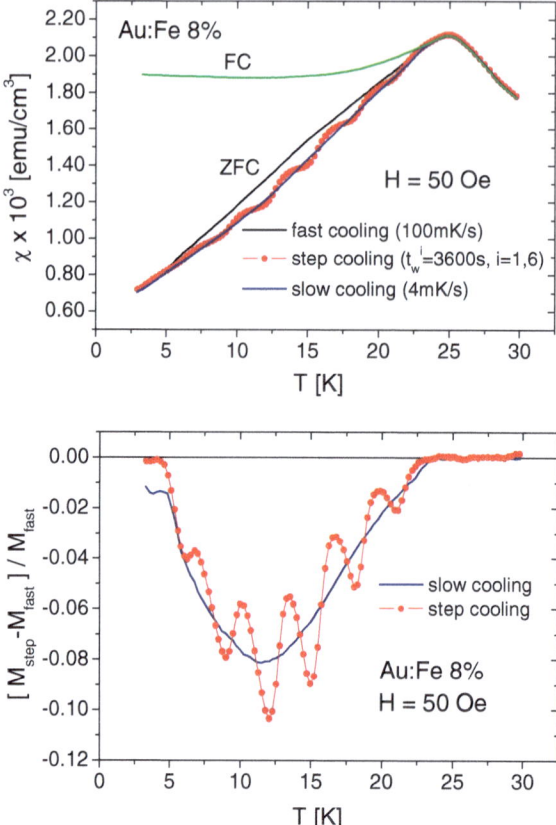

Fig. 2.27. Effect of various cooling procedures on the ZFC magnetization of the Au:Fe$_{8\%}$ spin-glass (*top part*). Comparison of fast and slow coolings, with and without stops. *Bottom part*: difference with the magnetization obtained after fast cooling. From [48]

have also plotted (bottom part of Fig. 2.27) the difference between the curves obtained after a specific cooling history and the reference one obtained after a fast cooling. Fast oscillations (memory dips) show up on top of a wide bump (cumulative ageing).

Thus, the spin-glass should not be considered as exempt of cooling rate effects, but rather as being able to show rejuvenation and memory effects in addition to cooling rate effects. How can we now compare the spin-glass with "normal" glasses, which are considered to be dominated by cooling rate effects? [21] New experiments have been designed to search for rejuvenation and memory effects in such systems. And these effects have been found, as is shown in the experiment of Fig. 2.28 by the ENS Lyon group [64].

This experiment uses the same procedure (and the same representation of the results) as in Fig. 2.27, but the cooling is only interrupted by one

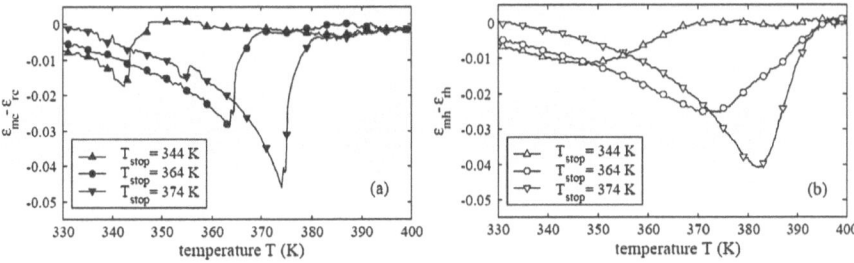

Fig. 2.28. Memory effect in the ageing of the PMMA polymer glass, from [64]. The plots show the difference in the dielectric constant between an experiment with a normal cooling and another one with a stop of 10 hours at T_{stop}, for 3 values of T_{stop}. *Left part*: cooling. *Right part*: re-heating, showing a dip at a temperature corresponding to T_{stop}

stop at one given temperature. Upon re-heating, the dielectric constant of the PMMA polymer indeed shows a dip, centred at a temperature slightly higher than that of the stop, and the comparison of 3 experiments with stops at 3 different temperatures shows very clearly that the position of the dip follows the temperature of the stop (Fig. 2.28). The range of temperatures in which the ageing effects are important in PMMA is much narrower than in spin glasses, and the width of the dip may appear to be larger because it spreads over the whole explored temperature range. However, the temperature-dependence of the dip position is very clearly evidenced, signing up the occurrence of ageing processes which are strongly temperature specific, as is the case in spin-glasses.

An even more dramatic example of rejuvenation and memory effects in a structural glass (Fig. 2.29) has been obtained in a study of the mechanical response of gelatine by a group of the food company Firmenich SA (Switzerland) [65].

Gelatine is a complex protein made of folded helices, and it has indeed many degrees of freedom related to helix unfolding in the vicinity of room temperature. This experiment is an *ac* measurement of the elastic modulus G', and is again comparable with the *ac* experiment of Fig. 2.22. During ageing at fixed temperature, G' relaxes upwards (ageing, the gelatine stiffens), and upon further cooling some rejuvenation can be seen. When re-heating, G' shows a dip at the ageing temperature, and the authors could even realize a double memory experiment in which two memory dips can be distinguished (Fig. 2.29).

Thus, it appears that ageing effects in glasses in general can be considered as showing both "T-cumulative" and "rejuvenation and memory" contributions. The specificity of spin-glasses might then be their ability to show sharp memory effects. However, the next section shows that the sharpness of these memory effects may be different in different spin glasses, and the further investigation of memory effects in *structural* glasses may bring other surprises.

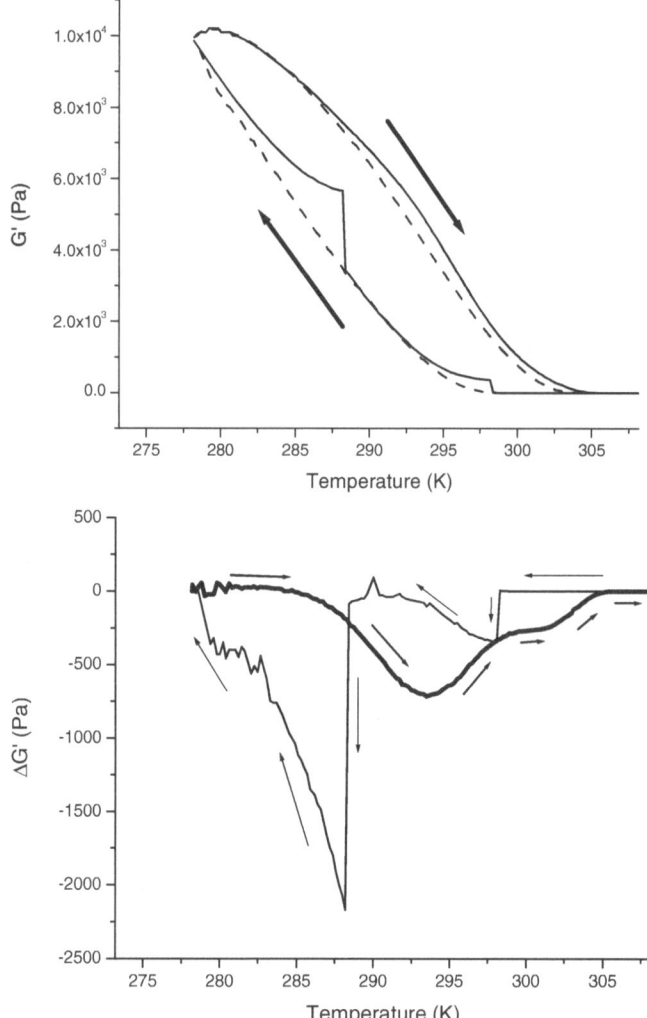

Fig. 2.29. Memory effect in the ageing of a gelatine gel (measurement of the elastic modulus G', from [65]). Two stops of 2 hours were made at 25 and 15°C during cooling. The *upper part* shows G' (*solid line*: with stops, *dashed line*: without stops) as a function of temperature. During the stops, G' increases slowly (ageing, the gelatine gel stiffens). Upon re-heating, a wide-spread excess of G' is seen when compared with the curve obtained without stops. But, in the lower part of the figure which shows the difference plot, the memory of both stops is clearly revealed on re-heating

2.4 Characteristic Length Scales for Ageing

As ageing goes on, the dynamical response of the spin-glass becomes slower. We have seen (Fig. 2.5) that the time derivative of the magnetization relaxation after a field change gives access to an effective distribution of relaxation times, which shows a wide peak centred in the $\log t = \log t_w$ region [22]. For longer t'_ws, this distribution shifts towards the longer time region. We have no direct access to the spin configurations which correspond to these longer and longer response times, but it is reasonable to assume that longer response times are associated with flipping a larger number of correlated spins. This is the point of view that we have adopted above in this paper, discussing the multiple memory experiments (Fig. 2.22) in terms of a hierarchy of embedded dynamical length scales selected by temperature (Fig. 2.25) [53]. No simple symmetry allows an easy observation of these dynamical correlation lengths, but, considering that such characteristic dynamical lengths are underlying the ageing phenomena, we have designed experiments which bring rather strong constraints on their properties. These experiments can be grouped in two classes: field variation and temperature variation experiments.

2.4.1 Length Scales from Field Variation Experiments

The idea of these experiments, based on [28], has been developed by R. Orbach and his group (UCLA and Riverside) [66]. It starts from the observation that the magnetization relaxation following a field change (as well in TRM as in ZFC procedure) becomes faster when a higher field amplitude is used, going beyond the linear response regime. An example is shown in Fig. 2.30.

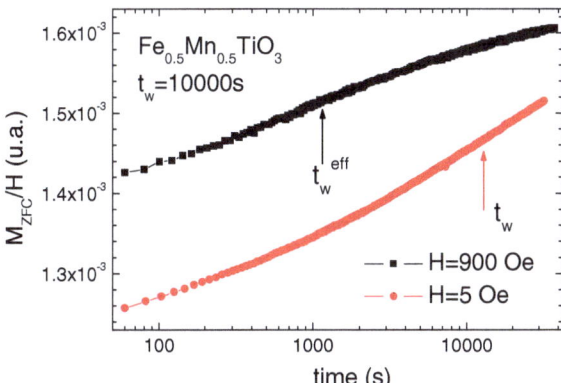

Fig. 2.30. ZFC relaxations of the $Fe_{0.5}Mn_{0.5}TiO_3$ Ising sample, for $t_w = 10000\,s$ and 2 different values of the field H (data from [67]). The *lower curve*, taken with a low-field H = 5 Oe, shows an inflection point in the t_w region. In the *upper curve*, taken with a much higher field H = 900 Oe, the inflection point is found at a shorter time $t_w^{eff} \sim 1000\,s$

The lower curve in Fig. 2.30 shows the ZFC relaxation obtained after applying a (small) 5 Oe field after $t_w = 10000$ s. Its inflection point is located as usual around $t \sim t_w$. The upper curve is obtained with a much higher field of 900 Oe, applied after the same t_w. This relaxation is faster than the first one in two respects: the initial rise up of the magnetization is higher, and the inflection point is found at shorter times, indicating that the distribution of the relaxation times now peaks at $t_w^{\text{eff}} \sim 1000$ s, one order of magnitude smaller than $t_w = 10000$ s. We propose to characterize the relaxation curves by their inflection point t_i (time at which the relaxation rate is maximum), defining a typical free-energy barrier U which can be overcome by thermal activation at temperature T after a time t_i with an attempt time $\tau_0 (\tau_0 \sim 10^{-12}$ s is a paramagnetic fluctuation time):

$$U = k_B T \ln(t_i/\tau_0) . \tag{2.7}$$

In the case of the low-field experiment with low-field H_0, $t_i \cong t_w$, which defines a barrier Δ as

$$\Delta(H_0) = k_B T \ln(t_w/\tau_0) . \tag{2.8}$$

In the experiment with a higher field H, the barrier $\Delta(H) = k_B T \ln(t_w^{\text{eff}}/\tau_0)$ is smaller since $t_w^{\text{eff}} < t_w$. Assuming that, in a relaxation experiment performed after a given t_w, the spin correlations extend up to a typical number of spins $N_s(t_w)$, we propose to ascribe the free-energy reduction $\Delta(H_0) - \Delta(H) = E_Z(H)$ to the Zeeman energy of coupling of the magnetic field to the typical number of correlated spins $N_s(t_w)$ that must be flipped in the relaxation process [28, 66]. In a low-field experiment this Zeeman energy is negligible, and we have $t_i \cong t_w$, but for a higher field H $E_Z(H)$ becomes significant, and we obtain it as the result of the measurement:

$$E_Z(H) = k_B T \ln(t_w/t_w^{\text{eff}}) . \tag{2.9}$$

The Zeeman energy is $E_Z = M.H$, M being the magnetization of the N_s spins. At this stage, we need to write explicitly the dependence of M on N_s, which is not completely obvious for a disordered system. For a small number of spins N_s in a random configuration, the magnetization is proportional to the typical fluctuation $N_s^{1/2}$, and is independent of the field: $E_Z = N_s^{1/2}\mu H$, where μ stands for the magnetic moment of 1 spin in the compound. On the other hand, at the *macroscopic* scale, the magnetization is an extensive quantity, proportional to the number of spins, and (to first order) proportional to the field via the susceptibility χ of 1 spin: $E_Z = N_s \chi H^2$.

It is likely that the general dependence of E_Z on N_s is a crossover shape from H to H^2 dependence, but this would mean too many free parameters to interpret the results. In principle, the experiment should tell us which one is the dominant regime in the conditions of the measurement, since we can measure $E_z(H) \propto \ln t_w^{\text{eff}}(H)$ for various values of H, and conclude whether

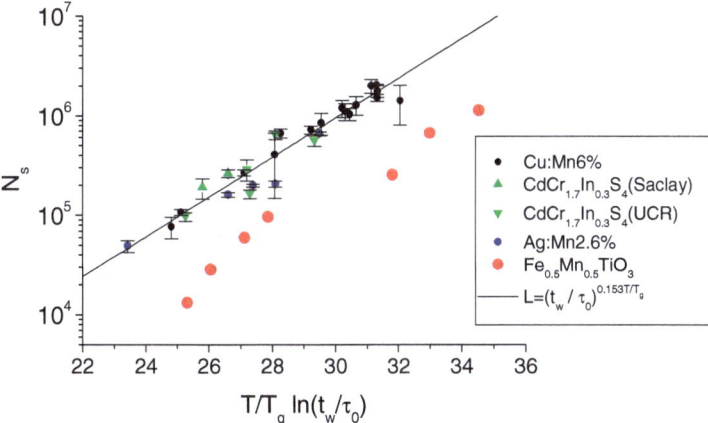

Fig. 2.31. Number of correlated spins extracted from field change experiments, as a function of the reduced variable $T/T_g \ln(t_w/\tau_0)$. The points with error bars correspond to Heisenberg-like spin-glasses [66], they are well fitted by the straight line $N_S \sim (t_w/\tau_0)^{0.45T/Tg}$. The full circles lying below the others are from the $Fe_{0.5}Mn_{0.5}TiO_3$ Ising sample [67]

$E_z(H)$ has an H or H^2 dependence. However, as in all fitting procedures, the result may depend on the range of fields explored, and the response is not completely unambiguous. Let us present now the experimental results that we obtained from various spin-glass samples.

In an early series of experiments [66], we explored several spin-glasses of different chemical nature: the insulating thiospinel $CdCr_{1.7}In_{0.3}S_4$, and the metallic alloys $Cu{:}Mn_{6\%}$ and $Ag{:}Mn_{2.6\%}$. With respect to spin anisotropy, these compounds are all Heisenberg-like [63]. For each sample, we measured ZFC relaxation curves for various amplitudes of the field H, at different temperatures T and for various waiting times t_w. For fixed t_w and T, the dependence of $\ln t_w^{\text{eff}}(H)$ versus H^2 was found to be significantly more linear than as a function of H, and we determined N_s from the observed slope of $E_Z = N_s \chi H^2$ versus H^2 [66]. The results are shown in Fig. 2.31.

In this plot, which is presented as a function of the reduced variable $T/T_g \ln(t_w/\tau_0)$, the results from the 3 Heisenberg-like samples at 2 different temperatures do all fall on the same line. The number of correlated spins is, as expected, an increasing function of t_w, and the numbers reached in the experimental times are $\sim 10^4 - 10^6$, which means a range of 10–100 lattice units for the correlation length (assuming $L \sim N^{1/3}$). The 3 samples have a common (universal for Heisenberg-like?) behaviour, which is well fitted by a unique straight line. The solid line shown in the graph corresponds to the power law dependence $N_S = (t_w/\tau_0)^{0.45T/Tg}$. This was a rather big surprise because, soon after these experiments, numerical simulations of the Ising spin-glass (Edwards-Anderson model) were performed in the ageing regime by several groups, who could compute the four point correlation function and "directly obtain" an estimate of the correlation length $L(T, t_w)$ [13]. The numerical result, common to the different groups, is $L \cong (t_w/\tau_0)^{0.15T/Tg}$ (recovering dynamic scaling $L \sim t^z$ of the equilibrium correlation length at $T_g, z = 1/0.15 \cong 6$ being the usual dynamic exponent). This is the same result as in the experiments (if $N \propto L^3$). At this stage, the difference between Heisenberg-like (in experiments) and Ising (simulations) spins was not really discussed, and what was emphasized was the striking similarity between the simulations of the Edwards-Anderson model [13], performed up to $t_w/\tau_0 \sim 10^5$, and the experiments [66], which are performed 10 orders further in time in the $t_w/\tau_0 \sim 10^{12-17}$ regime.

This comparison motivated a second series of experiments [67], in which the properties of the strongly anisotropic system $Fe_{0.5}Mn_{0.5}TiO_3$ [68], considered a representative example of an Ising spin-glass, were investigated using the same technique. We show in Fig. 2.32 the measured $Ln\, t_w^{\mathrm{eff}}$ as a function of H^2 and also H.

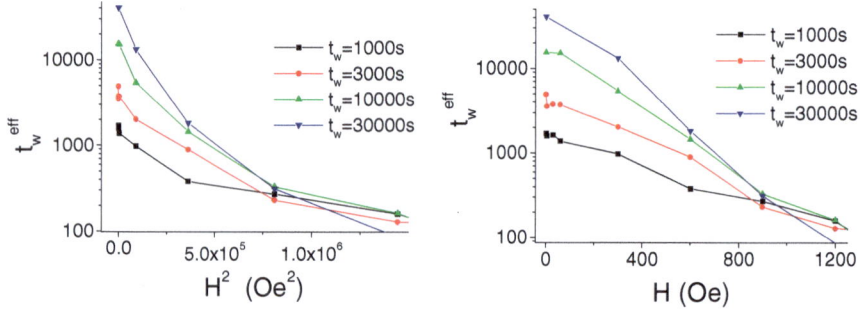

Fig. 2.32. Effective waiting times (log scale) obtained from the field change experiments on the $Fe_{0.5}Mn_{0.5}TiO_3$ Ising spin-glass, as a function of H^2 (*left*) and H (*right*) [67]. Four values of the waiting time t_w have been explored

In this Ising case, this is the linear behaviour of $Ln\, t_w^{\mathrm{eff}}$ as a function of H rather than H^2 which is favoured. Therefore we decided to analyse the Ising results in terms of $E_Z = N_s^{1/2}\mu H$. Checking afterwards the results of an analysis using $E_Z = N_s\chi H^2$ (only possible in the small field range), we found that it does not yield very different conclusions anyway. The results from the Ising sample are plotted in the same graph as those from the other samples in Fig. 2.31 [67]. They lie – by almost a factor of 10 – lower than the others: in the Ising sample, after a given t_w, the number of correlated spins is smaller than in the Heisenberg-like samples. But, at fixed temperature, the t_w dependence of N_s is faster in the Ising case. The overall conclusion of this comparison is that the same simple power law, of the type $N \sim t_w^{aT/Tg}$, is not able to reproduce both sets of results from Heisenberg and Ising samples.

The progress in computer simulations has finally allowed the numerical study of the Heisenberg spin-glass, which is still more greedy in computer time. In [15], the comparison of the numerical results is presented using the same variables as in Fig. 2.31. The correlation length $L(T, t_w)$ is found smaller in the Ising case, as in experiments. The extrapolation of the numerical data to the time regime of the experiments is rather hazardous, but it is possible that the long time slope of $L(T, t_w)$ versus t_w at low T becomes weaker for Heisenberg than for Ising (however, this is not the case in the numerical time range).

2.4.2 Length Scales from Temperature Variation Experiments

The temperature variation experiments bring a lot of information on the time-temperature relation in ageing phenomena. In a negative temperature-cycle experiment [19, 44], the spin glass is aged during t_1 at T, then during t_2 at $T - \Delta T$ and finally during t_3 at T. A specific state of ageing is established by this temperature history. If after this a field change is applied (like in the TRM procedure), the relaxation curve that is obtained reflects the properties of the state that has been prepared. For small ΔT values, it is possible to obtain the same relaxation curve after ageing at constant temperature T during a total waiting time $t_1 + t_2^{\mathrm{eff}} + t_3$, in such a way that the effect of waiting t_2 at $T - \Delta T$ is the same as waiting t_2^{eff} at T. The identity of the relaxation curves tells us that the same state of ageing has been established in both histories, at least for the ageing processes whose time scales are probed in a dc relaxation experiment ($\sim 10^0$ to 10^5 s) [19]. Now, the idea of is to consider that this same ageing state corresponds to the same dynamical length L up to which correlations are established. Hence, from a couple of experiments as described, we constrain the time and temperature dependence of $L(t, T)$:

$$L(t_2, T - \Delta T) = L(t_2^{\mathrm{eff}}, T) . \tag{2.10}$$

We have performed TRM experiments with negative temperature cyclings on a series of representative spin-glass samples, in order to better understand the differences between Ising and Heisenberg systems [67]. For this purpose, we have used a series of spin glasses which have also been studied in Orsay by torque measurements [63]. The torque measurements allowed sorting these spin glasses by their measured spin anisotropy (random anisotropy arising from Dzyaloshinsky-Moriya interactions). D. Petit and I. Campbell found [63] that the critical exponents at the spin-glass transition present a systematic dependence on the spin anisotropy, ranging from Edwards-Anderson type exponents for the Ising example to chiral ordering exponents [69] in the most isotropic case. These samples are, K_r being the relative anisotropy constant $K_r = (K/T_g)/(K/T_g)_{\mathrm{AgMn}}$, normalized to the AgMn value [63]:

(1) $Fe_{0.5}Mn_{0.5}TiO_3$, $T_g = 20.7\,K$, strongly anisotropic single crystal [68] (no K_r estimate, but large)

(2) $(Fe_{0.1}Ni_{0.9})P_{16}B_6Al_3$, amorphous alloy with $T_g = 13.4\,K$ and $K_r = 16.5$

(3) $Au:Fe_{8\%}$, diluted magnetic alloy with $T_g = 23.9\,K$ and $K_r = 8.25$

(4) $CdCr_{1.7}In_{0.3}S_4$, insulating thiospinel with $T_g = 16.7\,K$ and $K_r = 5.0$

(5) $Ag:Mn_{2.7\%}$, diluted magnetic alloy with $T_g = 10.4\,K$ and, by construction, $K_r = 1$ (in the particular case of $Ag:Mn_{2.7\%}$, we use former data from [70]).

The experimental procedure is sketched in Fig. 2.33a, and a set of results with the thiospinel sample (#4, Heisenberg-like) is presented in Fig. 2.33b.

In Fig. 2.33b, relaxation curves obtained after temperature cycling of amplitude ΔT are compared with those obtained after isothermal ageing at $T = 12K = 0.7\,T_g$ during $t_w = t_1 + t_3 = 1000\,s$ (bottom solid curve) and $t_w = t_1 + t_2 + t_3 = 10000\,s$ (top solid curve). If we look for instance at the curve obtained after temperature cycling $\Delta T = 0.5\,K$ (full circles), we see that it almost lies on the isothermal $t_w = t_1 + t_3 = 1000\,s$ reference, far below the $t_w = t_1 + t_2 + t_3 = 10000\,s$ reference. That is, in this case we have $t_2^{\mathrm{eff}} \sim 0$, which means that ageing during $t_2 = 9000\,s$ at $T - \Delta T$ is almost of no influence on ageing at T, even though ΔT is only of 0.5 K: this is the "temperature microscope effect" that was invoked above to explain the possibility of multiple memories (Fig. 2.22, with the same thiospinel sample). The comparison with the Ising sample is rather interesting (Fig. 2.33c).

For the Ising spin-glass, T_g is slightly different, but the temperatures are the same in units of T_g. The relaxations are performed at $T = 15\,K = 0.7\,T_g$, and we can look at the curve resulting from a temperature cycle with $\Delta T = 0.6\,K = 0.03\,T_g$ with solid squares (same fraction of T_g as for the solid circles for the Heisenberg case in Fig. 2.33b). This curve lies in the middle region between the $t_w = t_1 + t_3 = 1000\,s$ and $t_w = t_1 + t_2 + t_3 = 10000\,s$ references, which means that there is a significant effect of ageing at $T - \Delta T$ on ageing at T: in the Ising case, the T-microscope effect with temperature is not so strong as it is in a Heisenberg spin-glass.

This visual appreciation of the curves can be expressed in quantitative terms. Using the scaling procedure described in Sect. 2.2, we can ascribe an effective waiting time $t_1 + t_2^{\mathrm{eff}} + t_3$ to each of the temperature cycled curves, adjusting precisely the value of t_2^{eff} which allows the superposition of each of the T-cycled curves with a set of isothermally aged references. The result of each temperature cycling experiment is a value of t_{eff} for a given ΔT. In Fig. 2.34, we present in the same graph the results obtained from the 5 samples for $T = 0.85\,T_g$ (similar results have been obtained for $T = 0.7\,T_g$) [67].

Of course t_{eff} is a decreasing function of ΔT: for larger values of ΔT, the contribution of ageing at $T - \Delta T$ to ageing at T becomes weaker. Remarkably, we find that the slope of $t_2^{\mathrm{eff}}(\Delta T)$ varies systematically with the spin anisotropy of the sample. The slope is weaker for the Ising sample than for the thiospinel (Heisenberg-like) sample #4, as expected from the trend observed in Fig. 2.33bc, but the effect is systematic over the 5 samples stud-

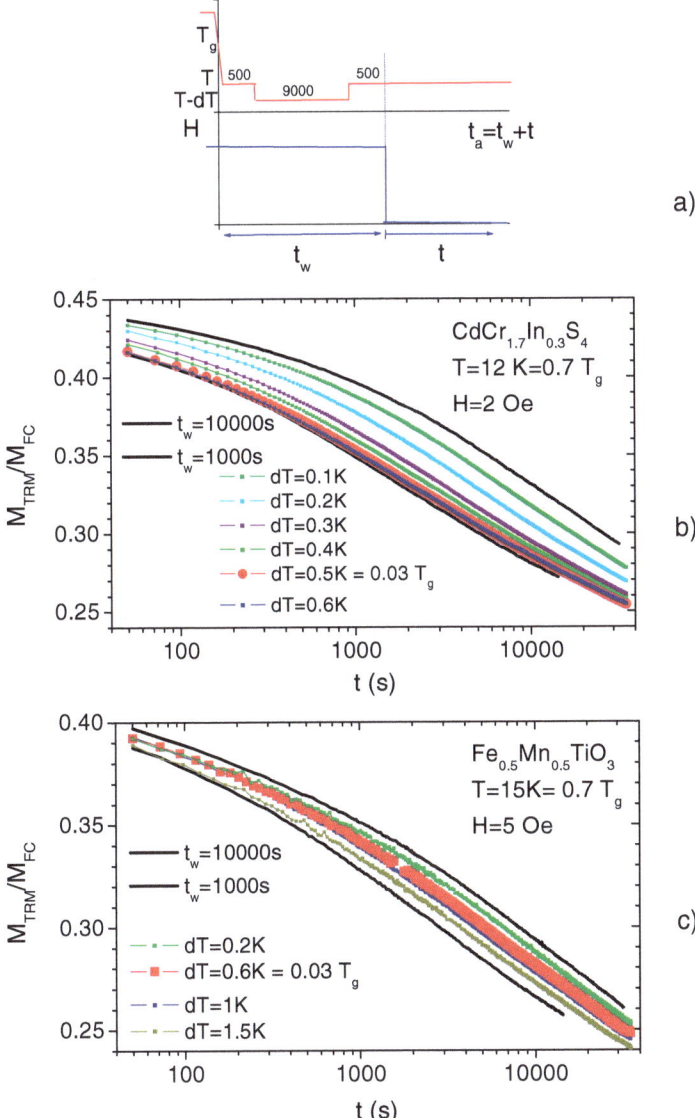

Fig. 2.33. TRM experiments with a negative temperature cycle from T to T-dT during the waiting time [67]. The extreme solid lines are reference curves, obtained after isothermal ageing during $t_w = 1000$ and 10000 s. The thick full circles are obtained after a negative temperature cycling with $dT = 0.03\,T_g$. Results from other dT values are also presented. *Top part*: CdCr$_{1.7}$In$_{0.3}$ thiospinel (Heisenberg-like) spin-glass. *Bottom part*: Fe$_{0.5}$Mn$_{0.5}$TiO$_3$ Ising spin-glass

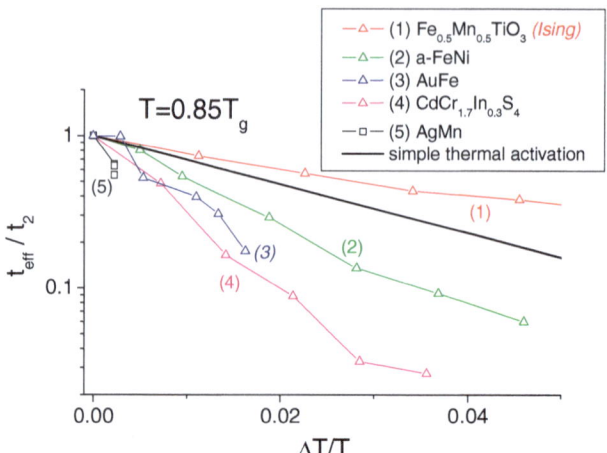

Fig. 2.34. Effective waiting times deduced from the temperature cycle experiments performed around $T = 0.85\,T_g$, for the 5 samples investigated (ranked by decreasing anisotropy from #1 to #5) [67]. The *straight line* stands for usual thermal slowing down (constant energy barriers) with $\tau_0 = 10^{-12}\,$s

ied. Going from Ising to Heisenberg situation, the weaker the spin anisotropy, the steeper the decrease of $t_2^{\mathrm{eff}}(\Delta T)$, which means a stronger and stronger T-microscope effect.

We can compare the steepness of this decrease with usual thermal slowing down. A free-energy barrier $U(T - \Delta T)$ can be defined corresponding to the ageing process during t_2 at $T - \Delta T$, and for usual thermal slowing down this barrier is the same as $U(T)$ corresponding to ageing during t_2^{eff} at T, which reads

$$U(T - \Delta T) = k_B(T - \Delta T)\ln(t_2/\tau_0),$$
$$U(T) = k_B T \ln(t_2^{\mathrm{eff}}/\tau_0), \tag{2.11}$$
$$U(T) = U(T - \Delta T).$$

From Eq. (2.11), we obtain

$$\ln(t_2^{\mathrm{eff}}/t_2) = -\Delta T/T \ln(t_2/\tau_0), \tag{2.12}$$

which is a straight line of slope $-\ln(t_2/\tau_0)$ in the log-log plot of t_2^{eff}/t_2 versus $\Delta T/T$ in Fig. 2.34 (for $\tau_0 = 10^{-12}\,$s, solid line in the figure). For samples #2 − 3 − 4 − 5, the slowing down is *stronger* than for usual thermal activation, a behaviour that was already observed in early experiments [70], and has been interpreted as the signature of a "super-activated" behaviour: the free-energy barriers U increase as the temperature decreases, i.e. $U(T - \Delta T) > U(T)$. These results cannot be ascribed to a decrease of τ_0, which would then take unphysical small values (for the thisopinel sample #4, one would have $\tau_0 = 3.10^{-27}$ s at $0.7 T_g$, and even $\tau_0 = 6.10^{-48}$ s at $0.85 T_g$). On the other hand, an increase of $U(T)$ for decreasing T is indeed what is expected from the hierarchical picture [44] sketched in Fig. 2.23; as the temperature is lowered, free-energy barriers grow up, subdividing the valleys into new sub-valleys. Early temperature-cycling experiments on Heisenberg-like spin-glasses [70] were already analyzed in terms of a barrier growth towards low temperatures, but the conclusions were somewhat different, since the rapid barrier growth was interpreted as an indication of divergences at all temperatures below T_g.

The behaviour of the Ising sample is rather surprising; the thermal slowing down is less steep than expected from usual thermal activation, and corresponds to an inverse temperature dependence of effective barriers, of the type $U(T - \Delta T) < U(T)$, which seems quite unlikely. A way to understand this result is to consider that the hypothesis of a paramagnetic attempt time $\tau_0 = 10^{-12}$ s is not valid in this case. The weak slope of the Ising results in Fig. 2.34 means a smaller value of $\ln(t_2/\tau_0)$, implying a longer value for $\tau_0, \tau_0 \sim 2.10^{-7}$ s. This renormalization of the microscopic attempt time can be due to critical fluctuations of the type encountered in the vicinity of T_g, which would have a stronger influence in the Ising case. Following this idea, we propose a common quantitative analysis of the 5 samples in the next section.

At this stage, an important remark should still be done. If ageing corresponds to establishing correlations up to a typical length $L(t_w, T)$, our results in Fig. 2.34 have clear-cut consequences on the possible time and temperature dependence of L, since in these experiments the same stage of ageing (and hence the same L) can be obtained in 2 different temperature histories. If the L dependence is of a simple power law type $L \sim (t_w/\tau_0)^{aT/Tg}$, as suggested earlier from the first (Heisenberg-like) experiments [66] and from the Ising simulations [13], then we should have

$$(t_2^{\mathrm{eff}}/\tau_0)^{aT/Tg} = (t_2^{\mathrm{eff}}/\tau_0)^{a(T-\Delta T)/Tg} , \tag{2.13}$$

which is identical to $U(T) = U(T - \Delta T)$ in Eq. (2.11). In other words, a power law behaviour of L would entail that, in a graph like Fig. 2.34, all results from all samples lie on straight lines of slopes determined by the value of τ_0. For the Ising sample this is not excluded, but for the Heisenberg-like spin-glasses τ_0 would then reach unphysical small values, smaller and smaller when approaching T_g. The conclusion from our temperature cycle experiments is that $L(t_w, T) \sim (t_w/\tau_0)^{aT/Tg}$ cannot account for all results, and that one has to go beyond a power law behaviour for L, as already concluded from the field variation experiments described above (Fig. 2.31).

Recent numerical simulations of Ising and XY spin-glasses [71], using a new method for determining L, obtain results which are compatible with a power law behaviour of L for both classes. However, in another set of simulations [16] following [15], Berthier and Young compare Heisenberg Ising spin-glasses using the same procedure as in our temperature cycle experiments, that is, comparing the $t_{\mathrm{eff}}(\Delta T)$ behaviours in both cases. The comparison with the experiments is rather puzzling. In [16], the $t_{\text{eff}}(\Delta T)$ line for the Ising case lies slightly above the line corresponding to simple thermal activation with constant barriers, as is the case in the experiments. But, at variance with the experiments, the numerical results for the Heisenberg case lie above the Ising ones. The authors [16] emphasize that this may be related to the difference in time scales. Therefore, they have explored the influence of increasing the time t_2 spent at $T - \Delta T$, and they do find that $t_{\text{eff}}(\Delta T)$ becomes steeper for increasing t_2, an effect which is much stronger in the Heisenberg than in the Ising case. The time scales explored experimentally and numerically remain very far from each other, but it is not completely excluded that, in the distant limit of experimental times, the numerical $t_{\text{eff}}(\Delta T)$ line for Heisenberg becomes lower than for the Ising case, in the same way as in the experiments.

2.4.3 The Dynamical Correlation Length from Both Temperature and Field Variation Experiments

In temperature variation experiments, a super-activated behaviour is observed for Heisenberg-like spin-glasses, and the Ising results point towards a renormalization of the microscopic attempt time τ_0 to a longer time scale τ_0'. We propose to express τ_0' as a fluctuation time related to the correlation length with a usual dynamic scaling relation

$$\tau_0' = \tau_0 L^z , \qquad (2.14)$$

z being the dynamic critical exponent which is measured above T_g, assuming that dynamic critical scaling may hold below T_g in the same way as above T_g. To express the dependence on L of the barrier Δ which must be crossed for flipping an ensemble of spins of size L, we follow the idea developed in the context of the droplet model for spin-glasses [56] and write

$$\Delta(L, T) = \Upsilon(T) L^\psi \qquad (2.15)$$

(L being dimensionless, in units of lattice spacing), where the "stiffness" energy $\Upsilon(T)$ of the barrier

$$\Upsilon(T) = \Upsilon_0 (1 - T/T_g)^{\psi \nu} \qquad (2.16)$$

vanishes at T_g with the same critical exponent ν which governs the divergence of the equilibrium correlation length $\xi \sim (T/T_g - 1)^{-\nu}$ above T_g. In other words, we assume that (like in pinned ferromagnets) ξ behaves in the same way below and above T_g, and that the barrier related to objects of size L is

$$\Delta(L, T) = \Upsilon_0 [L/\xi(T)]^\psi . \qquad (2.17)$$

Thermal activation over the barrier time $\Delta(L, T)$ yields the time t needed for a rearrangement of spins at scale L as $t = \tau_0' \exp[\Delta(L, T)]$, which reads explicitly [53]

$$t = \tau_0 L^z \exp\{\Upsilon_0(1 - T/T_g)^{\psi\nu} L^\psi\} \, . \tag{2.18}$$

This is a crossover expression between a purely critical regime $t = \tau_0 L^z$, obtained in the limit $L \ll \xi(T)$, and a superactivated regime in which the barriers grow as $(1 - T/T_g)^{\psi\nu}$ when the temperature is decreased. It is clearly different from the power law $L \sim (t/\tau_0)^{aT/T_g}$ that was considered earlier, however Eq. (2.18) can also be written $t \sim L^{z_{\text{eff}}(T)}$ by defining

$$z_{\text{eff}}(T) = d \log t/d \log L = z + (\psi\Upsilon(T)L^\psi)/k_B T \, , \tag{2.19}$$

where $z_{\text{eff}}(T)$ is now an effective temperature (and length) dependent exponent, which is equal to the dynamic exponent z at T_g [53].

We have fitted the $t(L, T)$ expression (Eq. (2.18)) to both our *field* and *temperature* variation experiments, using the data of the 3 samples for which both kinds of measurements have been performed: Ising sample (#1), thiospinel (#3) and Ag:Mn$_{2.7\%}$ (#5) [67]. We have fixed $z\nu$ from published dynamic critical scaling data. We also fixed, to improve the global fit of all data, a geometrical factor $\alpha = 2$ in the relation $N = \alpha L^3$ between the length L and the number of spins N. Apart from α, which is the same for all samples, there are only 2 free parameters per sample in the adjustment of the whole set of data: Υ_0 and ψ. A unique set of parameters is able to account for all the properties of each of the 3 samples (see Table 2.1). The fits are presented in Figs. 2.35 and 2.36 for both sets of results.

Figure 2.35 shows the fit to the $N_s(t_w, T)$ results obtained from the relaxation experiments with various field amplitudes (Subsect. 2.4.1). In this representation, the simple power law behaviour of $L(\sim N^{1/3})$ was represented by a straight line, but for the more complex crossover behaviour of Eq. (2.18), the time/temperature reduced variable $(T/T_g) \ln(t_w/\tau_0)$ in the abscissa is no more relevant. Therefore, we have presented the results of the fit as curve segments, each segment representing, for one sample, the variation of N_s as a function of t_w at fixed temperature.

Figure 2.36 shows the fits to the $t_{\text{eff}}(\Delta T)$ results from temperature cycle experiments (performed around $T = 0.7$ and $0.85 \, T_g$).

In each Figs. 2.35 and 2.36 taken separately the quality of the fits is not excellent for all points, but it is important to remember that the results in both figures are fitted to Eq. (2.18) with a unique set of 2 parameters, and that the number of data points per sample is $\sim$15 (see Table 2.1).

Actually, the parameters are not defined with a great quantitative accuracy, since their effects on the fit are strongly correlated. However, a consistent qualitative picture emerges. The main tendency is an *increasing* value of the barrier stiffness parameter Υ_0 and a *decreasing* barrier exponent ψ for *increasing* values of the anisotropy. This behaviour of the exponent is similar to that found in the analysis of previous *ac* temperature cycling experiments [72]. It contrasts with that derived from the time/frequency scaling of χ'' relaxations proposed in [73], which however is based on a less constrained analysis.

Table 2.1. Free parameters used in Eq. (2.18) to fit the temperature cycle and field variation experiments [67]. The "data points" column indicates the total number of data points that are fitted for each sample

	Υ_0	ψ	Data Points
$Fe_{0.5}Mn_{0.5}TiO_3$ (#1)	14.5	0.03	16
$CdCr_{1.7}In_{0.3}S_4$ (#4)	1.2	1.1	17
$Ag:Mn_{2.7\%}$ (#5)	0.7	1.55	13

The extracted coherence length is noticeably smaller in the Ising sample (large Υ_0) but grows faster with time (small ψ). At present, it is not clear how the strong single spin anisotropy in the Ising sample gives rise to both a high value of energy barriers and a very small value of the barrier exponent. Within a droplet description, $\psi \cong 0$ would imply that the droplet energy exponent θ is also zero, in agreement with recent numerical works on excitations in Ising spin glasses [74]. The case $\psi = 0$ also corresponds to barriers growing as the logarithm of the domain size. This behaviour has been argued by Rieger to hold in many disordered systems [75], including spin-glasses. In this case, the "effective exponent" z_{eff} defined above becomes a true, temperature dependent, dynamical exponent.

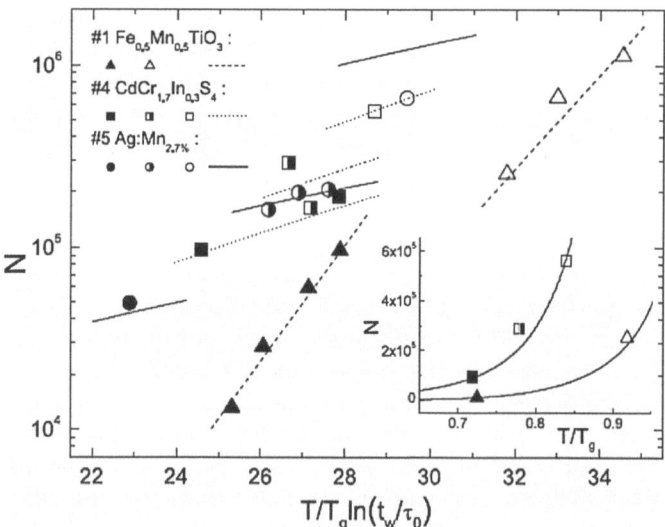

Fig. 2.35. Number of correlated spins from field change experiments (same data as in Fig. 2.31), with the results of the *common* fit to both *field change* and *temperature variation* experiments [67]. Each curve segment is obtained at fixed temperature as a function of t_w. The inset shows the number of correlated spins N as a function of temperature after $t_w = 1000$ s for samples #1 and #4, emphasizing their different behaviours

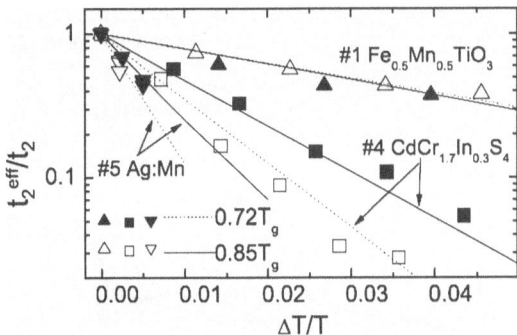

Fig. 2.36. Effective waiting times deduced from the temperature cycle experiments (like in Fig. 2.34), for 3 samples and 2 temperatures, together with the results (lines) of the common fit to both field change and temperature variation experiments [67]

Beyond the detailed values of the fitting parameters, the overall difference between an Heisenberg-like (thiospinel, #4) and the Ising samples is emphasized in the inset of Fig. 2.35, which displays, at fixed $t_w = 1000\,\mathrm{s}$, the temperature dependence of N for samples #1 and 4.

N is larger for the Heisenberg-like sample, which leaves more space for building independent embedded active length scales at different temperatures, and also the temperature variation of N is faster in the Heisenberg case, which signs up a faster separation of the active length scales with temperature (stronger temperature microscope effect). This should correspond to an increased sharpness of the memory dips in an experiment like that of Fig. 2.22. This is indeed what has been observed in the very first experiments of comparison between an Ising and an Heisenberg spin-glass [72]. Figure 2.37 shows the results of this ac memory dip experiment on the $Fe_{0.5}Mn_{0.5}TiO_3$ Ising spin-glass, in which it was already visible (in comparison with Fig. 2.22) that the memory effects are more spread out in temperature, in a way that we now understand in terms of a weaker T-microscope effect in the Ising case.

2.4.4 Separation of Time and Length Scales with Temperature: How Much?

In a spin-glass, as the temperature is decreased, some ageing processes become frozen (memory), while new ones are activated (rejuvenation). By "frozen", we mean that the time scale of a given relaxation process has become extremely large with respect to the experimental time window. In this sense, it is clear that there is a "separation of time scales" as a function of temperature in the spin-glass, but it is interesting to see more precisely how far this time separation maps onto a "separation of length scales", as discussed by Berthier and Young in [16], of which we extract a characteristic figure as our last Fig. 2.38.

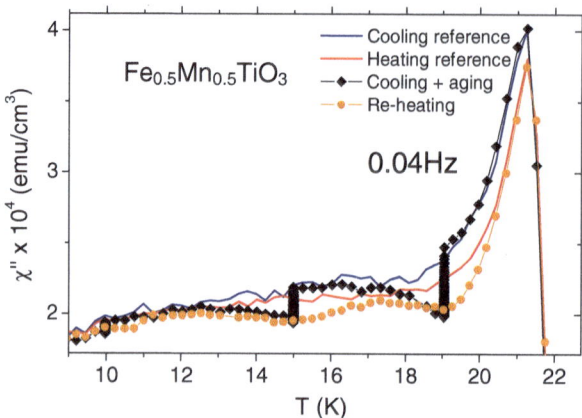

Fig. 2.37. Multiple rejuvenation and memory experiment with the $Fe_{0.5}Mn_{0.5}TiO_3$ Ising sample, from [72] (same as Fig. 2.22, which is for the thiospinel Heisenberg-like sample). The *solid lines* show a reference behaviour for continuous cooling and reheating at 0.001 K/s (the reheating curve is slightly lower than the cooling curve). *Diamonds*: cooling with stops at 19, 15, and 10 K, during the stops χ'' relaxes due to ageing, and when cooling resumes χ'' merges with the reference curve (rejuvenation). *Circles*: when reheating after cooling with stops for ageing, the memory of ageing is retrieved. The memory dips are not so sharply peaked in temperature as in the thiospinel (Heisenberg-like) sample (Fig. 2.22)

In Fig. 2.38, the authors have plotted the time variation of $L(T,t)$ using our parameterization (Eq. (2.18)) for an Heisenberg-like spin-glass. This figure gives a precise idea of the length scales which are play in typical (Heisenberg) experiments and numerical simulations. In an experiment with ageing during 10000 s at $T_1 = 0.825\,T_g$, the active length scale grows up to ~25 lattice units. The time separation with temperature is brutal, since 3.10^{21} years would be needed to obtain $L = 25$ at $T_2 = 0.7\,T_g$. However, the active length that is reached after 10000 s at T_2 (starting from zero) is not that different, of the order of 15: the "separation of *length* scales" from T_1 to T_2 takes place between 25 and 15, which is not spectacular, but enough to produce rejuvenation and memory effects, thanks to the fast separation of *time* scales.

Of course it is also very interesting to compare with the length scales that are reached at the time scale of the simulations, ~10^5 Monte Carlo steps. They are $L(T_1) \sim 6$ and $L(T_2) \sim 4.5$. This is not a powerful microscope in this case. Yet, due to the fast separation of the corresponding time scales rejuvenation and memory effects exist at the time scale of the simulations, and are now seen in the Heisenberg spin-glass at $d = 3$. They have not been found in the simulations of the Ising spin-glass, probably because of a still weaker temperature microscope effect.

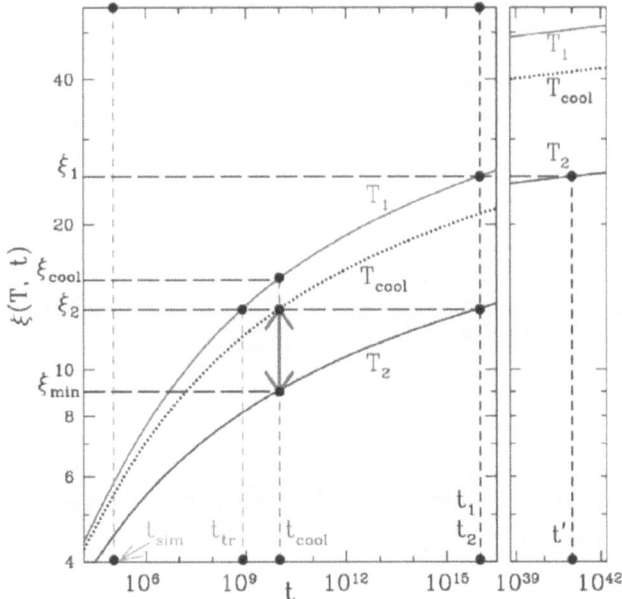

Fig. 2.38. From [16], growth of the dynamical correlation length as a function of time (in units of the elementary time $\tau_0, \tau_0 \sim 10^{-12}$ s for the experiments and $\tau_0 = 1$ for the simulations), as obtained from the parameterization of our experimental results [67] in a Heisenberg-like case. See text (Subsect. 2.4.4) and [16] for details

2.5 Conclusions

In this rather general paper extracted from a summer school course, we have tried to review the most important features of the slow, out-of-equilibrium, dynamics of spin-glasses. Perhaps the reader will have been convinced that, as stated in [53]:

"*Although spin-glasses are totally useless pieces of material, they constitute an exceptionally convenient laboratory frame for theoretical and experimental investigations. ... There are at least two reasons for this: (a) the theoretical models are conceptually simpler (although still highly nontrivial), and (b) the use of very sensitive magnetic detectors allows one to probe in detail the ac and dc spin dynamics of these systems down to very small external fields. The corresponding mechanical measurements in other glassy systems are much more difficult to control, although some recent progress has been made*".

The waiting time dependence of the dynamical response (ageing effect) is indeed a widely spread phenomenon observed in very different physical systems like polymer and structural glasses [20, 36, 42, 43, 64], disordered dielectrics [76, 77], colloids and gels [32–35, 65], foams, friction contacts [78], etc... Scaling laws of ageing have been established in the rheology of glassy polymers [20], which precisely apply to the case of spin-glasses [19]. In common with many different physical situations is also the subageing phenomenon, slight but systematic departure from pure t/t_w scaling, of which we do not still know whether it is intrinsic or related to experimental artefacts (finite size effects [30,31], too slow cooling rates compared with microscopic times... [24]).

The response measurements can now be completed by direct measurements of the spontaneous fluctuations. The experiments of Ocio and Hérisson [29] could, for the first time, reveal the crossover to a modified fluctuation-dissipation relation when entering the strongly ageing time regime of a spin-glass. Further such experiments in spin-glasses are needed, and an experimental way of normalizing the autocorrelation function has still to be found. In polymers and colloids, very interesting fluctuation dissipation studies could be performed these last years, which raise many new questions, among which the nature of the relationship between mechanical and dielectric properties of disordered systems [79].

The rejuvenation and memory experiments in spin-glasses show that the effect on ageing of the temperature history is highly non-trivial. The hierarchical structure of the numerous metastable states, proposed in the past [19,44], remains an efficient guideline to account for all details of the experiments, as discussed in various developments of Random Energy Models [47,50]. This "phase space" hierarchy can now be transcribed into a "real space" hierarchy of embedded length scales [53]. The basic ingredient is a strong separation of the time scales that govern the dynamics of the system on different length scales. Changing the temperature changes the length scale at which the system is observed, thereby allowing the coexistence of rejuvenation (that concerns short length scales) and memory (stored in long length scales). The relevance of "temperature-chaos" [56,61] for the occurrence of rejuvenation is still under debate [55]. In principle, rejuvenation may simply stem from the thermal variation of the equilibrium population rates of the metastable states, in the absence of any chaos effect [53], and in numerical simulations rejuvenation can indeed be observed without chaos [26]. However, it may well be that the experiments be influenced by chaos effects occurring at much larger length scales than can be directly explored [55].

A scenario of embedded active length scales is certainly at play in disordered ferromagnets, in which slow dynamics corresponds to hierarchical reconformations of elastic walls in a random pinning disorder [6,60]. The possible extension of this wall reconformation scenario to spin-glasses raises some puzzling questions such as the nature of domains and walls in a spin-glass.

The ageing length scales can be captured in experiments which determine the dynamical correlation length that is growing during ageing [67]. Several different sets of experiments can now be understood in terms of a unique form for the time and temperature dependence of the correlation length, which is a crossover between a critical regime and a super-activated regime, with energy barriers vanishing at T_g [53]. From the study of five representative spin-glass examples, we have found a clear trend to a stronger separation of active length scales with temperature when going from the Ising to the

Heisenberg case (corresponding to sharper memory effects) [67]. The origin of this systematic dependence on spin anisotropy remains mysterious. Having again in mind the comparison with ferromagnets, a clue may be that less anisotropy should make broader walls, hence providing the spin-glass with larger dynamical regions [80]. The comparison of Ising and Heisenberg spin-glasses is intensively investigated in numerical simulations, which are now able to attack the time-consuming computation of Heisenberg spin dynamics. But the gap between numerical and experimental time scales remains immense [15, 16, 71].

The concept of a slowly growing and strongly temperature dependent dynamical correlation length allows understanding on the same basis the rejuvenation and memory effects and the cooling rate effects. It is now likely that this scenario of ageing as a combination of "temperature specific" (rejuvenation and memory) and "temperature cumulative" processes [21], characterized in spin-glasses, is also relevant for polymer and structural glasses, which were previously thought as dominated by cooling rate effects. Memory effects have now been observed in some polymers and gels [64, 65]. It will be very interesting to see how far future experiments on various types of glasses may confirm the validity of a unique scenario for disordered systems which are made of such different building blocks.

References

1. See numerous references in "spin-glasses and random fields", A.P. Young Editor, Series on Directions in Condensed Matter Physics Vol. 12, World Scientific (1998).
2. M. Mezard, G. Parisi and M.A. Virasoro, "Spin-Glass Theory and Beyond", World scient. (Singapore) 1987.
3. J.A. Mydosh, "spin-glasses, an experimental introduction", Taylor & Francis, London (1993).
4. V.S. Dotsenko, M.V. Feigel'man and L.B. Ioffe, Sov. Sci. Rev. A. Phys. Vol. 15 (1990).
5. M. Alba, J. Hammann, M. Noguès, J. Phys. C **15**, 5441 (1982); J.L. Dormann, A. Saifi, V. Cagan, M. Noguès, Phys. Stat. Sol. (b) **131**, 573 (1985).
6. V. Dupuis, E. Vincent, M. Alba, J. Hammann, Eur. Phys. J. B **29**, 19 (2002).
7. J.L. Dormann, D. Fiorani, E. Tronc, Advances in Chemical Physics, Vol. XCVIII, I. Prigogine and A. Rice Eds, J. Wiley & Sons, Inc. (1997).
8. P. Jonsson, Adv. Chem. Phys. **128**, 191 (2004), and references therein.
9. H. Mamiya, I. Nakatani, and T. Furubayashi, Phys. Rev. Lett. **82**, 4332 (1999).
10. D. Parker, F. Ladieu, E . Vincent, G. Mériguet, E. Dubois, V. Dupuis, R. Perzynski, J. Appl. Phys. **97**, 10A502 (2005).
11. P. Nordblad, L. Lundgren, L. Sandlund, J. Magn. Magn. Mat. **92**, 228 (1990).
12. H.G. Katzgraber, M. Körner, and A.P. Young, Phys. Rev. B **73**, 224432 (2006)
13. J. Kisker, L. Santen, M. Schreckenberg, and H. Rieger, Phys. Rev. B **53**, 6418 (1996); E. Marinari, G. Parisi, F. Ricci-Tersenghi, and J.J. Ruiz-Lorenzo, J. Phys. A **31**, 2611 (1998); T. Komori, H. Yoshino, and H. Takayama, J. Phys. Soc. Jpn. **68**, 3387 (1999).

14. J. Houdayer, F. Krzakala, and O.C. Martin, Eur. Phys. J. B **18**, 467 (2000); F. Krzakala, and O.C. Martin, Phys. Rev. Lett. **85**, 3013 (2000).
15. L. Berthier and A.P. Young, Phys. Rev. B **69**, 184423 (2004).
16. L. Berthier and A.P. Young, Phys. Rev. B **71**, 314429 (2005).
17. L. Lundgren, P. Svedlindh, P. Nordblad and O. Beckman, *Phys. Rev. Lett.* **51**, 911 (1983).
18. M. Ocio, M. Alba and J. Hammann, *J. Phys. Lett. (France)*, **46,** L1101 (1985); M. Alba, M. Ocio and J. Hammann, *Europhys Lett.* **2**, 45 (1986).
19. E. Vincent, J. Hammann, M. Ocio, J.-P. Bouchaud and L.F. Cugliandolo, in *Complex behavior of glassy systems*, ed. by M. Rubi, Springer-Verlag Lecture Notes in Physics, Berlin, Vol. 492 p. 184, 1997 *(also available as cond-mat/9607224)*.
20. L.C.E. Struik, "Physical ageing in amorphous polymers and other materials", (Elsevier, Houston, 1978).
21. J. Hammann, E. Vincent, V. Dupuis, M. Alba, M. Ocio and J.-P. Bouchaud, J. Phys. Soc. Jpn. **69** Suppl. A, 206 (2000).
22. L. Lundgren, P. Svedlindh, and O. Beckman, Phys. Rev. B **26**, 3990 (1982).
23. J.P . Bouchaud, J. Phys. I (France) **5**, 265 (1995).
24. V. Dupuis, F. Bert, J.P. Bouchaud, J. Hammann, F. Ladieu, D. Parker and E. Vincent, Proceedings of Stat Phys 22 (Bangalore, India, 2004), Pramana J. of Physics **64**, 1109 (2005), *preprint available as cond-mat/0406721*; D. Parker, F. Ladieu, J. Hammann, E. Vincent, Phys. Rev. B **74**, 184432 (2006).
25. G.F. Rodriguez, G.G. Kenning, R. Orbach, Phys. Rev. Lett. **91**, 037203 (2003).
26. L. Berthier and J.-P. Bouchaud, Phys. Rev. B**66**, 054404 (2002).
27. V.S. Zotev, G.F. Rodriguez, G.G. Kenning, R. Orbach, E. Vincent and J. Hammann, Phys. Rev. B **67**, 184422 (2003).
28. E. Vincent, J.P. Bouchaud, D.S. Dean, J. Hammann, Phys. Rev. B **52**, 1050 (1995).
29. D. Hérisson and M. Ocio, Phys. Rev. Lett. **88**, 257202 (2002); D. Hérisson and M. Ocio, Eur. Phys. J. B **40**, 283 (2004).
30. J.-P. Bouchaud, E. Vincent, J. Hammann, J. Phys. I (France) **4**, 139 (1994).
31. Y.G. Joh, R. Orbach, G.G. Wood, J. Hammann, E. Vincent, J. Phys. Soc. Jpn. **69** Suppl. A, 215 (2000).
32. M. Cloitre, R. Borrega, and L. Leibler, Phys. Rev. Lett. **85**, 4819 (2000).
33. A. Knaebel, M. Bellour, J.-P. Munch, V. Viasnoff, F. Lequeux and J.L. Harden, Europhys. Lett. **52**, 73 (2000).
34. L. Cipelletti, S. Manley, R.C. Ball, and D.A. Weitz, Phys. Rev. Lett. **84**, 2275 (2000).
35. D. Bonn, H. Tanaka, G. Wegdam, H. Kellay and J. Meunier, Europhys. Lett. **45**, 52 (1998).
36. R.L. Leheny and S.R. Nagel, Phys. Rev. B **57**, 5154 (1998).
37. M. Ocio, H. Bouchiat, and P. Monod, J. Phys. Lett. **46**, 647 (1985); M. Alba, J. Hammann, M. Ocio, Ph. Refregier, and H. Bouchiat, J. Appl. Phys. **61**, 3683 (1987); Ph. Refregier and M. Ocio, Rev. Phys. Appl. **22**, 367 (1987).
38. L.F. Cugliandolo and J. Kurchan, J. Phys. A **27**, 5749 (1994); L.F. Cugliandolo, J. Kurchan, and L. Peliti, Phys. Rev. E **55**, 3898 (1997).
39. S. Franz, M. Mézard, G. Parisi, and L. Peliti, Phys. Rev. Lett. **81**, 1758 (1998).
40. E. Marinari, G. Parisi, F. Ricci-Tersenghi, J.J. Ruiz-Lorenzo, J. Phys. A **31**, 2611 (1998).

41. M. Alba, in progress.
42. W.K. Waldron, Jr. and G.B. McKenna, M.M. Santore, J. Rheol. **39**, 471 (1995).
43. S.L. Simon and G.B. McKenna, J. Chem. Phys. **107**, 8678 (1997).
44. Ph. Refregier, E. Vincent, J. Hammann and M. Ocio, J. Phys. (France) 48, 1533 (1987); E. Vincent, J.-P. Bouchaud, J. Hammann, and F. Lefloch, Philos. Mag. B **71**, 489 (1995).
45. K. Jonason, E. Vincent, J.P. Bouchaud, and P. Nordblad, Phys. Rev. Lett. 81, 3243 (1998); K. Jonason, P. Nordblad, E. Vincent, J. Hammann, and J.-P. Bouchaud, Eur. Phys. J. B **13**, 99 (2000).
46. F. Lefloch, J. Hammann, M. Ocio and E. Vincent, Europhys. Lett. **18**, 647 (1992).
47. M. Sasaki, V. Dupuis, J.-P. Bouchaud, and E. Vincent, Eur. Phys. J. B **29**, 469 (2002).
48. V. Dupuis, PhD Thesis, Orsay University (France) 2002.
49. J.-P. Bouchaud, V. Dupuis, J. Hammann, and E. Vincent, Phys. Rev. B **65**, 024439 (2001).
50. J.P. Bouchaud and D.S. Dean, J. Phys. I (France) **5**, 265 (1995).
51. L.F. Cugliandolo and J. Kurchan, Phys. Rev. B **60**, 922 (1999).
52. S. Miyashita and E. Vincent, Eur. Phys. J. B **22**, 203 (2001).
53. J.-P. Bouchaud, V. Dupuis, J. Hammann, and E. Vincent, Phys. Rev. B **65**, 024439 (2001).
54. H. Yoshino, A. Lemaître and J.-P. Bouchaud, Eur. Phys. J. B **20**, 367–395 (2001).
55. P.E. Jönsson, R. Mathieu, P. Nordblad, H. Yoshino, H. Aruga Katori, A. Ito, Phys. Rev. B 70, 174402 (2004).
56. D.S. Fisher and D.A. Huse, Phys. Rev. B **38**, 373 (1988); **38**, 386 (1988).
57. G.J.M. Koper and H.J. Hilhorst, J. Phys. (Paris) **49**, 429 (1988).
58. L. Balents, J.-P. Bouchaud, and M. Mézard, J. Phys. I (France) **6**, 1007 (1996).
59. J.-P. Bouchaud, in *Soft and Fragile Matter*, edited by M.E. Cates and M.R. Evans, Institute of Physics Publishing, Bristol (2000), accessible as a preprint as cond-mat/9910387.
60. E. Vincent, V. Dupuis, M. Alba, J. Hammann and J.-P. Bouchaud, Europhys. Lett. **50**, 674 (2000).
61. A.J. Bray and M.A. Moore, Phys. Rev. Lett. **58**, 57 (1987).
62. R. Mathieu, P. Jönsson, D.N.H. Nam, and P. Nordblad, Phys. Rev. B **63**, 092401 (2001).
63. D. Petit, L. Fruchter, and I.A. Campbell, Phys. Rev. Lett. **88**, 207206 (2002).
64. L. Bellon, S. Ciliberto and C. Laroche, Europhys. Lett., **51**, 551 (2000); Eur. Phys. J. B **25**, 223 (2002).
65. Alan Parker and Valéry Normand, preprint cond-mat/0306056.
66. Y.G. Joh, R. Orbach, J.J. Wood, J. Hammann, and E. Vincent, Phys. Rev. Lett. **82**, 438 (1999); Y.G. Joh, R. Orbach, G.G. Wood, J. Hammann and E. Vincent, J. Phys. Soc. Jpn. **69** Suppl. A, 215 (2000).
67. F. Bert, V. Dupuis, E. Vincent, J. Hammann, and J.-P. Bouchaud, Phys. Rev. Lett. **92**, 167203 (2004).
68. H. Aruga Katori and A. Ito, J. Phys. Soc. Jpn. **63**, 3122 (1994).
69. H. Kawamura, Phys. Rev. Lett. **68**, 3785 (1992); H. Kawamura, Phys. Rev. Lett. **80**, 5421 (1998); H. Kawamura and D. Imagawa, Phys. Rev. Lett. **87**, 207203 (2001).

70. J. Hammann, M. Lederman, M. Ocio, R. Orbach, and E. Vincent, Physica (Amsterdam) **185A**, 278 (1992).
71. H.G. Katzgraber and I.A. Campbell, Phys. Rev. B **72**, 014462 (2005).
72. V. Dupuis, E. Vincent, J.-P. Bouchaud, J. Hammann, A. Ito, and H. A. Katori, Phys. Rev. B **64**, 174204 (2001).
73. P.E. Jönsson, H. Yoshino, P. Nordblad, H. Aruga Katori, and A. Ito, Phys. Rev. Lett. **88**, 257204 (2002).
74. See, e.g., M. Palassini, F. Liers, M. Juenger, and A. Young, Phys. Rev. B **68**, 064413 (2003), and references therein.
75. R. Paul, S. Puri, H. Rieger, Phys. Rev. E **71**, 061109 (2005).
76. F. Alberici, J.-P. Bouchaud, L. Cugliandolo, J. Doussineau, and A. Levelut, Phys. Rev. Lett. **81**, 4987 (1998).
77. E.V. Colla, L.K. Chao, and M.B. Weissman, Phys. Rev. B **63**, 134107 (2001); E.V. Colla, L.K. Chao, M.B. Weissman, D.D. Viehland, Phys. Rev. B **72**, 134105 (2005).
78. L. Bureau, T. Baumberger, C. Caroli, Eur. Phys. J. E **8**, 331 (2002).
79. L. Bellon, S. Ciliberto, Physica D. **168–169**, 325 (2002); L. Buisson, L. Bellon and S. Ciliberto, J. Phys.: Condens. Matter **15**, S1163 (2003).
80. P.G. de Gennes, suggestion.

3

About the Nature of the Structural Glass Transition: An Experimental Approach

J. K. Krüger[1,2], P. Alnot[1,3], J. Baller[1,2], R. Bactavatchalou[1,2,3,4],
S. Dorosz[1,3], M. Henkel[1,3], M. Kolle[1,4], S. P. Krüger[1,3], U. Müller[1,2,4],
M. Philipp[1,2,4], W. Possart[1,5], R. Sanctuary[1,2], Ch. Vergnat[1,4]

[1] Laboratoire Européen de Recherche, Universitaire Sarre-Lorraine-(Luxembourg)
   jan.krueger@uni.lu
[2] Université du Luxembourg, Laboratoire de Physique des Matériaux, 162a,
   avenue de la Faïencerie, L-1511 Luxembourg, Luxembourg
[3] Université Henri Poincaré, Nancy 1, Boulevard des Aiguillettes, Nancy, France
[4] Universität des Saarlandes, Experimentalphysik, POB 151150, D-66041
   Saarbrücken, Germany
[5] Universität des Saarlandes, Werkstoffwissenschaften, POB 151150, D-66041
   Saarbrücken, Germany

Abstract. The nature of the glassy state and of the glass transition of structural glasses is still a matter of debate. This debate stems predominantly from the kinetic features of the thermal glass transition. However the glass transition has at least two faces: the kinetic one which becomes apparent in the regime of low relaxation frequencies and a static one observed in static or frequency-clamped linear and non-linear susceptibilities. New results concerning the so-called α-relaxation process show that the historical view of an unavoidable cross-over of this relaxation time with the experimental time scale is probably wrong and support instead the existence of an intrinsic glass transition. In order to prove this, three different experimental strategies have been applied: studying the glass transition at extremely long time scales, the investigation of properties which are not sensitive to the kinetics of the glass transition and studying glass transitions which do not depend at all on a forced external time scale.

3.1 Introduction

Synthetic glassy materials are known for more than 8000 years. Nevertheless, the question about the nature of the glassy state and of the thermal glass transition (TGT) of structural (or canonical) glasses is still open [1–26]. Usually, synthetic glasses belong to the class of structural (canonical) glasses which behave mechanically as solids, but which have an amorphous, that means a liquid-like, structure. In other words, glasses are hybrids which have similarities as well with solids as with liquids. This hybrid nature also becomes obvious in the course of the transformation from the liquid to the

J.K. Krüger et al.: *About the Nature of the Structural Glass Transition: An Experimental Approach*, Lect. Notes Phys. **716**, 61–159 (2007)
DOI 10.1007/3-540-69684-9_3   © Springer-Verlag Berlin Heidelberg 2007

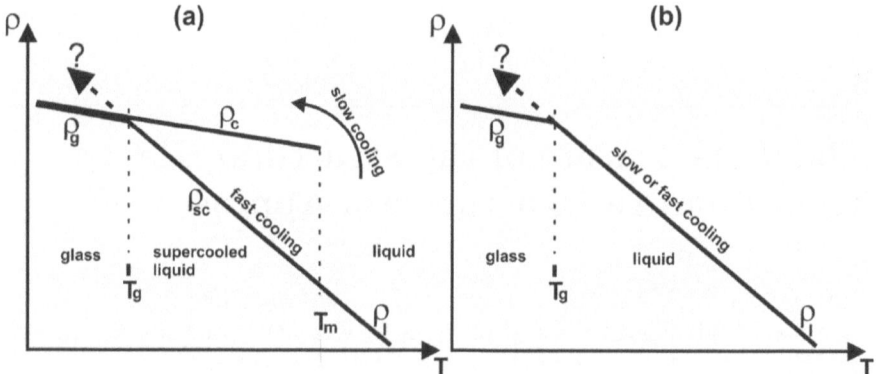

Fig. 3.1. Schematic drawing of the effect of slow and fast cooling on the mass density ρ of (**a**) a crystallisable and (**b**) a non-crystallisable liquid. ρ_l: mass density of the equilibrium liquid, ρ_c: mass density of the crystalline state, ρ_{sc}: mass density of the under-cooled liquid, ρ_g: mass density of the glassy state

glassy state. On cooling the glass-forming liquid, the transition from the liquid to the glassy state is accompanied by experimental features which may be attributed to phase transitions as well as to under-cooling effects (Figs. 3.1a,b, 3.2).

Under-cooling means that the temperature of the liquid sample can be decreased faster than certain of its physical properties do respond. For instance, crystallization of a liquid may be prevented by fast cooling if the viscosity of the liquid increases much faster than nucleation takes place. As a result a super-cooled liquid is obtained. Figure 3.1a shows schematically this situation for the mass density ρ for slow and fast cooling. Below the melting temperature T_m the liquid becomes super-cooled and is therefore out of equilibrium. The degree of metastability of the super-cooled state depends amongst other parameters on the glass-forming liquid itself and on the temperature deviation $\Delta T = (T_m - T)$. As a matter of fact the mass density curve of the super-cooled state is a direct continuation of the equilibrium liquid state. At still lower temperatures around the so-called thermal glass transition temperature T_g the mass density curve $\rho(T)$ shows a kink-like anomaly (Fig. 3.1a). Depending on the sharpness of this kink the volume expansion coefficient α_V shows a step-like anomaly at T_g. Figure 3.1b schematically shows for comparison the temperature behaviour of the mass density of a non-crystallisable liquid (e.g. atactic polymers). The only temperature induced anomaly which remains is the kink-like anomaly of $\rho(T)$ at the glass transition temperature T_g. The question marks in Fig. 3.1a,b stress the open question about the origin of this kink: is it purely due to a super-cooling effect (Fig. 3.1b) or to a further super-cooling effect (Fig. 3.1a) or does there exist an intrinsic event which causes this kink-like anomaly of the mass density? Assuming that the density-kink is purely kinetically conditioned ("kinetic hypothesis")

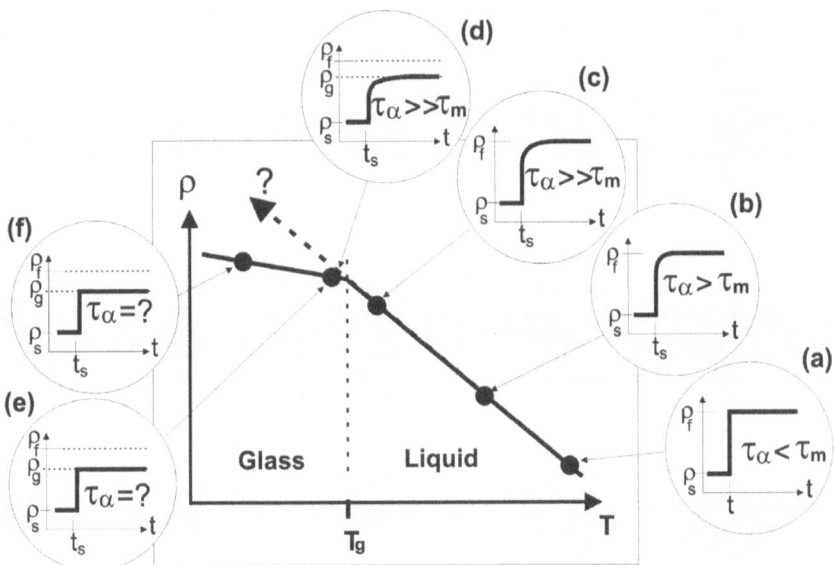

Fig. 3.2. Schematic drawing of the evolution of a structural glass transition. Mass density $\rho = \rho(T)$: the graphics (a) to (f) show the temporal evolution of the mass density ρ at fixed temperature T

the temperature position of this kink should shift to lower temperatures if the cooling rate is slowed down. Time-domain investigations, as schematically drawn in Fig. 3.2a–d, help to elucidate the validity of the "kinetic hypothesis".

For the thought (time-domain) experiment of Fig. 3.2 we assume that the mass density ρ is measured in a static way by determining at every temperature T the mass and the volume of the sample. At high temperatures, far above the freezing temperature T_g, mass-density and volume-changes equilibrate as fast as the temperature of the sample does. After a temperature step $-\Delta T$ the mass density responds almost instantaneously (Fig. 3.2a). As a matter of fact, on approaching the glass transition temperature T_g from above by temperature steps $-\Delta T$, the recovery of the mass density ρ takes more and more time t, i.e. $\rho(t)$ is less and less able to follow a temperature step $-\Delta T$ (Fig. 3.2b,c). This is due to the fact that glass-forming liquids exhibit internal degrees of freedom with relaxation times τ_α, which become very large on approaching T_g from above and which even show a certain tendency to diverge in a temperature interval below T_g (Fig. 3.2e–f). The experimental time needed to equilibrate the new temperature after a temperature jump $-\Delta T$ is assumed to be τ_m. If at sufficiently high temperatures $\tau_m \gg \tau_\alpha$ then the mass density ρ is in internal equilibrium. If the recovery time of the mass density τ_α becomes larger than the experimental time constant of the temperature adjustment τ_m, the mass density $\rho(t)$ shows relaxation behaviour (Fig. 3.2a–d). This relaxation behaviour has been interpreted as the result of cooperative interaction of the molecules close the glass transition (e.g. [2]) and is called the α-relaxation process.

In order to bring the mass density $\rho(T)$ still to equilibrium the experimental time constant τ_m has to be increased sufficiently in order to realize $\tau_m \gg \tau_\alpha$ or at least $\tau_m > \tau_\alpha$. For the understanding of the thermal glass transition the exciting question arises how the process develops in an equilibrium experiment when decreasing the temperature as slowly as possible (Fig. 3.2a–f). If on cooling, at a certain temperature T_g the α-relaxation time τ_α increases so strongly that τ_α becomes comparable to τ_m and even exceeds τ_m in that case the "bending temperature" T_g is indicative for the fact that the mass density $\rho(T)$ falls out of "equilibrium" (Fig. 3.2a–3.2d–f). In the following this will be denoted as time trap [15]. If on the other hand, the mass density curve shows a kink at T_g although the time trap could be avoided, in that case T_g represents an intrinsic transition temperature (Fig. 3.2a–3.2c,e,f). The α-relaxation phenomenon is typical for the glass transition in liquids and will be discussed in Sect. 3.4 of this article as the dynamical aspect of the thermal glass transition.

Thus, as a consequence of the strongly increasing α-relaxation times in the vicinity of T_g, there exists an inherent risk for every experiment performed to measure non-equilibrium properties close to T_g. If the time for temperature equilibration and/or for the equilibration of the measured quantity exceeds the "patience of the experimentalist", non-equilibrium properties are measured. If, as argued in literature (see any text book on glasses, e.g. [2]), the relaxation time τ_α of the α-relaxation process diverges at a temperature $T_0 \cong (T_g - 40)$ K then the cross-over between τ_α and the experimental time scale is unavoidable. If the experiment of interest was driven too fast and as a consequence captivated in the "time-trap" a so-called ageing process is possible on a long time scale. This process brings eventually the physical quantity of interest to its equilibrium value. This specific "kinetic" face of the glass transition and the question under which conditions quenched physical properties will age towards their equilibrium values will be discussed in Sect. 3.3. Whether $\tau_\alpha(T)$ really diverges at a temperature T_0 is rather questionable.

Section 3.5 deals with the problem whether static or quasi-static properties can be measured in principal around T_g, and what they can tell us about the mechanisms behind the TGT. Experimental tools are presented which are at the same time sensitive to large and very small relaxation times and which are therefore sensitive to test the "time-trap" argument. Another experimental approach which has the potential to give some insight into the nature of the thermal glass transition deals with the transition from the dynamically frozen to the solid state. The fact, that this transition is experimentally observable gives a strong hint for the existence of an intrinsic glass or ideal glass transition possibly hidden behind dynamic and kinetic features. Non-linear elastic properties measured at T_g yield a hint to the development of unexpected structural changes at the thermal glass transition. An intrinsic glass transition would be iso-structural and solid in nature and is for fixed external

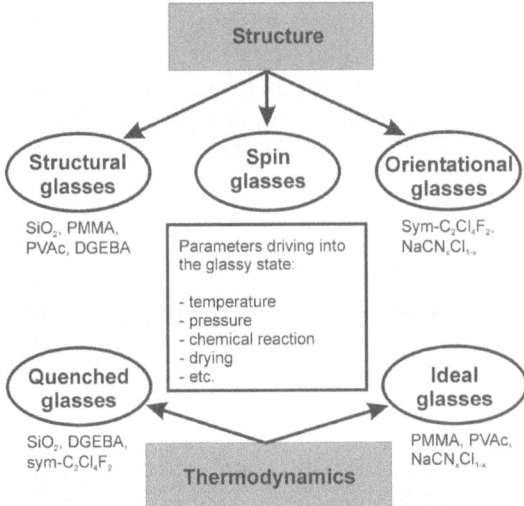

Fig. 3.3. Scheme of the most important glass-forming materials as classified by structure and thermodynamics. SiO_2 silicon dioxide, PMMA polymethylmethacrylate, PVAc polyvinylacetate, DGEBA diglycidilether of bisphenol A, $C_2Cl_4F_2$ difluorotetrachloroethane, $NaCN_xCl_{1-x}$ sodium cyanide chloride mixed crystal

thermodynamic variables a property of the material. In agreement with daily observations, the static shear modulus is expected to be larger than zero.

An alternative approach to the understanding of the nature of the glass transition is to avoid in principle the time trap by choosing a type of glass transition where no external variables are involved in the transition process. In Sect. 3.6 the scenario of the "chemical glass transition" is proposed as a scientific vehicle to get an alternative view on the glass transition.

In recent years, in addition to structural glasses further classes of materials with inherent freezing features have been discovered. Figure 3.3 gives a schematic drawing about the most important classes of glass forming materials currently discussed in physics.

Structural glasses are glasses like polymethylmethacrylate (PMMA) or silicon dioxide (SiO_2) which are completely amorphous but behave mechanically like a solid. Spin glasses are usually crystals where the magnetic spins are randomly oriented but dynamically frozen. Orientational glasses are crystalline materials with frozen orientational disorder. Classifying glasses from the thermodynamic point of view, we distinguish between quenched and ideal glass-formers. Quenched glasses are glassy systems where the disorder of the liquid state is conserved due to fast cooling. The fast cooling prevents the system from crystallization. On the contrary ideal glasses are systems which have no crystalline reference state with lower free energy.

As a matter of fact, the mechanisms which result in spin glass and/or orientational glass transitions are much more evident than those leading to the glassy state of structural glass formers [27] as there is only one well-known internal degree of freedom provoking the glass transition. For some time it was therefore believed that spin glasses and orientational glasses could be used as rather simple but elucidating model systems for the *thermal glass transition* (TGT) in structural glass formers. However, the results met only in part the expectations [28, 29], but they lead to a better insight into the main ingredients provoking the freezing process, which are: (i) disorder, (ii) frustration and (iii) non-linear molecular interactions [27, 30]. Therefore it makes sense to elucidate the open problems of the thermal glass transition in structural glasses on the background of those occurring in much simpler systems like orientational glass-formers.

Hence prior to the discussion of the nature of glass transitions in structural glasses, three introductory educational examples of orientational glass formers together with some shortcomings will be discussed: mixed crystals of the type $Na(CN)_xCl_{1-x}$, single crystals of sym-$C_2Cl_4F_2$ (difluorotetrachloroethane, DFTCE) and poly-siloxane side-chain liquid crystals.

Cyanide mixed crystals are molecular crystals of the type $M(CN)_xZ_{1-x}$ and $M_xX_{1-x}(CN)$ (where M and X stand for alkali metals and Z for halogenide ions). They are obtained by mixing different pure alkali cyanides or pure alkali cyanides with alkali halogenides. The concentration x very strongly influences the structural phase transition behaviour compared with the pure alkali cyanides. In the following we deal with $Z = Cl$ and $M = Na$. With decreasing temperature pure NaCN undergoes a strong first-order ferroelastic phase transition at $T_c \sim 285\,K$ from cubic to rhomboedric symmetry. Usually, the crystal does not mechanically survive this transition and breaks into pieces. Brillouin spectroscopy (see Sect. 3.2), however, is able to measure in the temperature regime below T_c by focussing the scattering volume into one of the intact domains. Figure 3.4 shows a decrease of the shear elastic stiffness on approaching T_c from higher temperatures followed by a huge jump into the low-temperature phase. Obviously c_{44} is the order parameter susceptibility related to this transition. By decreasing the CN-concentration x the transition temperature T_c from the cubic to the rhomboedric phase can be shifted continuously to lower temperatures.

In the case of $M(CN)_xZ_{1-x}$ crystals the average cubic symmetry observed in the high temperature phase remains unchanged for all temperatures if the concentration x is lower than a critical value x_c [28]. In this concentration range ($x < x_c$) there is no more structural phase transition observed in the temperature dependence of different physical parameters, but the system is believed to undergo an orientational glass transition [27,28,31–36] at a definite temperature characterized by the existence of a minimum value of the shear elastic stiffness $c_{44}(T)$.

Figure 3.5 demonstrates this behaviour for the critical concentration $x_c \sim 0.65$. Compared with pure NaCN (Fig. 3.4) the strong softening of $c_{44} = c_{44}(T)$ is maintained but the first-order character of the c_{44}-anomaly is completely lost. The transition temperature largely shifts from $T_c = 287\,K$

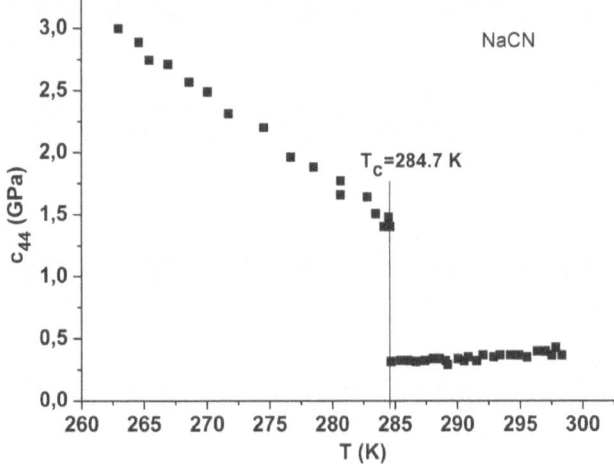

Fig. 3.4. Temperature dependence of the elastic shear stiffness c_{44} of NaCN around the phase transition from the cubic ($T > T_c$) to the rhomboedric phase ($T < T_c$)

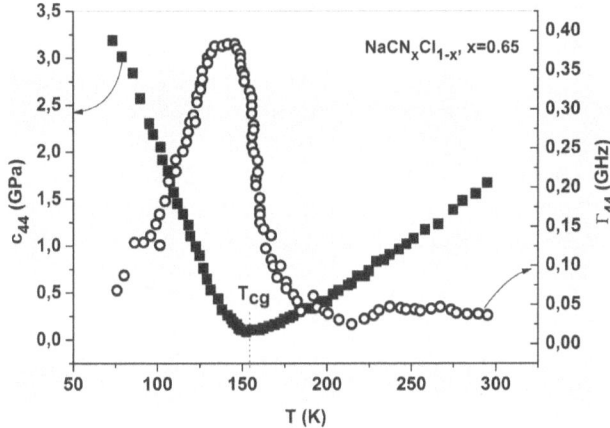

Fig. 3.5. Temperature dependence of c_{44} and the related acoustic attenuation Γ_{44} of the mixed molecular crystal NaCN$_x$Cl$_{(1-x)}$ with $x = x_c \sim 0.65$

to $T_{cg} = 153$ K. The identification of T_{cg} as the glass transition temperature originates from the fact that there is a continuous approach of the first-order phase transition anomaly with increasing Cl-concentration to this minimum of $c_{44} = c_{44}(T)$. In other words this orientational or quadrupolar glass transition emerges from a first order transition due to an increase of positional disorder (CN→Cl) and due to orientational frustration. This interpretation is in clear contradiction to the observations in "quenched structural glasses". In the latter glasses the glass transition temperature T_g is always well below the melting point T_m.

A striking feature of Fig. 3.5 is the observation that the c_{44}-minimum is accompanied by a maximum of hypersonic loss. This behaviour recalls a dynamic glass transition of structural glass formers rather than that of a static glass transition. That objection against the current interpretation [32] is supported by the fact that the temperature where this minimum takes place is frequency dependent, that also indicates its dynamical character.

The nature of this glass transition seems to be spin glass-like [27,30,37,38]. It is well known that pure NaCl does not show any glass-like behaviour. In so far it is interesting to know to which extent the glassy behaviour imposed by the disorder of the CN-dipoles is still active. According to Fig. 3.6 the minimum within the $c_{44} = c_{44}(T)$ curve is still present at the rather low CN-concentration of $x = 0.2$ which indicates that some relaxation processes are still present at hypersonic frequencies below $T = 150\,\mathrm{K}$ [37].

This result raises the question, whether the earlier interpretation of the orientational glass transition in alkali halides was not wrong and whether eventually a quasi-static glass transition does exist at even lower temperatures. If this view of this kind of "spin glass transition" was correct in that case, the similarities to the TGT of structural glass formers would be much more intimate.

The theoretical description of $c_{44}(T,x)$ has been subject of numerous papers [e.g.] [27, 31, 39, 40]. A consistent application of theoretical models to the experimental data has not given a satisfactory description of the behaviour, as a function of temperature and concentration, in the range $x < x_c$. The consistency of the theory requires that the concentration- and temperature-dependent properties of c_{44} should be described with a unique

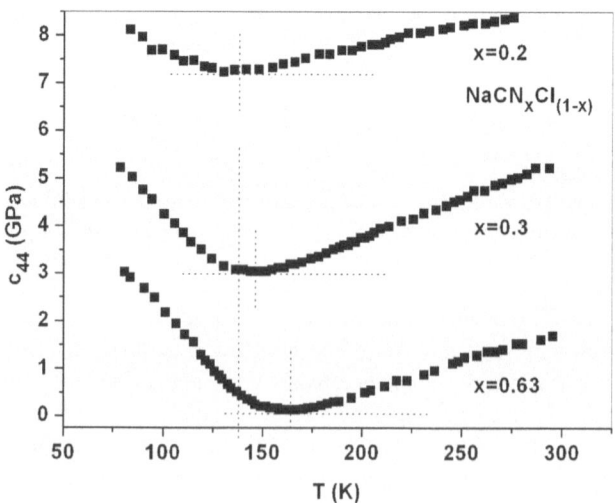

Fig. 3.6. Temperature dependence of the elastic shear stiffness c_{44} of $NaCN_xCl_{1-x}$ as a function of the concentration x of CN-dipoles

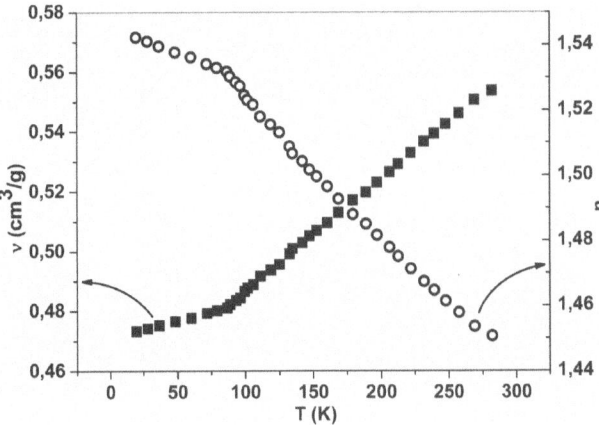

Fig. 3.7. Temperature dependence of the specific volume v and the refractive index n of DFTCE

set of parameters independent of concentration and temperature in both concentration ranges $x > x_c$ and $x < x_c$.

An orientational glass transition (spin glass transition) which is much more similar to the one in structural glass formers was found for symmetric difluorotetrachloroethane ($C_2Cl_4F_2$ = DFTCE) [41,42]. This material is at ambient temperature a plastic crystal with cubic symmetry (bcc) [29], which, even in the polycrystalline state, doesn't show optically any grain boundaries. Moreover the DFTCE molecules rotate and undergo permanently trans-gauche transitions according to Boltzmann statistics. Both dynamics interfere but are compatible with the bcc-symmetry. With decreasing temperature the molecular rotations slow down and freeze below the actual structural phase transition ($T_c = 130$ K) [42]. The trans-gauche transitions are still present and lead to some molecular disorder and frustration. At the glass transition temperature slightly below 90 K (Fig. 3.7) the super-cooled cubic DFTCE transforms from the dynamically disordered cubic state to the statically disordered cubic state. Therefore, the frozen disorder is predominantly an intramolecular disorder.

DFTCE single crystals have been grown on ultrathin mono-crystalline films of polytetrafluoroethylene (PTFE) [43,44]. Figure 3.8 shows the elastic indicatrix of a thin DFTCE crystal plate with a thickness of 50 μm as measured with Brillouin spectroscopy (see next section) at ambient temperature. The crystal orientation is believed to be such that the large faces of the crystal plate correspond to cubic faces.

Figure 3.7 shows the temperature dependence of the specific volume and of the refractive index of this DFTCE sample. Both curves show the typical kink-like behaviour usually observed at the TGT of structural glass formers. The specific volume v was measured by x-ray analysis of the cubic lattice. The refractive index n was calculated using the Lorentz-Lorenz relation [45, 46] (see Eq. (3.2)) and calibrating the specific refractivity by n-measurements on

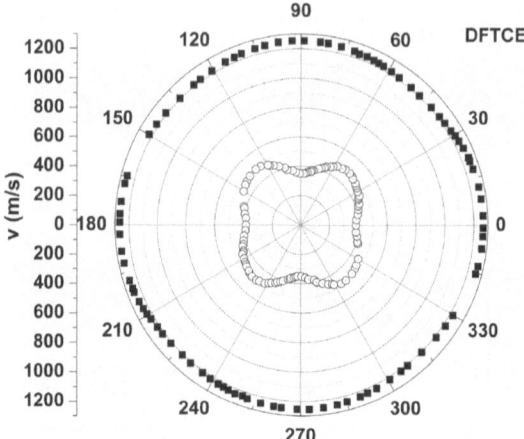

Fig. 3.8. Elastic indicatrix of an arbitrary crystal cut of cubic DFTCE. The *filled squares* represent the quasi-longitudinally polarized acoustic phonon branch and the *open circles* represent the quasi transversely polarized acoustic phonon branch

DFTCE at ambient temperature using an Abbé refractometer. The specific refractivity was taken as temperature independent. It should be stressed that the specific volume measurements based on x-ray analysis are much more reliable than dilatometer and pyknometer measurements as used for polymers. Of course the refractive index data depend on the chosen model of Lorenz-Lorentz. The kink-like behaviour of the specific volume curve around the TGT signifies a step-like behaviour of the thermal expansion coefficient at T_g whereby the specific volume is lower in the frozen than in the dynamically disordered state of the sym-$C_2Cl_4F_2$ molecules. For structural glass formers usually T_g is interpreted as the temperature at which the free volume becomes minimal [11,12]. Below T_g the glass forming material behaves solid-like. Since DFTCE is already a solid above T_g, the latter interpretation of the step-like change of the thermal expansion coefficient cannot be used. In other words, in a dense single crystal of cubic symmetry the classical concept of free volume becomes meaningless. For single crystals the concept of anharmonicity of the elastic interaction potential rather than the concept of free volume has to be taken into account. Applying this argument to the thermal glass transition of DFTCE it has to be concluded that the intermolecular interaction potential changes discontinuously at its TGT (s. a. below).

Using Brillouin spectroscopy (see Sect. 3.2) we have also studied the temperature dependence of the longitudinal hypersound velocity. It turns out (Fig. 3.9) that the hypersonic velocity also behaves kink-like at the glass transition temperature $T_g \sim 87\,\mathrm{K}$. Thus, at hypersonic frequencies the sound velocity is frequency-clamped near T_g and behaves qualitatively like the density and the refractive index. In contrast to the orientational glass transition in alkali halide mixed systems any elastic softening or elastic discontinuity

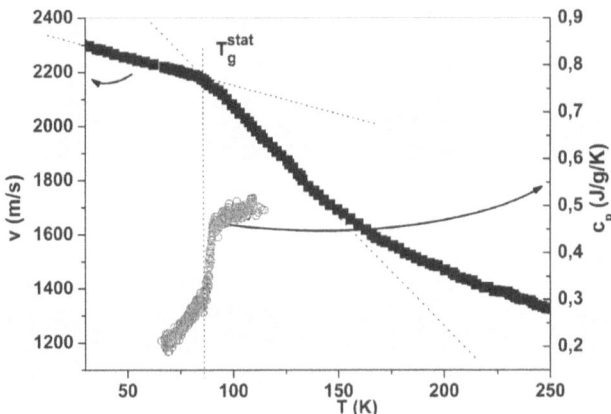

Fig. 3.9. Temperature dependence of the longitudinal sound velocity v and of the specific heat capacity c_p of DFTCE

is absent. Moreover, there is no pronounced acoustic attenuation in the vicinity of the TGT. Therefore the phenomenological properties of DFTCE at its thermal glass transition are much more similar to those of structural glass-formers than to those of orientational glasses as discussed above for $M(CN)_x Z_{1-x}$.

This point of view becomes strengthened by the temperature dependence of the specific heat capacity of DFTCE at T_g (Fig. 3.9). The specific heat capacity grows step-like with T at the TGT as it is characteristic for structural glass-formers. It has to be stressed that the specific heat capacity anomaly does not show the typical "Landau-behaviour" [47] usually observed at phase transitions of second order. Hence we conclude that the phenomenological properties of cubic DFTCE behave at about 87 K exactly in the same way as it is observed for structural glass formers at their TGT.

As already mentioned above, the kink-like anomaly observed for the specific volume and the sound velocity at T_g of DFTCE suggests a sudden change of the elastic interaction potential. For cubic crystals such changes can be probed by the so-called mode-Grüneisen parameters γ (s. a. Sect. 3.6) introduced by Mie and Grüneisen [48–51]. These parameters depict the relative change of the acoustic phonon frequency of a given acoustic phonon mode with the relative change of the density of the cubic crystal.

$$\gamma(\boldsymbol{p}, \boldsymbol{q}) = \frac{\rho}{\omega(\boldsymbol{p}, \boldsymbol{q})} \frac{\partial \omega(\boldsymbol{p}, \boldsymbol{q})}{\partial \rho} \tag{3.1}$$

where $\omega = 2\pi f$ is the phonon frequency, $\boldsymbol{p}$ is the polarisation of the phonon, $\boldsymbol{q}$ is the phonon wave vector and ρ is the mass density. It is worth noting that Brillouin spectroscopy meets exactly the measuring conditions for mode-Grüneisen parameters provided the specific volume of the crystal of interest is known: how do the phonon mode frequencies change if the crystal expands or shrinks? The origin of the expansion or shrinkage is not defined in Eq. (3.1), but one can think of pressure or temperature or any other parameter which can make a crystal expand or shrink. It is obvious that a kink in the sound velocity (or sound frequency) curve does not necessarily imply a discontinuity in the mode-Grüneisen parameters. At least in principle, the kink in the specific volume curve could compensate the effect in the acoustic phonon frequency (sound velocity).

For cubic DFTCE a very comfortable situation exists since precise acoustic phonon frequencies as well as precise specific volume data are available. Consequently, the longitudinal mode-Grüneisen parameter can be determined unambiguously. Figure 3.10 shows for DFTCE the longitudinal acoustic phonon frequency f_L as measured by Brillouin spectroscopy and the longitudinal mode-Grüneisen parameter γ_L. The mode-Grüneisen parameter shows a strong discontinuity at the thermal glass transition. Slightly below T_g the parameter γ_L is smaller than just above T_g. $\gamma_L \sim 4$ is a reasonable value for a frozen plastic crystal [52]. This discontinuity of γ_L at T_g signifies an abrupt change of the elastic interaction potential at the glass transition of cubic DFTCE (see Sect. 3.6). Such a discontinuous change of the elastic interaction potential in a crystal strongly suggests the existence of a phase transition. Taking into account that DFTCE remains cubic in average, this phase transition should be isostructural in nature. It should be stressed again that the kink-like behaviour of the longitudinal acoustic phonon frequency and of the specific volume data has nothing to do with a simple loss of free volume as it is found for liquids and structural glass formers. It is obvious, that the role of free volume will play a crucial role in the interpretation of the TGT of structural glass formers.

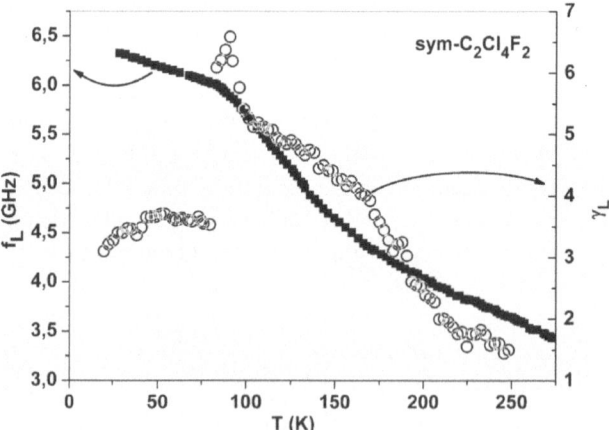

Fig. 3.10. Temperature dependence of the longitudinal sound frequency f_L and the longitudinal mode-Grüneisen parameter γ_L around the thermal glass transition of DFTCE

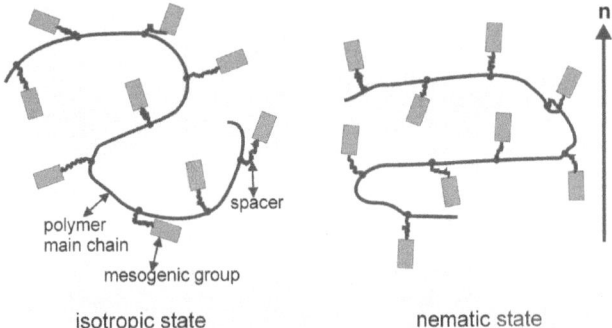

isotropic state   nematic state

Fig. 3.11. Schematic drawing of a polymer side-chain liquid crystal in its isotropic and its nematic state

Whereas the main feature of crystals is their translational order, the main feature of classical liquid crystals like 4-methoxybenzylidene-4'-butylaniline (MBBA) is their orientational order. Similar to crystals, for liquid crystals there exists a hierarchy of order which ranges from the nematic state to the different smectic states [53–56]. Classical liquid crystals are difficult to vitrify. The situation is different for polymer side-chain liquid crystals (PLC) [56–58]. Figure 3.11 shows schematically the isotropic and the nematic structure of PLC's. The polymer backbone is even in the nematic state distributed at random (random coil) provided the spacer molecules, usually n-alkane chains, are sufficiently long. In the so-called isotropic state the mesogenic groups also show random orientational order. In the nematic state an orientational order of the mesogenic groups forms along the director axis **n**. The transition from the isotropic to the nematic state is usually of weak first-order. This holds true e.g. for the refractive index. The material remains "liquid" in the isotropic as well as in the nematic state. Because of the continuously broken orientational symmetry the director can point in any direction of space [59]. Therefore liquid crystals are usually in a polydomain state. For our measurements we have homogeneously oriented the PLC on an ultrathin film of monocrystalline PTFE.

Figure 3.13 shows the sound velocity indicatrix of a poly-siloxane side-chain liquid crystal (Fig. 3.12) measured at ambient temperature which means in the nematic state. It is worth noting, that the nematic state has fibre-symmetry. In contrast to the elastic behaviour of classical nematic liquids,

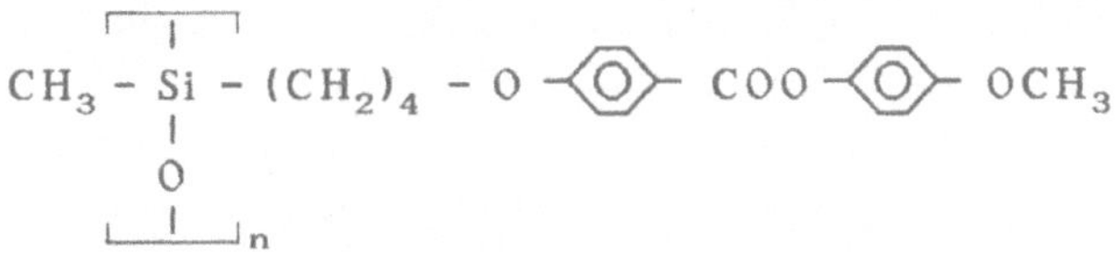

$$CH_3 - Si - (CH_2)_4 - O - \langle O \rangle - COO - \langle O \rangle - OCH_3$$

Fig. 3.12. Structural formula of poly-siloxane

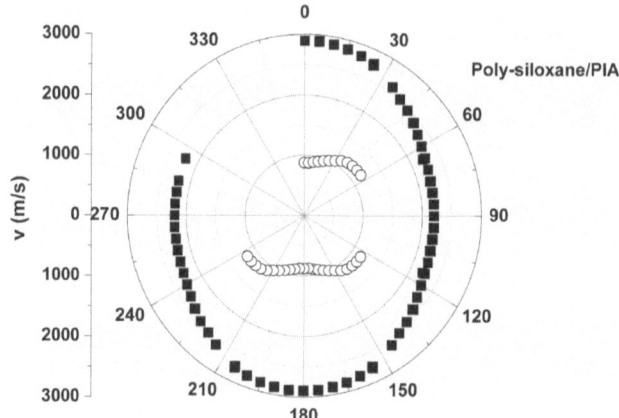

Fig. 3.13. Sound velocity v indicatrix (longitudinal: solid squares, transverse: open circles) of a poly-siloxane polymer side-chain liquid crystal measured within the nematic state. 0 degree indicates the direction of the director **n**

PLC's show a pronounced elastic anisotropy. The longitudiual sound velocity is maximal along the direction of the director **n** which indicates that in PLC's there exists a coupling between the nematic order parameter and the elastic deformation. This coupling is obviously provided by the randomly oriented polymer main chain when a solid-like behaviour, indicated by the existence of an acoustic shear mode, sets in (Fig. 3.13).

Figure 3.14 shows the temperature dependence of the sound velocities of the main acoustic modes propagating in the (3, 1)-plane. The 3-axis was chosen to be directed along the director **n** and the 1-axis is orthogonal to the 3-axis. As expected, within the high temperature phase, i.e. the isotropic phase, there is only the longitudinal acoustic mode. At the isotropic → nematic transition temperature T_{ni} the longitudinal phonon splits up into two modes displaying the change of symmetry. The acoustic shear mode is still not detectable. The reasons for that behaviour are the liquid nature of the material and its low viscosity at T_{ni}. So the shear mode is believed to be overdamped (no acoustic shear mode propagation).

The sound velocity curve v_{3L} shows an inflection point at about 325 K. At this inflection point a hypersonic loss maximum is absent. Therefore the inflection point cannot be interpreted in terms of a dynamical glass transition. In consequence the increase in slope below the inflection point needs an alternative explanation. Between T_{ni} and the inflection point the splitting of the longitudinal modes is probably dominated by the temperature evolution of the nematic order parameter (degree of alignment of the mesogenic side-groups along the director). Since the order parameter has to level with decreasing temperature, the ongoing freezing of the molecular ensemble starts to dominate the acoustic behaviour below the inflection point. The increasing elastic hardening of the polymer matrix emerges in the appearance of the

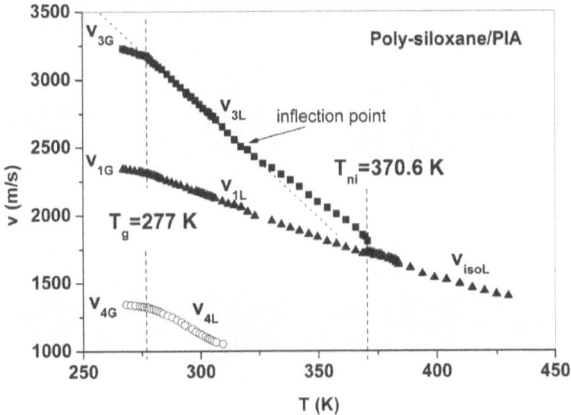

Fig. 3.14. Sound velocities v of the principle sound modes of poly-siloxane side-chain liquid crystals as a function of temperature

intrinsic stiffness of the oriented mesogenic side groups. This interpretation is based on the idea that the mechanical behaviour of the two component system (mesogenic groups and polymer matrix) can be approximated by the Reuss-model [57]. This view is supported by the appearance of the acoustic shear mode v_{4L}.

Thus Fig. 3.14 clearly shows the influence of anisotropy of the isotropic polymer main chain matrix on the hypersonic velocities. The dynamical freezing process influences predominantly the longitudinal acoustic mode propagating along the direction of the order parameter rather than orthogonal to this direction. That means, that similar to the freezing behaviour of the quadrupolar glass $NaCN_xCl_{(1-x)}$ and in contrast to the behaviour of DFTCE there is an intimate coupling between the order parameter of the nematic state and the dynamic mechanical properties of the matrix on approaching the thermal glass transition.

The common kink at $T_g = 277\,K$ in all three acoustic modes indicates the static glass transition. Below $T_g = 277\,K$ the poly-siloxane becomes an orientationally ordered solid. The elastic properties have to be described with an elastic tensor of full fibre symmetry now. Within the glassy state the sound velocities change only moderately with temperature. As in the case of DFTCE and in contrast to the transition behaviour of $NaCN_xCl_{1-x}$ the glass transition in PLC's is well below the structural phase transition. It is worth noting that the hypersonic velocities behave again kink-like at the static glass transition. Since the kinetic view of the glass transition does not know an intrinsic static transition temperature the v_{3L} branch of the liquid phase could be extrapolated to much lower temperatures if infinitely slow cooling is assumed. Because of the strong slope of v_{3L} this would lead to an unreasonably high elastic stiffness c_{33}. Therefore the appearance of the TGT is expected in order to reduce the strong slope of the acoustic mode v_{3L}. This glass transition is again accompanied by a step-like behaviour of the mode-Grüneisen parameters. The argument for that statement will become clear in Sect. 3.6, but bases on the fact that the change in slope for both longitudinal acoustic modes is different, whereas the one for mass density is the same for both modes.

In the following we shall definitively leave the ordered state and concentrate on purely amorphous, i.e. structural, glass formers but we shall use the concepts introduced above. In order to understand the nature of canonical glasses and of the TGT we shall concentrate our interest on structural glasses of the "quenched" and the "ideal type".

Quenched glasses are formed by freezing supercooled liquids. Thus quenched glasses usually exhibit a crystalline reference state and suffer the risk of recrystallization. In this respect this kind of glasses might be bad candidates for investigations of the relaxation behaviour in the long time limit at the TGT as the glassy state is metastable. However, the usefulness of quenched glasses for the investigation of the TGT depends on the rate of recrystallization.

In order to be sure that only glass transition properties are probed at the TGT *ideal glass formers* should be chosen as model substances. Ideal glasses are characterized by the fact that they do not exhibit a crystalline or any other reference state which is more stable than the liquid phase. Therefore these materials transform necessarily from the stable liquid state into the glassy state, which stability is the matter of debate.

Figure 3.15 shows for atactic polymer chains in a schematic way the frustration mechanism preventing these systems from crystallization. Allowing the monomers of type A and B to polymerize in an atactic, i.e. statistically disordered, manner, it is obvious that these macromolecules can hardly form a crystal with translational symmetry.

With increasing molecular chain length due to the chemical reaction, the molecular dynamics will slow down and may even result in a glassy state at

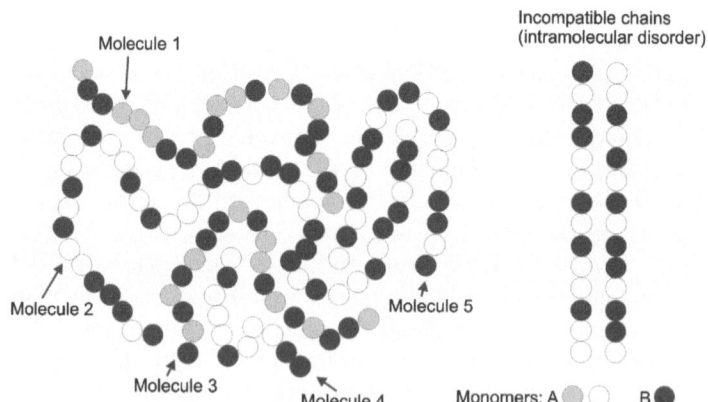

Fig. 3.15. Schematic drawing of an ideal glass-former based on the concept of atactic polymers

a given temperature (chemical glass transition see Sect. 3.6). The same effect can be observed for a molecular system of given chain length by decreasing the temperature of this system. In both cases the molecular translation diffusion, a basic feature of a liquid, will be hindered. Whereas there exists no debate on chemical freezing, thermal freezing is usually interpreted as the competition of the intrinsic time scale (α-relaxation process) with the patience of the experimentalist [21,60]. This means that the experimentalist does not wait long enough after a temperature perturbation for the relaxed equilibrium value of the measured susceptibility to be reached. This cross-over problem, sometimes called "time trap" [49,50], is complicated by the fact that the α-relaxation time does not behave Arrhenius-like but increases stronger than exponential with decreasing temperature. It even seems that the α-relaxation time diverges at a finite temperature T_0, called the *Vogel-Fulcher-Tammann* (VFT) temperature [61,62]. Equation (3.2) displays the phenomenological relation between temperature and the related α-relaxation time τ_α

$$\tau_\alpha = \omega_\alpha^{-1} = \tau_{\alpha 0} \cdot \exp\left[\Delta G/\left(R\left(T - T_o\right)\right)\right] \tag{3.2}$$

ω_α is the relaxation frequency of the α-relaxation process, $\tau_{\alpha 0}^{-1}$ is the so-called attempt frequency. ΔG is the free activation enthalpy and R is the general gas constant.

However, this relation has never been verified experimentally close to the VFT-temperature T_0. Keeping in mind that T_0 indicates the existence of an intrinsic glass transition at which all relevant relaxations are frozen, it is clear that the above-mentioned cross-over of the α-relaxation with any experimental time scale cannot be avoided. This is a strong argument of the kinetic view of the TGT. However this view is based on the divergence of the α-relaxation time which has never been proven experimentally.

The experimentally found thermal glass transition temperature is usually about 30–40 K above T_0. At such a distance from the temperature T_0 where the time constant diverges it should be at least in principle possible to verify to which extent the operative glass transition temperature can be shifted or whether there is a definite low-temperature boundary for this transition [15].

Thus the main questions concerning the understanding of the TGT are whether there exists, at sufficiently slow cooling, an underlying phase transition [15] to an ideal glassy state, or whether the TGT simply signifies the cross-over of the α-process with typical time constraints of the experimental technique (including the patience of the scientist) [15].

If the TGT is a purely kinetic effect, there will be only limited interest in this event. If on the other hand there is an underlying phase transition, in that case the question will arise about the nature of the glassy state of matter. In the latter case it will be necessary to introduce in addition to the three classical states of matter (gaseous, fluid and crystalline solid state) a forth state of matter: the structural glassy state. The importance of the problem makes it necessary not only to present and discuss new results but also to reconsider some older data which are often used as witnesses for the kinetic view of the TGT.

The α-process discussed here should not be confused with the high temperature α-process discussed in mode-coupling theories [3]. Mode-coupling theory predicts the possibility of an ideal glass transition at temperatures typically 30–40 K above the TGT. This glass transition is believed to be prevented due to so-called hopping processes which finally die out at the TGT.

3.2 The Method of Brillouin Spectroscopy

For the present work *Brillouin spectroscopy* (BS) is the central experimental technique. Although this technique is well-established [63–65], its power is not sufficiently known. Moreover, in recent years BS has been not only applied in new fields of physics but it has been developed further in order to give new physical information in addition to the hypersonic properties. Therefore, this method deserves an introduction.

BS is an optical technique which is predominantly used to investigate acoustic properties at hypersonic frequencies. Hence, BS can be applied only to transparent materials, at best to translucent materials. The acoustic wavelengths involved are in the range of 200 nm to several µm. Figure. 3.16 shows a modern Brillouin set-up being able to measure simultaneously at different scattering vectors or in other words in different scattering geometries. For the

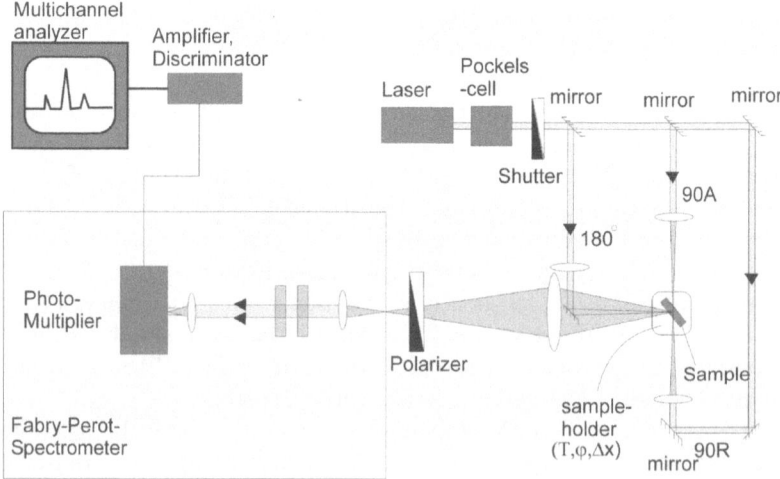

Fig. 3.16. Schematic presentation of a Brillouin spectrometer which can be used to measure within three different scattering geometries (even simultaneously): backscattering (180°), 90R-scattering and 90A-scattering (see Fig. 3.18). The sample holder allows for change of temperature (T), translational movement of the sample (Δx) and rotation of the sample (φ)

sake of simplicity, the sketched Fabry-Pérot interferometer is of the single-pass type. In modern BS multi-pass instruments or even Tandem spectrometers are used [66].

The spectral light intensity distribution measured by the Fabry-Pérot interferometer is converted by a photomultiplier into a suitable electronic pulse sequence which is then accumulated with a photon-counting system, usually an amplifier and discriminator combined with a so-called *multichannel analyzer* (MCA) (see Fig. 3.16). An example for a typical Brillouin spectrum is shown on the screen of the MCA.

Generally the scattering intensity $I(q, \omega)$ per solid angle and frequency interval is proportional to the space and time Fourier transform $\mathbb{F}_{r,t}\{A\}$ of the autocorrelation function A [67,68]

$$A_{klmn} = \langle \delta \alpha_{kl}^*(r, t) \cdot \delta \alpha_{kl}(r', t') \rangle \tag{3.3}$$

of the optical polarisability fluctuations $\delta \alpha_{kl}(r, t)$ in the scattering volume [67]

$$I_{si}(q, \omega) \propto F_{r,t}\{A\} \tag{3.4}$$

The subscripts s and i mean scattered and incident respectively, and refer to the appropriate polarization directions of the light waves here. Omitting, for simplicity, the tensor properties of $\delta \alpha$ we see that the spectral power density $I_{si}(q, \omega)$ is proportional to the mean square fluctuation component at frequency ω:

$$I_{si}(q, \omega) \propto \overline{(\delta \alpha (q))_\omega^2} \tag{3.5}$$

All kind of excitations like phonons, excitons, spin waves and so on may contribute to the spectral power density. Also higher-order processes as multi-phonon interactions may be included formally in this phenomenological treatment by expanding $\delta \alpha(q)$ into a power series in terms of symmetry coordinates [69].

Assuming a certain excitation mode characterized by an extensive parameter $\Psi(q, \omega)$ a conjugated force $F(q, \omega)$ and susceptibility $\chi(q, \omega)$, the linear law reads

$$\psi(q, \omega) = \chi^*(q, \omega) \cdot F(q, \omega) \tag{3.6}$$

The spectral power density $I(q, \omega)$ can be related to the imaginary part of $\chi^*(q, \omega)$ by the fluctuation-dissipation theorem [68]

$$I(q, \omega) \propto (n(\omega) + 1) \, Im(\chi^*(q, \omega)) \tag{3.7}$$

where

$$\frac{1}{n(\omega)} = \exp\left(\frac{\hbar \omega}{kT}\right) - 1 \tag{3.8}$$

The relation may be extended to include several, say s, coupled modes Ψ_j. To obtain the resulting field of generalized forces one has to add the contributions of the s modes

$$F_i = \sum_{j=1}^{s} \gamma_{ij} \cdot \psi_j \tag{3.9}$$

Where the diagonal elements of the matrix γ represent the inverse complex susceptibilities of the uncoupled modes, the off diagonal elements characterize the complex mode-mode coupling strength. Inverting Eq. (3.9) one gets the mode amplitudes

$$\psi_k^* = \sum_{i=1}^{s} \chi_{ki}^* \cdot F_i \tag{3.10}$$

The susceptibilities χ_{ki}^* determine the scattered light spectrum

$$I(\boldsymbol{q}, \omega) \propto (n(\omega) + 1) \, Im \left[\sum_{i,j=1}^{s} p_i p_j \chi_{ij}^* (\boldsymbol{q}, \omega) \right] \tag{3.11}$$

The coefficients p_i are related to the light scattering cross sections of the various modes and can be described for example by appropriate components of the elasto-optic tensor [70].

Considering only two modes and abbreviating the sum in Eq. (3.11) by $\overline{\chi^*}(\boldsymbol{q}, \omega)$ the result is

$$\overline{\chi}^* (\boldsymbol{q}, \omega) = \frac{\gamma_{22}^* p_1^2 - 2\gamma_{12}^* p_1 p_2 + \gamma_{11}^* p_2^2}{\gamma_{11}^* \gamma_{22}^* - (\gamma_{11}^*)^2} \tag{3.12}$$

or

$$\overline{\chi}^* (\boldsymbol{q}, \omega) = \frac{p_1^2}{\gamma_{11}^* - (\gamma_{12}^*)^2 \chi_2^{0*}} + \frac{p_2^2}{\gamma_{22}^* - (\gamma_{12}^*)^2 \chi_1^{0*}} - \frac{2p_1 p_2}{\gamma_{11}^* \gamma_{22}^* - (\gamma_{12}^*)^2} \tag{3.13}$$

Here $\chi_i^{0*} = \gamma_{ii}^{*-1}$ means the susceptibility of the uncoupled mode i and $(\gamma_{12}^*)^2 \chi_i^{0*}$ is related to its self energy. It may happen that the mode Ψ_j does not produce any change of the polarizability, so $p_j = 0$ and only one term of Eq. (3.13) will be left. In order to give an example, the elasto-optic coupling for transversely polarized acoustic phonons is often extremely small.

The specific form of the light scattering spectrum will depend mainly on the characteristics of the inverse susceptibilities γ_{jj}^* of the uncoupled modes. Taking for a damped harmonic oscillator with the damping constant Γ and an oscillator strength γ_{j0}

$$\gamma_{jj}^* = \gamma_{j0} \left(\omega_j^2 - \omega^2 + i \cdot \omega \cdot \Gamma_j \right) \tag{3.14}$$

and for a relaxator with a relaxation time τ

$$\gamma_{jj}^* = \gamma_{j0} \left(1 + i \cdot \omega \cdot \tau_j \right) \tag{3.15}$$

it is easy to predict the form of the scattering spectra. However, experiments do not give sufficient information for an unambiguous analysis. Already in the two-mode case the complex factor γ_{12} cannot be deduced definitely if coupling appears.

From (3.14), (3.15) together with (3.11) the spectral power density of one relaxator and one oscillator being uncoupled can be deduced. The imaginary part of the inverse of (3.14) is:

$$Im\,[\chi*]_{Osc} = \left(\gamma_0^{-1}\right)_{Osc} \frac{\omega \cdot \Gamma_{Osc}}{\left(\omega_{Osc}^2 - \omega^2\right)^2 + \omega^2 \cdot \Gamma_{Osc}^2} \tag{3.16}$$

For the relaxator the imaginary part of the inverse of (3.15) is:

$$Im\,[\chi*]_{Rel} = Im\,\left[\gamma^{*-1}\right]_{Rel} = \left(\gamma_0^{-1}\right)_{Rel} \frac{\omega \cdot \Gamma_{Rel}}{\omega^2 + \Gamma_{Rel}^2} \tag{3.17}$$

with $\Gamma_{Rel} = \frac{1}{\tau_{Rel}}$.

Combining (3.16) and (3.17) with (3.11) and ignoring the coupling yields:

$$I\,(\boldsymbol{q}, \omega) \propto (n\,(\omega) + 1) \left[\left(\gamma_0^{-1}\right)_{Rel} \cdot \frac{p_1 \cdot \omega \cdot \Gamma_{Rel}}{\omega^2 + \Gamma_{Rel}^2} \right.$$

$$\left. + \left(\gamma_0^{-1}\right)_{Osc} \cdot \frac{p_2 \cdot \omega \cdot \Gamma_{Osc}}{\left(\omega_{Osc}^2 - \omega^2\right)^2 + \omega^2 \cdot \Gamma_{Osc}^2} \right]. \tag{3.18}$$

Equation (3.18) gives the shape of a spectrum typical for a liquid of small viscosity consisting of a central peak due to entropy fluctuations and frequency shifted Stokes- and anti-Stokes lines related to density fluctuations resulting in a longitudinally polarized bulk phonon. It is worth noting that Eq. (3.18) gives the physical spectrum of the scattering processes involved, it does not contain the filter properties of the spectrometer. In order to get the information of Eq. (3.18) from a measured spectrum the latter one has to be deconvoluted. Generally, the deconvolution is a difficult task and is done by numerical techniques.

The Kinematic View of Brillouin Spectroscopy

The kinematic view of BS couples energy and momentum of the interacting photons and phonons involved in the scattering process. As usual in inelastic scattering processes energy and momentum conservation holds [63, 64]:

$$\hbar\omega_s = \hbar\omega_i \pm \hbar\Omega \tag{3.19}$$

$$\hbar\boldsymbol{k}_s = \hbar\boldsymbol{k}_i \pm \hbar\boldsymbol{k} \tag{3.20}$$

where ω_i, ω_s and Ω are the frequencies of the incident light, of the scattered light and of the phonon, $\boldsymbol{k}_i$, $\boldsymbol{k}_s$ and $\boldsymbol{q}$ are the respective wave vectors and Θ is the scattering angle within the sample (see Fig. 3.17).

Focussing still on a simple liquid and taking into account, that the energy transfer between photons and phonons is in BS extremely small

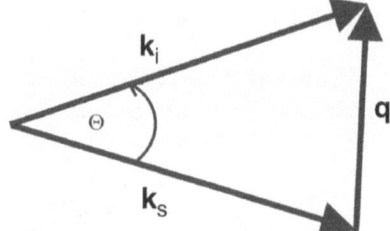

Fig. 3.17. Typical scattering diagram for wave vectors: k_i wave vector of the incident light, k_s wave vector of the scattered light, q scattering vector, Θ scattering angle

$(5 \cdot 10^9 / 5 \cdot 10^{14} = 10^{-5})$ the vectors k_i and k_s have almost the same length. Provided, acoustic attenuation is negligible ($\Gamma_{Osc} \ll \Omega$) the phase sound velocity can simply be calculated from geometric arguments. Using Eq. (3.19) and Eq. (3.20), and considering the refractive index n of the sample one can easily determine the sound velocity v related to a specific mode (longitudinal, transverse, etc.):

$$v = \frac{\Omega}{q} = \frac{\Omega}{2\pi} \frac{\lambda_{\text{Laser}}}{2n \sin\left[\frac{\theta}{2}\right]} \tag{3.21}$$

Knowing the mass density ρ of the sample under investigation, the elastic modulus related to the wave vector q can be calculated:

$$c(q) = \rho \cdot v^2(q) \tag{3.22}$$

Scattering Geometries and Pitfalls

As is seen from Eq. (3.21) the acoustic wavelength

$$\Lambda = \frac{2\pi}{q} = \frac{\lambda_{\text{Laser}}}{2n \sin\left[\frac{\theta}{2}\right]} \tag{3.23}$$

depends on the vacuum wavelength λ of the laser, on the scattering angle θ within the sample and on the refractive index n of the sample. As a consequence of latter influence the acoustic wavelength is e.g. not invariant under changes of temperature which may be a real problem in interpreting Brillouin data.

Fortunately, one of the scattering geometries shown in Fig. 3.18 yields an acoustic wave vector which is independent of the refractive index: the 90A-scattering geometry.

The relations between sound velocity, sound frequency and sound wavelength are as follows for the different scattering geometries [66, 71]:

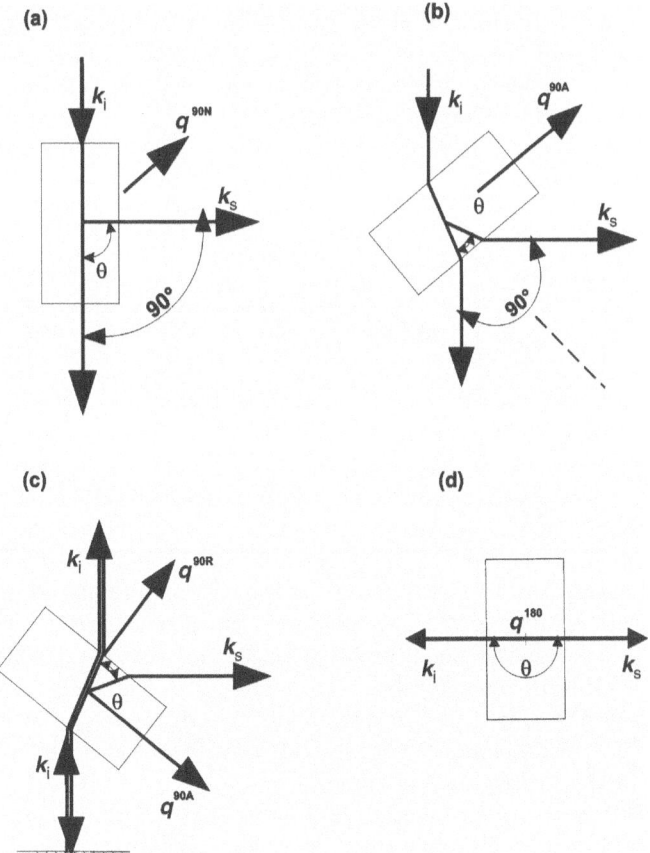

Fig. 3.18. Typical scattering geometries used for Brillouin spectroscopy. (**a**) 90N-scattering geometry, (**b**) 90A-scattering geometry, (**c**) combination of 90A- and 90R-scattering geometry, (**d**) back scattering geometry

$$v^{90N}(T) = \frac{\Omega^{90N}(T)}{2\pi} \frac{\lambda_{\text{Laser}}}{2n(T)\sin\left(\frac{90°}{2}\right)} \tag{3.24}$$

$$v^{90A}(T) = \frac{\Omega^{90A}(T)}{2\pi} \frac{\lambda_{\text{Laser}}}{2\sin\left(\frac{90°}{2}\right)} \tag{3.25}$$

$$v^{90R}(T) = \frac{\Omega^{90R}(T)}{2\pi} \frac{\lambda_{\text{Laser}}}{2n(T)\sin\left(\frac{90°}{2}\right)} \tag{3.26}$$

and

$$v^{180}(T) = \frac{\Omega^{180}(T)}{2\pi} \frac{\lambda_{\text{Laser}}}{2n(T)} \tag{3.27}$$

Besides Eq. (3.25) all other equations depend on the refractive index n of the sample, which means, that the acoustic wavelength depends on the refractive index n. In the case of Eq. (3.25) the refractive index enters the refraction process and the phase velocity in a compensating manner and therefore the sound wavelength becomes independent of n.

Provided sound dispersion is absent, Eq. (3.25) in combination with Eqs. (2.22, 2.24) and (3.27) can be used to calculate the refractive index from Brillouin data

$$n(T) = \frac{1}{\sqrt{2}} \frac{\Omega^{180}(T)}{\Omega^{90A}(T)} \tag{3.28}$$

$$n(T) = \frac{\Omega^{90N}(T)}{\Omega^{90A}(T)} \tag{3.29}$$

If the samples have an isotropic symmetry Eqs. (3.28) and (3.29) can be used to determine the refractive index of the samples.

If for the frequencies of interest acoustic relaxation processes are active, the right-hand sides of Eqs. (3.28) and (3.29) no more represent the refractive index but the so-called opto-acoustic dispersion functions (D-function) [66,72]. The difference $D(T) - n(T)$ gives a measure for dispersion respectively the relaxation strength if present. Further information can be found in Sect. 3.5.

In the case of film- or plate-like samples which have reflecting substrates on one side there exists an alternative scattering geometry, which we call the $RI\theta A$-scattering geometry [72]. It measures the scattered light from the laser beam at the substrate and combines the advantages of the 90A- and the backscattering technique. Figure 3.19 shows schematically the specific properties of this scattering technique.

The $RI\theta A$-scattering geometry [72] is not restricted to an outer scattering angle of 90° but can be applied for any outer scattering angle θ. Thus the $RI\theta A$-scattering geometry results in an a priori backscattering situation with

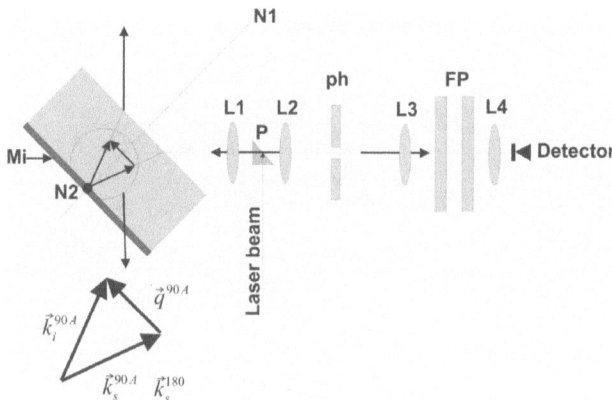

Fig. 3.19. Schematic drawing of the so-called $RI\theta A$-scattering geometry. N1, N2 rotation axes, L1–L4 lenses, P prism, ph pinhole, Mi mirror, FP Fabry-Pérot-Interferometer

the possibility to measure in addition a θA-phonon with a wave vector $q^{\theta A}$, which, because of the optical reflection law, is oriented in the film plane. Rotation around the N2-axis changes the magnitude of $q^{\theta A}$ but maintains its direction. Rotation of the sample around the N1-axis probes the in-plane acoustic symmetry.

3.3 The "Kinetic Face" of the Structural Glass Transition

In this section we will review some generally accepted results from literature on kinetic features of the glass transition and discuss some new decisive experimental data. By "kinetic aspects" we mean the influence of the cooling or heating procedure of the sample on the glass transition. So we will analyse the loss of thermodynamic equilibrium of the sample due to those temperature changes which are necessary to bring the sample closer to the glass transition. As already discussed briefly in the introduction, the main unsolved problem about the glass transition concerns the possibility of the existence of a phase transition possibly hidden by kinetic effects.

It is clear, that glasses are solid in the glassy state from an empirical point of view. From the structural point of view they are amorphous as confirmed by x-ray diffraction. Thus the symmetry of the glassy state is isotropic, which is obviously the same as for the liquid state. Consequently, if the TGT was a structural phase transition this transition would be of the iso-structural type.

In literature on glasses (e.g. [73–83]) the TGT is often discussed in terms of the temperature dependence of the shear viscosity η_{44} (we use the Voigt notation [84]). Figure 3.20a shows schematically the increase of this viscosity around the thermal glass transition temperature T_g. In literature on glasses one usually finds the argument that very close to T_g the shear viscosity takes

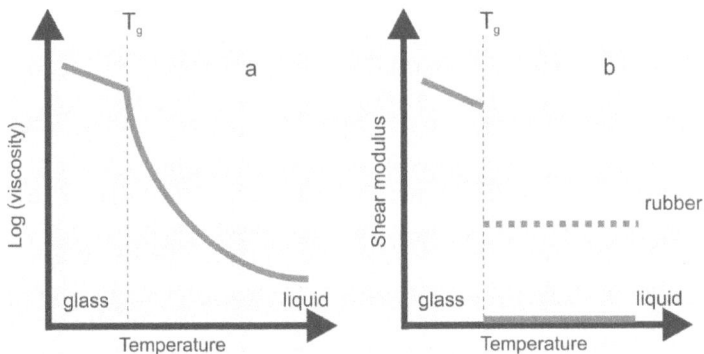

Fig. 3.20. Schematics of the temperature-dependence of (**a**) the visocisity and (**b**) the static shear modulus at the glass transition

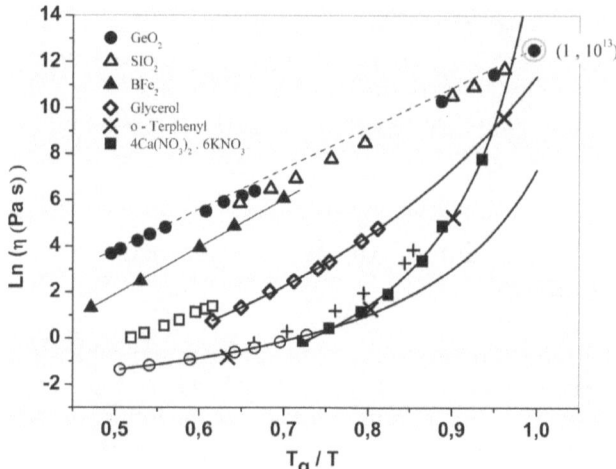

Fig. 3.21. Logarithmic viscosity versus the renormalized temperature T_g/T for several typical glass formers

values up to about 10^{12} Pa s. Sometimes a viscosity of $10^{12.3}$ Pa s [73,78,79,85] is used as a definition for the glass transition.

However, the temperature dependence of η_{44} is not the same for different glass-formers. *Strong glass-formers* (c.f. Fig. 3.21, data from [85]) show an Arrhenius-like temperature behaviour of η_{44} whereas *fragile glass-formers* behave VFT-like (c.f. Eq. (3.32)). In order to test the degree of fragility we have fitted some of the data (straight lines in Fig. 3.21) to the VFT-law using the Maxwell relation:

$$\eta_{44} = c_{44}^\infty \cdot \tau_\alpha \tag{3.30}$$

In Eq. (3.30) c_{44}^∞ is the dynamically clamped shear modulus which is assumed to show only a slight temperature dependence and $\tau_\alpha(T)$ is the relaxation time of the α-relaxation process according to Eq. (3.32). The fitted curves in Fig. 3.21 show, that different glass formers tend to very different viscosities close to T_g and that a limiting value for the shear viscosity of 10^{12} Pas is only a rough estimate. Thus, $\eta_{44}(T)$ seems not to be a good quantity in order to describe the TGT. In addition, the viscosity is a transport coefficient, usually defined for liquids. It seems that a susceptibility, like a mechanical modulus, defined for the liquid and the solid state, is more appropriate.

At least for fragile glass-formers the static shear stiffness seems to be more suitable to describe the transition from the liquid to the glassy state. By definition the static shear stiffness c_{44}^s is zero in the liquid state. Glasses, on the other hand, show a static shear stiffness $c_{44}^s > 0$ (at least on all accessible time scales) indicating an elastic stability comparable to that of crystals. The expected jump-like behaviour of the static shear modulus at T_g is schematically drawn in Fig. 3.20b.

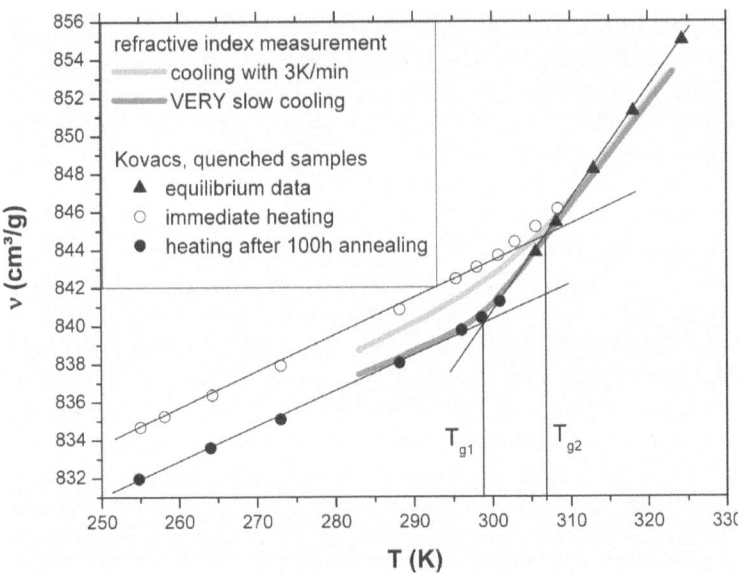

Fig. 3.22. Temperature dependence of the specific volume v of PVAc (dots taken from Kovacs [86]) around the glass transition after different cooling procedures

The kinetic face of the TGT in structural glasses is usually shown by demonstrating the ageing behaviour of volume data in the glassy state. In Fig. 3.22 specific volume data of polyvinylacetate (PVAc) given by Kovacs [86] are reconsidered and compared to recent data gained from refractive index measurements by means of the Lorenz-Lorentz equation

$$r \cdot \rho = \frac{n^2 - 1}{n^2 + 2} \tag{3.31}$$

where ρ is the mass density, r the specific refractivity (assumed to be constant) and n the refractive index. This equation is well established for polymers [87].

The dots in Fig. 3.22 denote data from Kovacs, the lines denote the new data. Both data sets agree very well in the high temperature branch ($T > 310$ K). The different behaviour below 310 K is due to different thermal treatments of the samples as explained in the following. The triangles represent equilibrium values ($T > T_g$), the circles represent specific volume data obtained on heating after quenching the sample. The values measured almost immediately after the quenching (open circles) are higher than those measured after annealing the sample for 100 h at $T = 255$ K (filled circles). The dark gray line was measured on cooling with a rate of 3 K/min, the light gray line on extremely slow cooling (step-wise with $\Delta T = 1$ K and an average rate of ca. 0.4 mK/min).

The smooth kink in the specific volume curves at the transition from the glassy to the liquid state is interpreted as the operative glass transition due to the cross-over of the intrinsic time scale of the α-relaxation process of PVAc with a time scale related to the thermal history of the sample. As indicated in Fig. 3.22 the intersection of the extrapolated high temperature straight line with those of the two glassy states are used to fix the operative glass transition temperatures T_{g1} and T_{g2}. This T_g shift, also reflected in the new data, is typical for the rate dependence of the TGT.

With the Kovacs data the annealing effect leads to a reduction of the specific volume of about -0.5% for 100 h annealing at $T \approx (T_g - 40\,\mathrm{K})$ with $T_g \approx 300\,\mathrm{K}$. This ageing has been attributed by Kovacs to the metastability of the glass with respect to its fluid phase after quenching. Apparently this annealed state coincides with that of our new data on extremely slow cooling, although the duration of our measurement exceeds the annealing time by at least a factor of 10. Accordingly this state seems not to be arbitrary at all, as different but slow thermal histories lead to the same state. This gives rise to the question: If the system is clearly out of equilibrium (due to a crossing of time scales) and if it relaxes even well below T_g towards its equilibrium, i.e. from the point of view of kinetics towards the extrapolated liquid branch, why should it stop relaxing at a certain point?

Indeed, the absolute values of the specific volume within the glassy state depend on the cooling history of the sample but the slope, which corresponds to the thermal expansion coefficient α, does not. So if the glassy state was metastable and strongly dependent on the thermal history, why should its temperature dependence be unique?

Figure 3.23 shows the effect of fast cooling as a source for metastability for the longitudinal elastic stiffness coefficient c_{11} (Voigt notation [84]) of polystyrene (PS) as obtained by Brillouin spectroscopy at sound frequencies in the GHz range. Both c_{11}-curves have been measured on step-wise heating with an average heating rate of 0.1 K/min. The curve of open circles was measured on a sample quenched in ice water from 400 K to 300 K. The black curve was also measured in the same way but after extremely slow cooling from above T_g to 300 K. The cooling was performed on step-wise cooling using the time domain method (see Sect. 3.4). The average cooling rate was extremely low (ca. 0.5 mK/min).

As shown before for the specific volume data (Fig. 3.22) the absolute values of the elastic stiffness modulus of the glassy state depend on the thermal history of the sample whereas the slopes of these curves do not. This result is not new but surprising from a physical point of view and will therefore be discussed in Sect. 3.5. The stiffness anomaly of the black curve in Fig. 3.23 around the TGT reflects an overheating effect: Since the sample was extremely slowly cooled to the glassy state, we may expect that the segment packing arrangement became so dense, that the random closed packed (rcp) state [78] was attained. Even on slow heating with $+0.1$ K/min the time needed for "liquifying" the glassy state was significantly larger than provided by that heating rate. The observed overheating temperature ΔT^{oh} is about $\Delta T^{oh} \approx 10$ K.

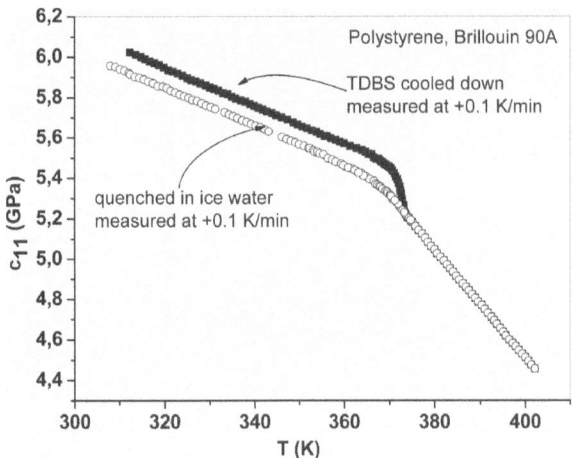

Fig. 3.23. Overheating effect at the glass transition temperature of polystyrene induced by strongly differing cooling and heating rates

The physical origin of the metastability demonstrated in Fig. 3.22 and Fig. 3.23 can be twofold at least:

1. Internal degrees of freedom which may be described in terms of irreversible thermodynamics by spatially homogeneous internal variables [88]. The equilibration of these internal variables is retarded on fast cooling or heating.
2. Spatial heterogeneities because of quenching.

The second effect is usually neglected in the discussion about the TGT but it deserves attention for the following reasons.

These spatial heterogeneities result from the low thermal conductivity which is typical for many glasses. Since usual heating and cooling processes transfer heat across the outer surface of the sample, specific volume gradients develop according to the temperature gradient. As an example one may think of a long polymer cylinder (Fig. 3.24) which is rapidly cooled down from high temperatures to below T_g. This quenching will freeze at first the outer skin of the sample (Fig. 3.24). Then this outer skin becomes a sort of rigid container which is filled with the same material in the fluid state. With ongoing quenching, the remaining fluid vitrifies layer by layer but with varying cooling rate and under the mechanical stress exerted by the outer solidified part. In the final state the sample consists of a layered structure with an inhomogeneous distribution of mechanical stresses and a radial distribution of glass transition temperatures. It is not surprising that such inhomogeneous samples show apparent ageing effects.

Figure 3.25 shows two specific heat curves measured with adiabatic calorimetry of a polymethylmethacrylate (PMMA) sample, which suffered different thermal histories prior to the two heating runs. The open circles were

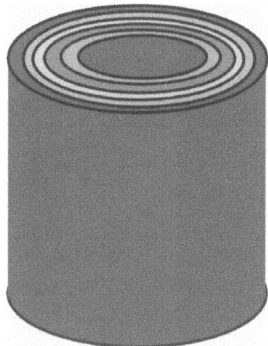

Fig. 3.24. Schematics of a polymer cylinder with shells representing different cooling rates and hence different glass transition temperatures due to fast freezing

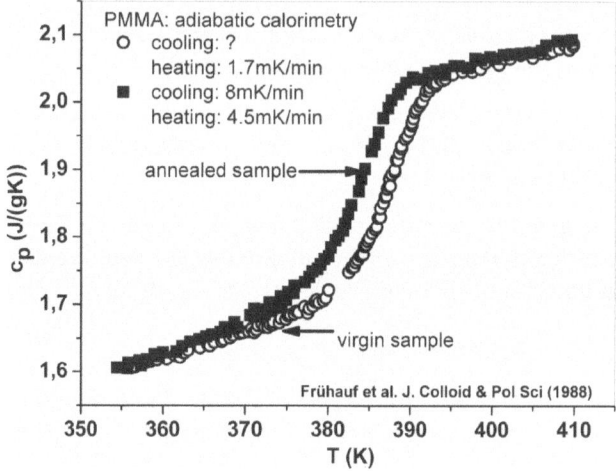

Fig. 3.25. Temperature dependence of the specific heat capacity of PMMA after different cooling and heating procedures

measured on a virgin (extruded) sample from ambient temperature to 410 K. Subsequently the sample was slowly cooled down again to ambient temperature by 8 mK/min and then measured again on heating by 4.5 mK/min (black dots). The superior experimental reliability and precision of these data measured with heat-pulse adiabatic calorimetry is illustrated in Fig. 3.26. The margin of error of these specific heat data is smaller than the diameter of data dots. The low- and high-temperature asymptotes of the $c_p(T)$-curve are very well described by straight lines with slightly different slopes being a little bit lower in the liquid state.

Based on this high data quality, the difference in T_g of about 3 K between the virgin and the slowly cooled state of the PMMA sample is real in Fig. 3.25.

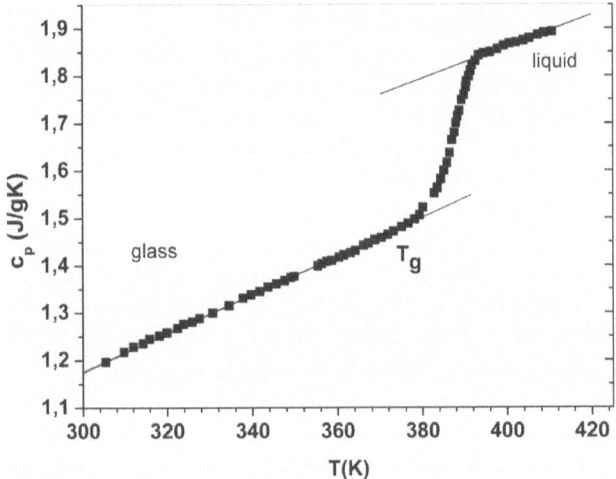

Fig. 3.26. Specific heat capacity around the TGT of PMMA, the ordinate is off by roughly 15% due to a calibration error [89]

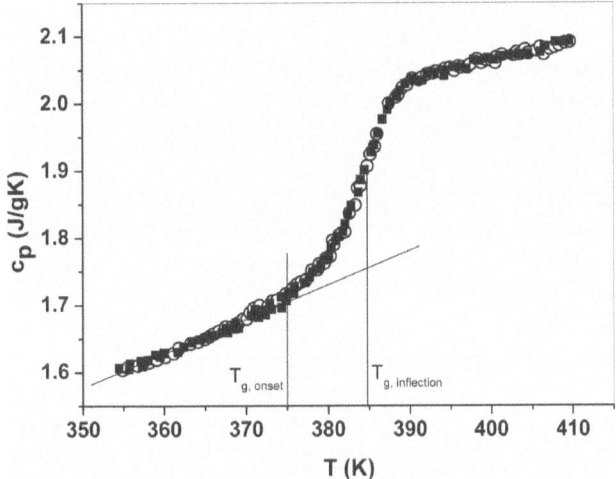

Fig. 3.27. Specific heat capacity around the TGT of PMMA for cooling rates differing by a factor of 60 (both of the order of magnitude: mK/min) [89]

Frühauf et al. [89] attributed this shift to the formation of internal stresses and a spatial distribution of glass transition temperatures.

In the following we will discuss data measured on two identical PMMA samples (Fig. 3.27), where very different cooling rates yield the same adiabatic specific heat capacity, indicating that at least in average the same glassy state is produced for both cooling scenarios. The samples have suffered cooling rates differing by a factor of more than 60.

The black curve in Fig. 3.27 was cooled very slowly with 8 mK/min and the circle curve was cooled faster with 519 mK/min which is still slow compared to rates applied in technological processes, however. Both samples were then measured by the same heating rate of about 4.5 mK/min. In the margin of error both curves are identical and show no cooling rate dependence at all. The reason for this complete independence of the TGT on the thermal history is only understood if the rate dependence of T_g can be neglected for the given range of cooling rates. This fact seems to be contradicted by the large change of the cooling rate in combination with the accuracy of the specific heat data.

An alternative interpretation for the observed behaviour could be connected to the observed cut-off of the relaxation frequencies, i.e. the relaxations stop at certain temperature with a finite relaxation frequency. This aspect will be discussed in Sect. 3.4.

If the conclusion can be made that the data of Fig. 3.27 are equilibrium or near equilibrium data, then the step-height of c_p at T_g represents the excess specific heat connected to the "liquefaction" of the glassy state. In physical terms, the precise location of T_g is not obvious. Many authors use the inflection point of the specific heat curve as a good measure to identify T_g. Since the temperature width of the thermal glass transition regime is almost 20 K an identification of the inflection point of the c_p-curves as a reliable glass transition temperature is questionable. It is more likely, that the onset of the excess specific heat on the background specific heat of the glassy state defines the glass transition temperature. This onset coincides quite well with the kink in the sound velocity and the refractive index.

A completely different view on the kinetic properties of the TGT is obtained having a look at the *"Generalized Cauchy Relation"* (gCR) of amorphous materials [90, 91]. From solid state physics it is well-known that the number of independent coefficients of the elastic stiffness tensor is deduced from the symmetry of the crystal [e.g. [92]]. Additional relations, which further reduce the number of independent elastic stiffness coefficients are known in solid state physics and are called *Cauchy relations* (CR) [93, 94]. Such relations hold only true if the crystal of interest obeys additional constraints about local symmetry (every lattice particle is a centre of inversion), molecular interaction forces (only central forces), and lattice anharmonicity (only harmonic potentials).

For cubic crystals these constraints lead to the relation:

$$c_{12} = c_{44} \qquad (3.32)$$

Equation (3.32) reduces the initially three independent elastic coefficients to only two, c_{11} and c_{44} [94]. Imagine now that we were able to produce a ceramic sample made of irregularly oriented cubic nano-crystals which obey a CR according to Eq. (3.32). In that case the ceramic body shows isotropic symmetry on the macroscopic scale. Hence the isotropy condition [92] holds for the orientational average (denoted by the brackets) of the elastic coefficients:

$$\langle c_{12} \rangle = \langle c_{11} \rangle - 2 \cdot \langle c_{44} \rangle \tag{3.33}$$

Combining Eqs. (3.32) and (3.33) we obtain a Cauchy relation for the isotropic state [94]:

$$\langle c_{11} \rangle = 3 \cdot \langle c_{44} \rangle \tag{3.34}$$

Equation (3.34) reduces the number of independent elastic coefficients of the isotropic nano-ceramic from two to one.

If the cubic crystal does not fulfil the requirements of the Cauchy relation, than you can write Eq. (3.32) in a more general way

$$c_{12} = c_{44} + A(T, p, \dots) \tag{3.35}$$

where A is a term which depends on temperature T, pressure p, etc.

The orientational average leads then to a generalized Cauchy relation of the isotropic state

$$\langle c_{11} \rangle = 3 \cdot \langle c_{44} \rangle + A(T, p, \dots) \tag{3.36}$$

According to Eq. (3.36) the deviation from the Cauchy relation of the cubic state measured by A is found as an additive term in the Cauchy relation of the isotropic state Eq. (3.35). This result sheds some interesting light on the possible mechanism leading to the generalized form of the Cauchy relation. Equation (3.35) together with Eq. (3.36) lead to the result that a ceramic based on nano-crystals with cubic symmetry can show an ideal Cauchy relation Eq. (3.35) provided that the cubic crystals follow a Cauchy-Relation ($A = 0$). As a consequence the additive constant A in the generalized Cauchy relation is not caused by the difference in local and global symmetry of the ceramic.

Amorphous glass-formers are not expected to meet the aforementioned conditions about local symmetry, the absence of defects and the absence of anharmonicity. Thus Eq. (3.34) should not apply to these materials.

However, theoretical work [90, 95] proposed a generalized Cauchy relation between the high-frequency elastic shear modulus $G^\infty = c_{44}^\infty$ and the compression modulus K^∞ for dynamically frozen liquid argon:

$$K^\infty = \frac{5}{3} \cdot G^\infty + 2 \cdot (P - n \cdot k_B T) \tag{3.37}$$

where P is the external pressure, T the temperature and n the particle density. Using

$$K^\infty = c_{11}^\infty - \frac{4}{3} c_{44}^\infty \tag{3.38}$$

one finally obtains

$$c_{11}^\infty = 3 \cdot c_{44}^\infty + 2(P - n \cdot k_B T) \tag{3.39}$$

Compared to the CR for the isotropic nano-crystalline case Eq. (3.35), this *"generalized" Cauchy relation* (*g*CR) for the amorphous argon adds a term, which depends on pressure and temperature and therefore much more equals Eq. (3.36). According to Eq. (3.37) the pre-factor 3 is reproduced for c_{44}, but the relation between c_{44} and c_{11} is not linear for changing pressure or temperature.

To check this point, the TGT of several glass formers (DGEBA, PVAc, Salol, DBP, etc.) were recently studied during slow cooling ($\dot{T} < 0.5\,\text{K}$ /min) by Brillouin spectroscopy at hypersonic frequencies [96]. Due to the high measuring frequency, the method provides the clamped stiffness coefficients $c_{11}^{\infty}(T)$ and $c_{44}^{\infty}(T)$ (Fig. 3.29). As the result, a *linear* relation

$$c_{11}^{\infty}(T) = A^0 + B \cdot c_{44}^{\infty}(T) \tag{3.40}$$

was obtained, A^0, B temperature-independent material constants and T being the temperature. Again, B = 3 was found within the margin of error for all materials under investigation [96]. As shown in Fig. 3.30, the TGT is completely hidden in the elastic data representation described by Eq. (3.40). Accordingly the influence of the TGT has to be identical on both the longitudinal and the transverse acoustic mode.

This observation raises the question whether (i) non-equilibrium processes do not violate the gCR at all or whether (ii) the TGT is not necessarily a non-equilibrium process? If hypothesis (i) holds true, the validity of relation (3.40) for the liquid state as well as for the solid state would be less surprising. If on the other hand (ii) holds true, the validity of the gCR could turn out to be a versatile tool to discriminate between mechanical equilibrium and non-equilibrium. With respect to this point of view it is highly interesting to elucidate the question whether fast cooling of the liquid to the non-equilibrium glassy states is able to violate the gCR. If it happens, whether these non-equilibrium states relax towards their equilibrium and if they do so, on which time scale and to which direction this happens.

All data reported in the following were measured with Brillouin spectroscopy using the $90A$-scattering geometry (c.f. Fig. 3.18b) detecting simultaneously the longitudinal sound frequency f_L^{90A}, and the transverse sound frequency f_T^{90A}. It should, however, be stressed that amongst all scattering geometries only the $90A$-scattering or more general the ΘA–scattering provides a phonon wave vector $q^{\Theta A}$ which is independent of the refractive index n of the sample (Sect. 3.2).

Having determined the sound velocities of the longitudinal and shear polarized phonon modes and assuming that the mass density ρ is known, the elastic constants $c_{11} = c_L$ and $c_{44} = c_T$ are given by

$$c_{ii}^{90A} = c_{L,T}^{90A} = \rho \cdot (V_{L,T}^{90A})^2, \qquad i = 1, 4 \tag{3.41}$$

In order to test whether the gCR Eq. (3.40) can be violated by fast quenching we have used the diglycidilether of bisphenol A (DGEBA, Fig. 3.28) as a glass forming organic liquid.

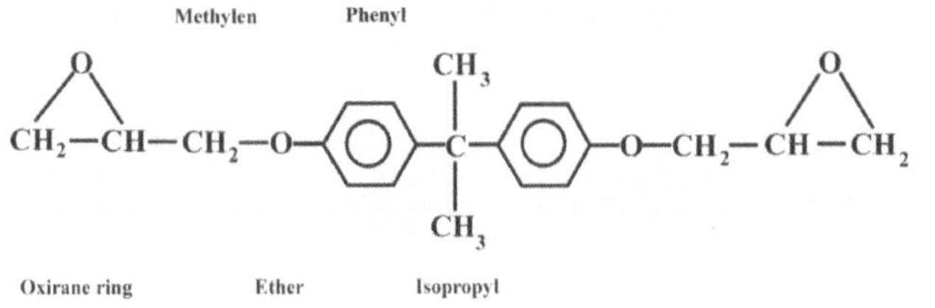

Methylen   Phenyl

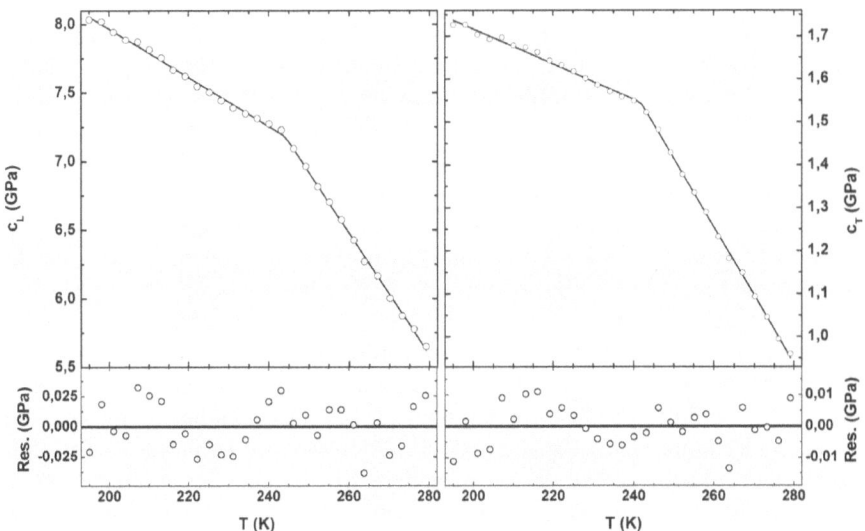

Fig. 3.28. Structural formula of DGEBA

It is well-known [96, 97] that bulk DGEBA is a fragile liquid that may crystallize, in special circumstances if nucleation is forced e.g. by rough surfaces, water droplets, etc. DGEBA shows at extremely slow cooling a static (intrinsic) TGT at $T_{gs} = 243$ K [97].

In Fig. 3.29 the elastic stiffness data $c_{11}(T)$ and $c_{44}(T)$ for an extremely slow cooling run are shown (~0.01K/min). Both elastic moduli versus temperature curves show a kink at about the static glass transition temperature T_g of 243 K [97]. According to the small residuals shown in Fig. 3.29 both frequency curves behave piecewise linearly in the investigated temperature interval.

According to Eq. (3.40) the related gCR is depicted in Fig. 3.30. The plot demonstrates that the elastic data measured under these conditions perfectly obey the gCR with $A = 2.6$ GPa and $B = 2.98 \pm 0.2$. Hence within the margin of error $B = 3$ holds true. The distribution of the residuals confirms the linear

Fig. 3.29. c_L and c_T of DGEBA as a function of the temperature T. The *straight lines* are fitted curves. The residuals in the lower part demonstrate the agreement of the data with a piece-wise straight line model

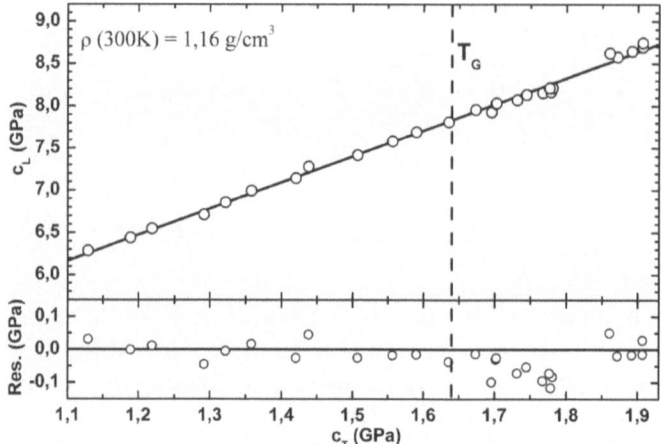

Fig. 3.30. gCR-representation for the Brillouin data of Fig. 3.29. In the lower part of the figure the corresponding residuals are given. The residuals indicate the good compatibility of the measured data with the gCR presented by the *straight line* according to Eq. (3.40)

behaviour of the data representation, and that the TGT is completely hidden in this parametrical representation.

If we tentatively accept a purely kinetic interpretation (see e.g. [98]) of the TGT, we had to assume that T_g could be shifted to higher temperatures, provided the cooling process was speeded up.

In other words, the gCR would remain a linear function but would become ambiguous in its physical meaning, i.e. the parameters A and B would have the same values in the equilibrium as in the non-equilibrium state. Consequently, the gCR would turn out to be rather insensitive to deviations from equilibrium. If, on the other hand, the slopes of the longitudinal and transverse frequency curves of the new hypothetical glassy state change their relation in comparison to that measured on slow cooling, the gCR would become violated for the new glassy state. Keeping in mind that within this approach the slower cooled sample is closer to equilibrium, such behaviour seems not to be very likely. In order to clarify this point we have tested whether the apparent insensitivity of the gCR with respect to equilibrium is a general feature of the gCR. For this purpose we have used quenching procedures in order to create forced non-equilibrium glassy states. During this fast cooling we are not able to record Brillouin data simultaneously. Therefore we cannot say anything about the evolution of the elastic behaviour during the fast cooling process. As will be shown below (Fig. 3.31c) we were able, however, to measure the sound frequency evolution of samples submitted to intermediate cooling rates (0.05 K/min).

Figure 3.31 shows the $c_{11} = c_{11}(c_{44})$-relation as obtained from Brillouin data derived from different quenching procedures. As indicated above, these quenching procedures are not really under control. Figures 3.31a,b,d show Brillouin data measured during heating the quenched sample from the glassy to the liquid state. Figure 3.31c shows the Cauchy-representation of Brillouin data measured on intermediate fast cooling of the DGEBA sample. The open circles in Fig. 3.31 give the measured data, the straight lines give the gCR. From Fig. 3.31 it is evident that non-equilibrium glassy states produced by quenching lead to a violation of the gCR. The quenched samples of Fig. 3.31 show significant positive deviations from the gCR. As a matter of fact, this positive deviation is related to a lag of the shear stiffness behind the longitudinal stiffness.

In other words, on quenching the DGEBA sample to low temperatures, the longitudinal modulus develops closer towards the increasing equilibrium value than the transverse modulus does. Moreover, these deviations are able to relax towards the gCR during heating and cooling procedures within the glassy state (c.f. Fig. 3.33).

Figure 3.31c demonstrates how the violation of the gCR takes place if the BS-measurements are performed during intermediate fast cooling (about 0.05 K/min). It is evident that even if the sample is moderately cooled into the glassy state, significant positive deviations from the gCR can be produced. The linear transformation properties implied by the gCR and deviations from this linear transformation can be evidenced in a different way. We show in Fig. 3.32 the temperature dependence of the longitudinal and shear phonon corresponding to the Cauchy representation of Fig. 3.31b.

As is seen in Fig. 3.32 we have spread the plot axis for both phonon frequency curves in such a way that the longitudinal and transverse phonon branches coincide within the liquid phase. In the glassy phase the relative slope of the transverse mode becomes smaller in comparison to that of the longitudinal mode, whereas the slope of the longitudinal mode is the same as for slow cooling. According to the related frequency-temperature plots the different ageing processes presented by Figs. 3.31a,b,d meet the "equilibrium glass branches" of the longitudinal and transverse polarized modes defined by Figs. 3.29, 3.30. The low-temperature branch of Fig. 3.31a merges at $T - T_g = 16$ K with the gCR. The slope of this low-temperature branch amounts to about $m = 4$ in comparison to $B = 3$. The low-temperature branch of Fig. 3.31b even merges at $T - T_g = 21$ K. The slope of this branch amounts to $m = 4.8$. In Fig. 3.31d the onset of the deviation from the gCR occurs only at the TGT. No excess of the slope ($m > B = 3$) was ever observed above T_g.

In order to elucidate the observed significant ageing behaviour within the glassy state of DGEBA we have performed different cooling and heating cycles within the glassy state of this material. The data are depicted in Fig. 3.33. After quenching the sample to 120 K (cooling rate ~ -200 K/min) we performed a first heating experiment (run 1) in Fig. 3.33, which we stopped at 195 K. This is more than 40 K below T_g.

The slope of the low-temperature branch is $m = 4.15$ in comparison to $B = 3$ in the liquid state. Subsequently we have cooled the sample again

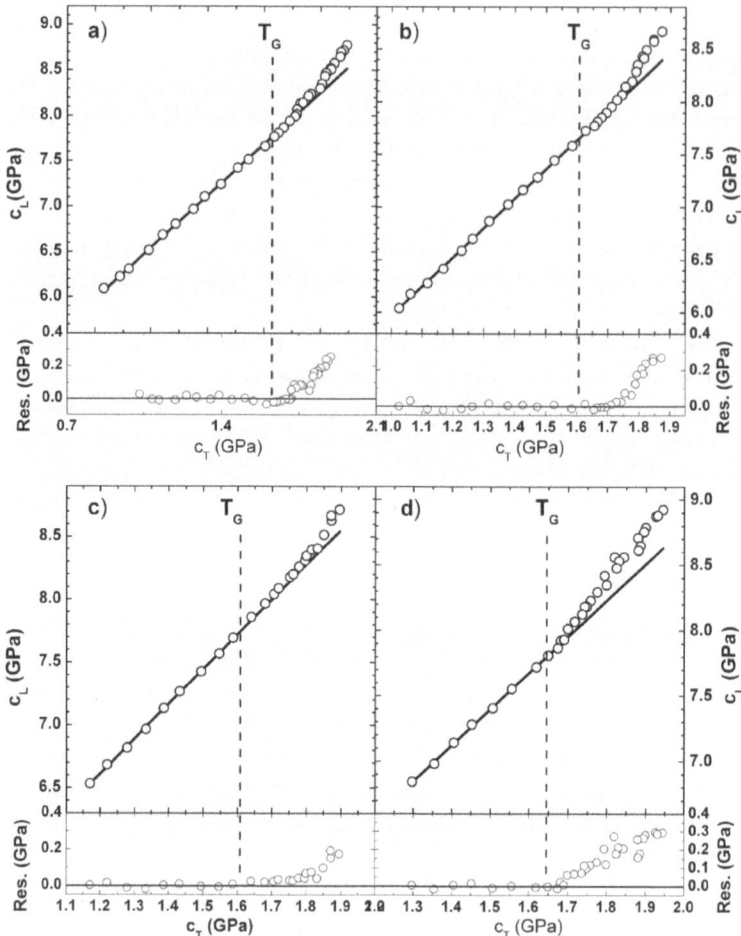

Fig. 3.31. gCR-representations for different quenching and heating cycles. The residuals indicate the deviations of the measured data from the gCR presented by the *straight lines*. Figures a, b, d represent Brillouin data measured on heating after different quenching procedures. Figure c shows a cooling run on intermediate fast cooling. See text for further explanations

to 120 K and then heated the sample to $T_g - 10$ K. The slope of the low temperature branch has now decreased to $m = 3.92$. A further cooling to 137 K and a subsequent heating to ambient temperature yields a slope of $m = 3.48$ for the glassy state. These results confirm the above anticipated ageing process within the glassy state. This thermal relaxation process converges in all our experiments versus a unique glassy state, which we have identified recently on the basis of time domain Brillouin measurements [97] as the equilibrium glassy state of DGEBA (c.f. Sect. 3.4). Definitely, we found no relaxations towards the liquid state, which exceed the sound frequency data

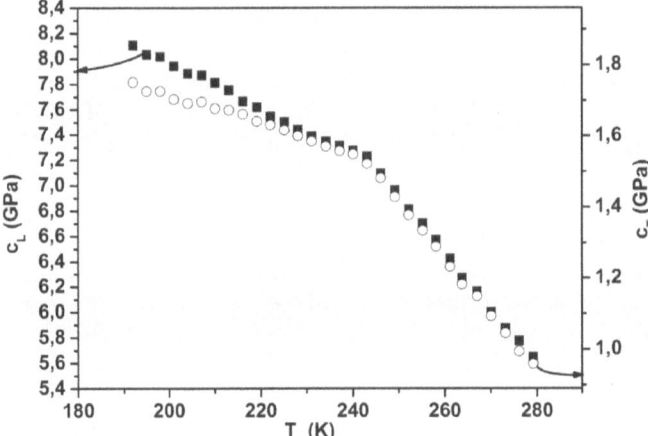

Fig. 3.32. Longitudinal and transverse elastic moduli (c_L and c_T) for DGEBA as a function of the temperature T, measured after quenching to about 77 K

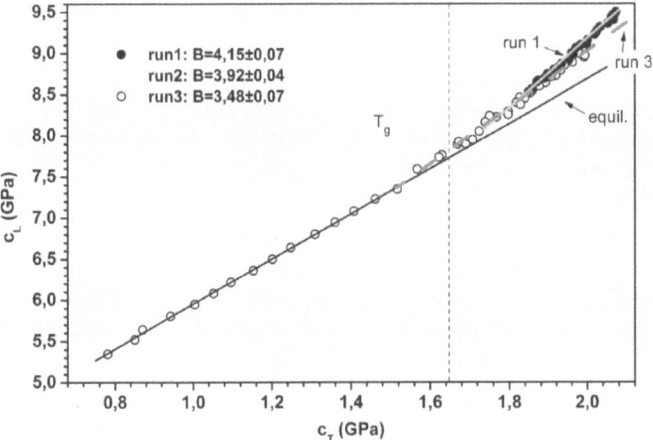

Fig. 3.33. gCR-representations for different consecutive quenching and heating cycles. All measurements were performed on heating. The full *straight line* represents the gCR for the equilibrium state. For further explanations, see text

within the temperature representation of the equilibrium glassy state. This result is based on at least 15 different quenching cycles. According to Ref. [97] the structural α-relaxation shows a cut-off at the glass transition temperature (c.f. Sect. 3.4). It is therefore likely that the dynamical processes involved in the ageing process have little in common with the structural α-relaxation process. We do not want to compare time constants found in the gCR representation during ageing in the glassy state with time constants reported for the hypersound velocities in the liquid state [97]. There is no hint for a direct connection between both processes.

Being convinced that the ageing process within the super-cooled liquid state discussed above has nothing to do with the α-relaxation process the question about the reason for the violation of the gCR has still to be elucidated. As a matter of fact, the violation of the gCR is only observed if the quenching process starts at temperatures more than a hundred Kelvin above T_g and is as fast as than -200 K/min. At that cooling speed usual glass structures cannot be formed. In our opinion, the supercooled liquid (with respect to the crystalline state) is driven into a further supercooled state sc2 but this time the super-cooling is related to the glassy state. The state sc2 behaves as a metastable glass. That the glassy state is the related preferential thermodynamic reference state to which the metastable state sc2 relaxes is evident from the experimental facts. The usual local molecular arrest at the intrinsic glass transition is impeded. The experimental data show, that the achieved metastable state sc2 has smaller shear stiffness than the glassy state, whereas the longitudinal modulus is, in comparison to that of the glassy state, not altered. In other words the sc2-state behaves "stiff" on compression but rather "soft" on bending the sample: the state sc2 behaves more liquid-like than a glass. As a matter of fact, heating up to the glass transition temperature erases any memory about the super-cooled state sc2.

In favour of this view are the facts that (i) the temporal evolution of the ageing process is much faster than the hypothetical α-process of DGEBA in the temperature range $T_g > T > T_{gs}$, but very close to T_g and that (ii) we found ageing well below the VFT temperature T_o. As a consequence, the liquid state of DGEBA seems to be a forbidden state below T_{gs} and the glassy state, which is approached during the thermal ageing appears as an "equilibrium glassy state" of DGEBA for $T < T_g$. Of course we cannot exclude the existence of another energetically more favourable state, but if it exists, this state must be separated from our equilibrium glassy state by such high potential barriers that it is virtually inaccessible. In order to be observable in a Brillouin experiment, the molecular non-equilibrium excitations have to couple to the transverse and longitudinal polarized sound modes and their relaxation times have to be in a time-window accessible to this kind of experiment. It is worth noting that, if this coupling is weak or if the relaxation times are too fast or too slow in comparison to our experimental time scale, Brillouin spectroscopy is ineffective with respect to the test of ageing phenomena.

Studying the experimental results presented in this chapter, we can conclude as follows:

(i)   Using appropriate thermal treatments, distinct glassy states with different macroscopic properties, like density or mechanical moduli can be realized.

(ii) In the glassy state there can exist relaxations (due to ageing) which are distinct from α-, β-, γ-, etc-relaxations (see Sect. 3.4).

(iii) It seems that in structural glass-formers there exists an intrinsic glassy state.

(iv) the ideal glass transition temperature marks the stability boundary for the intrinsic glassy state and the super-cooled glassy states.

3.4 The Dynamic View of the Thermal Glass Transition

In this section we will analyse the glass transition and related phenomena which are due to the cross-over of the inverse probe frequency with the intrinsic α-relaxation frequency of the glass-forming liquid.

It is usually believed by the glass community that the slowing down of molecular dynamics on approaching the thermal glass transition plays a crucial role for the physical changes appearing at the TGT [99–104], as well as at the glass transition predicted by mode-coupling theory [105, 106] etc. The dynamics associated with those molecular motions which are intimately related to the TGT are usually described by the VFT-law introduced in Sect. 3.1. This primary glass relaxation process is usually called α-relaxation process. According to the schematic drawing in Fig. 3.34 the activation plot of the α-relaxation frequency is strongly bent towards zero at the so-called VFT-temperature T_0 [103, 104, 107]:

$$\omega_\alpha = \tau_\alpha^{-1} = \omega_{\alpha 0} \cdot \exp\left[-\Delta G_\alpha / \left(R\left(T - T_o\right)\right)\right] \qquad (3.42)$$

This phenomenological law holds true at least for temperatures $T > T_g > T_0$ and implies a divergence of the α-relaxation time τ_α at T_0. In Eq. (3.42) the parameter $\omega_{\alpha 0}$ is the attempt frequency displaying the molecular mobility at high temperatures and ΔG_α is the activation free energy of the α-relaxation process.

It is the trust in this divergence which has stimulated many researchers in the field of glass transition to believe in the kinetic nature of this phenom-

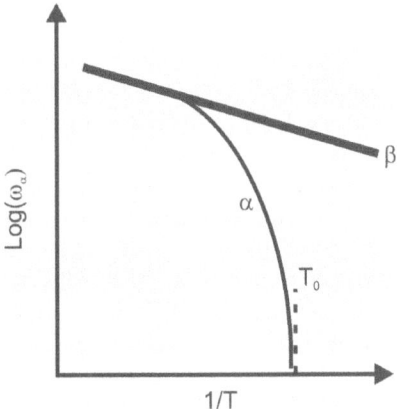

Fig. 3.34. Schematic representation of the VFT-law for the α-relaxation and the Arrhenius-like behaviour of the β-relaxation

enon. This belief was supported by manifold observations of ageing processes in the pre- and post-transition temperature intervals of the operative glass transition temperature T_g. It should however be stressed, that these observations in favour of a purely kinetic interpretation of the TGT are no proof for any divergence of τ_α but they are at best a proof for sufficiently large relaxation times in order to compete with usual experimental time constants τ_{exp}. Thus the real temperature behaviour of the α-relaxation time in the limit of large experimental time constants is the fundamental problem. The question, whether τ_α really diverges or remains finite, decides whether the "cross-over" problematic is unsolvable or just a quantitative problem of the experimentalists patience. This is the reason why it deserves huge interest and is the main topic of this section. Of course, besides the cooperative α-relaxation process there may exist other relaxation processes which are thermally activated. In Fig. 3.34 such an Arrhenius-process is designated as β-process. This kind of relaxation processes is usually not affected by the glass transition.

The deviation of the VFT-law from the Arrhenius behaviour is an experimentally evident precursor of the thermal glass transition not only observed well above T_g but which is found for quenched glass formers even clearly above the melting temperature T_m. That means that this precursor of the later TGT is already present in the glass forming material at temperatures at which the material has not decided whether it will densify by crystallization or densify by freezing. In so far the slowing down of the molecular dynamics described by a VFT-behaviour puts the question whether it can be interpreted as a general precursor for a transition into a dense solid state. Independently of the thermodynamic path the primary goal of a cooled liquid seems to be its densification and consequently, to find a state which is as densely packed as possible (close packing). Below the melting point two possibilities exist for this densification. (i) the crystallization and (ii) the thermal freezing. Since crystallization is usually a strong first order transition, the α-relaxation is clearly no dynamical precursor for the crystallization process. This argument is supported by the fact that during the undercooling of a glass forming liquid, the crystallisation process is suppressed and the potential melting/crystallization process leaves no mark on the VFT-law. Since in a crystal the average position of the lattice molecules is fixed, the α-relaxation process of the related liquid state usually disappears or is quantitatively different from that in the crystalline state. An example of the latter behaviour is DFTCE which has been discussed in the first section. Thus, the first order transition to the crystalline state can be interpreted as one efficient way to achieve the goal of dense packing. The remaining alternative, which is necessarily disordered, is the glassy state with random closed packing (rcp). Then the question occurs, if densification by crystallization changes fundamentally the α-relaxation behaviour or even stops it, why should the α-process be unaffected by the glass transition and in turn produce the "time-trap" discussed in the first section?

A further comparison between the glass transitions and structural phase transitions might be elucidating. So-called soft mode theories predict that the lattice instability occurring at a structural phase transition is accompanied by a slowing down of the soft mode frequency to zero [108]. In reality this slowing down of the soft mode frequency to zero never happens, instead the phase transition takes place at a soft mode frequency well above zero. Assigning the experiences with soft modes of structural phase transitions in crystals to the potential behaviour of the α-relaxation time the divergence of the α-relaxation time $\tau_\alpha = 1/\omega_\alpha$ at the glass transition of a glass forming liquid is little probable. This argument is a further counter-indication against the "cross-over" argument but not a proof yet.

In order to elucidate the true α-relaxation behaviour and its influence on phenomenological physical properties at the TGT, the temperature dependence of the latter within the pre-transitional phase close to T_g has to be revisited. It is usually accepted that every liquid can be quenched to the glassy state. According to Sect. 3.1 "quenched glass formers" become supercooled below their melting temperatures whereas ideal glass formers can stay in their thermodynamic equilibrium state until the TGT. Fast cooling (quenching) of a glass forming liquid is accompanied by a decrease of its specific volume $\nu_l(T)$ according to the volume expansion coefficient $\alpha_{lsc}(T) = \alpha_l(T)$ (sc supercooled) of the equilibrium liquid. The specific volume in the supercooled liquid state $\nu_{lsc}(T)$ is in general higher than in the crystalline reference state $\nu_{lsc}(T) > \nu_{\text{cryst}}(T)$. Thus, supercooling provides at first a deviation of the specific volume from that of the crystalline reference state but continues that of the thermodynamic equilibrium liquid into the undercooled liquid state. The linear temperature dependence of the specific volume found in the pre-transition temperature interval does not reflect the VFT-behaviour of the main glass relaxations. Only at lower temperatures, in a limited temperature interval, called glass transition interval, the thermal expansion coefficient changes rather abruptly from that of the equilibrium liquid state $\alpha_l(T)$ to that of the so-called glassy state $\alpha_g(T < T_g) < \alpha_l$ [109–111] (see Sect. 3.3). Ideal glass-formers behave in exactly the same way but transform from the equilibrium liquid state immediately into the glassy state without passing through a metastable pre-transition temperature interval. This means, that from a phenomenological point of view quenched- and ideal glass-formers do not differ from each other although they are in different thermodynamic states.

As a matter of fact, transport properties like the shear viscosity of fragile liquids behave in the same way. Quenched glass forming fragile liquids and ideal glass forming fragile liquids are accompanied by an increase of viscosity stronger than exponential (see Fig. 3.21) and do not reflect the behaviour of the equilibrium liquid at much higher temperatures [112–114].

It appears that the pre-transitional temperature interval as seen by the shear viscosity is much broader than that seen by mass density. From the viewpoint of shear viscosity one has to differentiate between two temperature regions: the rather large pre-transitional temperature interval in the liquid phase and the rather small glass transition zone (kink). The question arises about the reason why the transport of molecules should feel hindered much more in the pre-transitional temperature interval than by usual thermal activation at much higher temperatures? Geszti [115] has introduced a phenomenological relation between the shear viscosity and the molecular dynamics

$$\eta_{44} = c_{44}^{\infty} \cdot \tau_{\alpha} \tag{3.43}$$

where τ_{α} is again the relaxation time related to the TGT (Sects. 3.1, 3.3) and c_{44}^{∞} is the frequency clamped shear stiffness of a liquid. Equation (3.43) couples in some way the imaginary part ($\sim \eta_{44}$) of the complex shear stiffness of a mechanical continuum to a relaxation time τ_{α}. In the case of glass forming liquids it couples the solid state like behaviour c_{44}^{∞} of a liquid observed under frequency clamped conditions to its acoustic attenuation. Keeping in mind, that c_{44}^{∞} depends linearly on temperature in the pre-transitional liquid phase [116, 117], the more than exponential increase of η_{44} with decreasing temperature must be due to molecular dynamics, i.e. the temperature behaviour of τ_{α}. From a formal point of view the observed saturation (kink) of η_{44} at the glass transition might be due to either a saturation of the relaxation time τ_{α} and/or to a saturation of c_{44}^{∞}.

Combining Eq. (3.42) with Eq. (3.43) leads to

$$\eta_{44} = c_{44}^{\infty} \cdot \tau_{\alpha 0} \cdot \exp\left[\Delta G_{\alpha} / \left(R\left(T - T_{o}\right)\right)\right] \tag{3.44}$$

The strong bending of the activation plots of τ_{α} and η_{44} has been related to an increasing cooperativity [102] of the molecular motions in the pre-transition interval on approaching the TGT. That means, that in order to put one molecule or a molecular group in a better packed environment the molecules in the shell of next neighbours, in the shell of over-next neighbours and so on have to move cooperatively in order to allow for this improved packing. It is this cooperative molecular motion which allows even in the very dense packed state the combined rotational and translational molecular motion necessary for a further densification. Equation (3.44) implies that in a very dense state a further improvement of the density is more time consuming than in a less dense state.

The increases of the bending strength of τ_{α} and η_{44} also signalize an increase of fragility of the glass former. The magnitude of the temperature difference $|T_g - T_0|$ is a measure for the fragility of the glass former and amounts e.g. to about 40 K for polymers [101, 112, 113].

At this state of discussion the reader is confronted with the situation that the shear viscosity of a fragile glass former increases stronger than exponential on approaching T_g and shows a pronounced kink (saturation) at this temperature. The mass density and the frequency clamped elastic shear stiffness behave linearly in the pre-transitional temperature interval but show a kink at T_g. Finally, in contradiction to the temperature dependence of the shear viscosity η_{44} the relaxation frequency ω_{α} is predicted to follow a VFT-law Eq. (3.42) across the operative glass transition temperature T_g remaining continuous and differentiable at T_g but with a divergence at $T_0 < T_g$. Thus, a saturation of τ_α at T_g is not predicted by the VFT-law and obviously, it would contradict the "cross-over argument" given in Sect. 3.1. In other words, the unavoidable "time-trap" introduced in Sect. 3.1 as an argument in favour of the kinetic nature of the glass transition would become redundant if the VFT-law breaks down at T_g (see also the discussion below).

Insisting for the moment on the validity of the VFT-law Eq. (3.42), the only way to force η_{44} to saturate at T_g is to saturate c_{44}^∞. In this kinetic picture the saturation of c_{44}^∞ is synonymous to reaching a non-equilibrium state. But there is also the possibility of structural changes at T_g.

The bridge to the "structural changes" eventually might be found in a phenomenological relation between the shear viscosity η_{44}, the free volume V_f and the occupied volume V_o given by Doolittle [100]

$$\eta_{44} = A \cdot \exp\left[\frac{b \cdot V_o}{V_f}\right] \tag{3.45}$$

where A and b are constants. Following the idea of free volume [118, 119] a volume element V from the glass-forming liquid is then composed of $V = V_f + V_o$. According to Eq. (3.45) the temperature dependence of η_{44} stems from the temperature dependencies of V_f and V_o. Defining the related volume expansion coefficients α_f and α_o respectively

$$\alpha_f = \frac{1}{V_f}\frac{dV_f}{dT} \tag{3.46a}$$

and

$$\alpha_o = \frac{1}{V_o}\frac{dV_o}{dT} \tag{4.46b}$$

The volume expansion coefficient α and the relative temperature derivative of η_{44} can be calculated yielding

$$\alpha = \frac{1}{V}\frac{dV}{dT} = \frac{1}{V}\frac{d\,(V_f + V_o)}{dT} = \frac{V_f}{V}\alpha_f + \frac{V_o}{V}\alpha_o \tag{3.47}$$

and

$$\frac{1}{\eta_{44}}\cdot\frac{d\eta_{44}}{dT} = \frac{b\cdot V_o}{V_f}\{\alpha_o - \alpha_f\} = \frac{1}{c_{44}}\frac{dc_{44}}{dT} + \frac{1}{\tau_\alpha}\frac{d\tau_\alpha}{dT} = \frac{1}{c_{44}}\frac{dc_{44}}{dT} - \frac{\Delta G}{R}\frac{1}{(T - T_o)^2} \tag{3.48}$$

In Eq. (3.47) the coefficient α_o measures the thermal expansion of the molecules due to thermal molecular excitations in an anharmonic potential and α_f measures the decrease of the free volume with decreasing temperature. Having again a look at the saturation of η_{44} at T_g either α_o or α_f or both in Eq. (3.48) have to change at the TGT.

At least for α_o a discontinuous change is only possible, if structural changes, preferably intramolecular structural changes, occur at the TGT. According to the "free volume theory" [118, 119] the free volume V_f goes to a constant minimum value or even zero at $T^* > 0$ K which means, that the corresponding volume expansion coefficient α_f goes to zero (random closed packing). A spontaneous disappearance of α_f at T^* suggests a volume expansion coefficient α which changes discontinuously at T^*. A discontinuous change of the volume expansion coefficient is synonymous to the fact that the molecular interaction potential changes discontinuously at T^*. In any case T^* has to be lower than the usual operative glass transition temperature T_g. In some theories T^* is brought in connection with T_{VFT} but without any proof.

In addition, the right hand part of Eq. (3.48) relates the saturation of the shear viscosity to the saturation of the frequency-clamped shear stiffness. It remains the essential question: why does the free volume becomes constant at an operative glass transition temperature T_g while the cooperative motions of the molecules still follow the VFT relation (see Eq. 3.48)? To answer this question a precise reinvestigation of the α-relaxation process and of the shear stiffness c_{44}^∞ in the pre-transitional temperature interval is imperative. Provided the generalized Cauchy relation (discussed in the previous section) holds true it suffices to investigate the longitudinal elastic modulus instead of the shear stiffness as the latter one is often difficult to measure.

For the sake of simplicity, the following discussion is reserved to ideal or almost ideal glass-formers. The α-process can be observed with different experimental techniques. Dielectric-, mechanical- and thermal spectroscopy are amongst the most common techniques, as the complex dielectric susceptibility $\varepsilon^*(\omega)$, the complex elastic stiffness tensor $c_{ij}^*(\omega)$ or the compliance tensor $s_{ij}^*(\omega)$ with $(i, j = 1, 2 \ldots, 6)$ and the complex specific heat capacity $c_p^*(\omega)$ couple to the α-process. According to irreversible thermodynamics the α-relaxation time τ_α depends on the thermodynamic boundary conditions and therefore on the measurement technique, at least in principle. In reality these different α-relaxation times are often very close to each other.

As indicated in Fig. 3.35 the α-relaxation frequency of polymethylmethacrylate (PMMA) follows in the pre-transitional temperature interval a VFT-law [120]. However, according to time domain Brillouin spectroscopy (TDBS, discussed in detail below) close to T_g deviations from the VFT-law occur (open circles in Fig. 3.35) [121]. Just below T_g a strong jump of ω_α versus $1/T$ is observed (large open circle in Fig. 3.35). The Brillouin results from Fig. 3.35 yield the first relaxation data in a frequency range not attainable so far. Below the temperature related to this jump no further low frequency relaxations are observed even though they should be observable with TDBS if the VFT-law was still applicable. There seems to exist a cut-off of the α-relaxation process. A β-relaxation process accompanying the α-process behaves, as expected, Arrhenius-like.

According to the experimental technique in use (TDBS) all relaxation frequencies gathered until the cut-off are equilibrium data. Now there are two possibilities (see Fig. 3.36): (i) No further α-relaxations exist. In that case the Brillouin data obtained at lower temperatures are equilibrium data (solid line in Fig. 3.36). (ii) the α-relaxation frequency slows down much steeper than

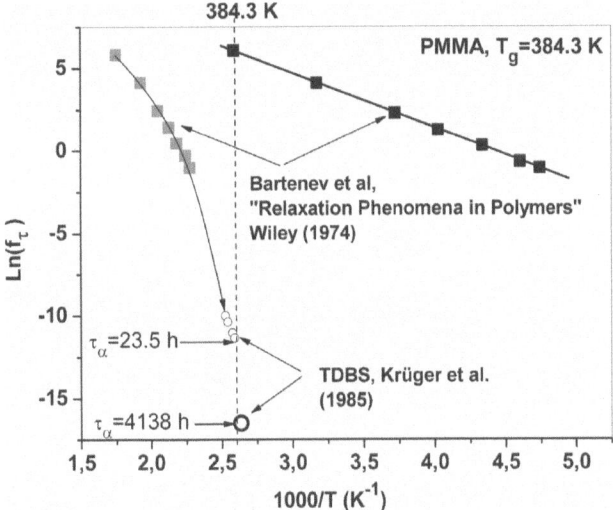

Fig. 3.35. Logarithmic α- and β-relaxation frequencies of atactic PMMA as obtained by dielectric spectroscopy and TDBS

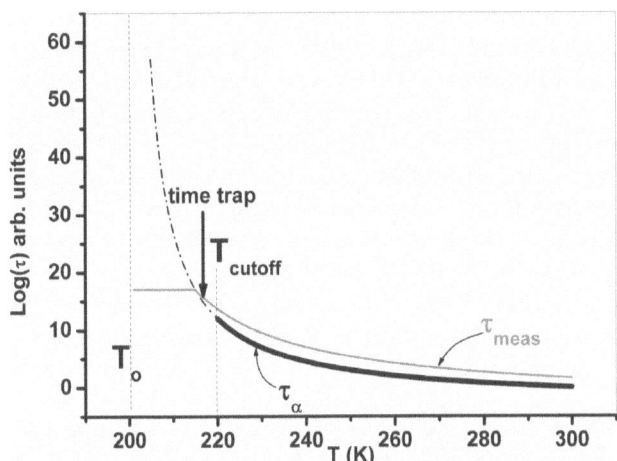

Fig. 3.36. Schematics of the evolution of the relaxation time of the α-process τ_α with an assumed cutoff at $T_{\text{cutoff}} > T_0$. The time scale of the measurement τ_{meas} is chosen in such a way that it is always larger than τ_α

predicted by the VFT-law. In that case measurements performed at lower temperatures are performed out of equilibrium not appearing in the figure.

Independently whether the α-relaxation time really diverges or just becomes a large quantity, the kinetic view of the TGT implies that the operative thermal glass transition temperature reflects a cross-over between the relaxation time τ_α with the time constant of the measurement τ_{meas} (see also

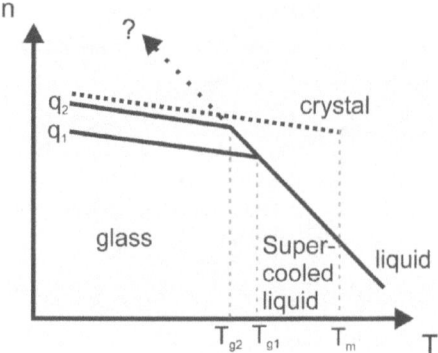

Fig. 3.37. Schematics of the cooling rate dependent behaviour of the TGT. q_1 q_2 cooling rates ($q_1 > q_2$); T_{g1}, T_{g2} related glass transition temperatures; T_m melting temperature of the crystalline state

Sect. 3.1). The operative glass transition temperature T_g is defined (Fig. 3.36) implicitly by $\tau_{meas} \approx \tau_\alpha (T_g)$. This cross-over is the "time trap" mentioned in Sect. 3.1. In that case static or high-frequency clamped phenomenological properties like the mass density $\rho(T)$ or the refractive index $n(T)$ start to deviate from equilibrium (Fig. 3.37). When the cooling/heating rate q of the experiment conflicts with the α-relaxation times of the material the operative glass transition temperatures $T_g(q)$ depend on the cooling rate or more generally on the thermal history of the sample. In terms of the refractive index n (Fig. 3.37) the operative glass transition is reached if during cooling with a rate q_1 the glass forming material is no more able to realize the equilibrium refractive index. If the cooling process develops too fast in comparison to the time needed to establish the equilibrium, a refractive index *inferior to the latter* is realized. If one even increases the cooling rate from q_1 to q_2, the operative T_g shifts according to Fig. 3.37 to higher temperatures. It is worth noting that these bending features within the $n(T)$-curve are kinetic in nature.

It should be stressed again that there exist no experimental proofs for a divergence of τ_α. On adjusting τ_{meas} always in such a way that $\tau_{meas} > \tau_\alpha$ it will emerge from the experiment whether an intrinsic glass transition at a temperature T_{gs} occurs or not and whether T_g coincides with T_0. Only if the condition $\tau_{meas} > \tau_\alpha$ is always fulfilled a kinetically induced glass transition is avoided and the *time trap* is under control. In that case, a kinetically induced slope change of the temperature dependent refractive index is avoided and the question appears about the inherent temperature dependence of $n(T)$ below T_g.

In order to clarify this situation a convenient experimental scenario has to be chosen to recognize a non-diverging time trap so far it exists. A suitable technique is the above mentioned method of "**time domain Brillouin**

spectroscopy" (TDBS) [121, 122]. This technique probes the evolution of the hypersonic velocity stimulated by a step-like change of temperature.

Because of the high probe frequency this technique measures frequency-clamped elastic moduli in the vicinity of the TGT. Using TDBS the only relaxation process which can impose on the sound frequency response can originate in a time lag between the establishment of the sample temperature on one hand and the equilibrium hypersonic velocity on the other. It is worth noting that high-performance TDBS is sufficiently sensitive to resolve temperature jumps $-\Delta T$ as small as 2 K.

Using this technique the two main ingredients necessary to discover an actual cut-off of the *α-process* in the glass transition region are: *a.) hypersonic probe frequencies which are sufficiently high in order to yield high-frequency clamped sound velocities in the pre-transitional temperature regime (frequency domain) and b.) the access to extremely low-frequent acoustic relaxations caused by instantaneous temperature changes (time domain).* For clarity Fig. 3.38 shows a schematic diagram displaying the development of the longitudinal elastic stiffness modulus $c_{11}(T)$ on cooling the sample stepwise from the liquid to the glassy state. At high temperatures the elastic properties are determined by the static longitudinal modulus c_{11}^s. At low temperatures $T < T_g$ the longitudinal modulus of the glassy state c_{11}^{glass} is measured. Above $T \geq T_g$ adjacent to T_g (the pre-transitional temperature regime) frequency clamped elastic moduli c_{11}^∞ are measured.

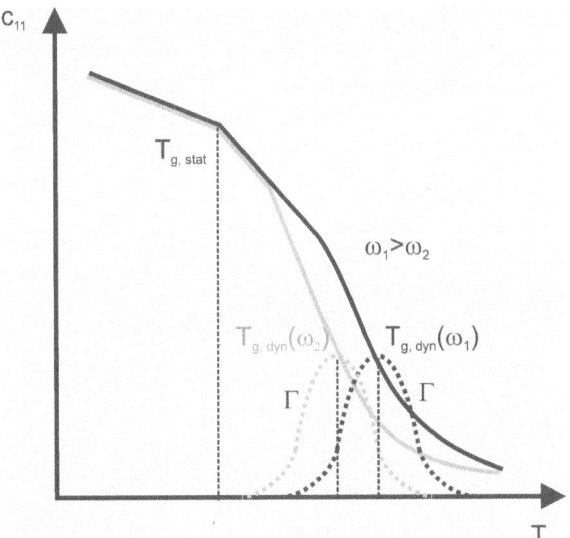

Fig. 3.38. Schematic representation of the static and dynamic glass transition. $T_{g,\,stat}$ static glass transition temperature, $T_{g,dyn}$ dynamic glass transition temperature (dependent on the measurement frequency ω), Γ related acoustic attenuation

The regime of elastic constants between c_{11}^s and c_{11}^∞ is a dispersion region. Within this region the elastic constant $c_{11}(T)$ is strongly temperature dependent and this is accompanied by strong acoustic attenuation (Γ). The attenuation maximum at the so-called dynamical glass transition temperature T_g^{dyn} depends on the probe frequency $\omega = 2\pi f$(e.g. ultrasonics) or alternatively on the acoustic wave vector q (e.g. Brillouin spectroscopy). This maximum determines the α-relaxation time τ_α by the relation

$$\tau_\alpha = 1/\omega\left(T_g^{\mathrm{dyn}}\right) \tag{3.49}$$

At Brillouin frequencies ($f \approx 10\,\mathrm{GHz}$) the hypersonic loss-maximum ($T = T_g^{\mathrm{dyn}}$) is often but not always located about 150 K above the TGT and the lower wing of that maximum ends well above T_g. In some cases the low-temperature wing of the hypersonic loss peak shows a cut-off directly at T_g (s.a. Fig. 3.39) which means that hypersonic relaxation processes persist just until T_g. The fact that these relaxation processes are stopped by the TGT indicates the close relation with the primary glass transition, since secondary relaxations would not be affected by the TGT. *Therefore it seems that the TGT is not caused by the α-process but rather that the TGT stops the freezing dynamics.* From Fig. 3.39 we find that in case of DGEBA the temperature difference between T_g and T_g^{dyn} is only about 100 K. A further discussion of this feature will be given in the next section.

A central question concerns the temporal behaviour of static or frequency-clamped susceptibilities in the vicinity of the TGT. In terms of high frequency elastic stiffness data the question appears whether the kink in the sound velocity-temperature curves or in the elastic moduli-temperature curves is maintained at extremely slow cooling or whether it can be shifted succes-

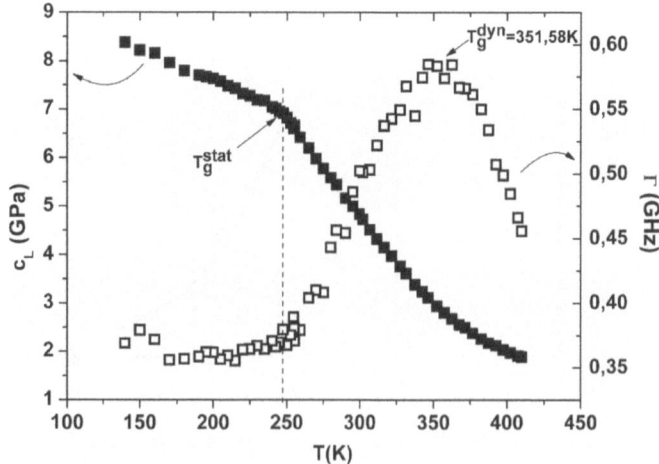

Fig. 3.39. Longitudinal elastic modulus c_L and hypersonic attenuation Γ versus temperature T for the fragile glass former DGEBA

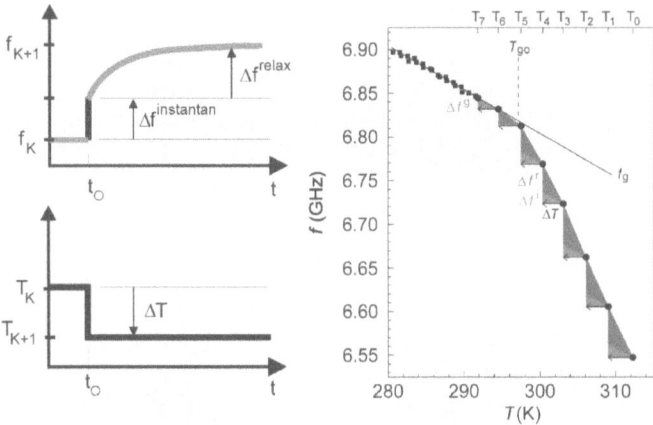

Fig. 3.40. Schematic drawing of the time domain BS. Left hand side: temperature step $T_k \rightarrow T_{k+1}$, related sound frequency response $f_k \rightarrow f_{k+1}$, separated in an instantaneous and a relaxing part. Right hand side: sound frequency f vs. temperature T, dots representing measured values while the triangles symbolize the instantaneous and the relaxing sound frequency response after each temperature step. Below the operative glass transition temperature T_{go} only an instantaneous response is measured

sively to lower temperatures with a temperature limit determined only by the patience of the experimentalist.

Figure 3.40 shows schematically the approach of TDBS in case of polyvinylacetate (PVAc) [124–129]. According to Fig. 3.40 a temperature jump of $\Delta T = T_k - T_{k+1}$ is imposed on the sample. Simultaneously the sound frequency response is recorded as a function of time t at the fixed temperature T_{k+1}. Every record is performed to that end that the sound frequency response has reached its asymptotic value f_{k+1}. As indicated in Fig. 3.40 the complete sound frequency response consists of two parts, an instantaneous response Δf^i and a relaxing response Δf^r. The instantaneous response is almost independent of the temperature T_{k+1} and corresponds to the entire sound frequency response within the glassy state. This result indicates, that the instantaneous response of the liquid state reflects already the temperature coefficient of the longitudinal elastic stiffness of the glassy state. This means that the temperature coefficient of the instantaneous hypersonic response reflects already on the anharmonicity of the frozen state and supports the observation that the temperature coefficient of the longitudinal elastic constant within the glassy state does not depend on the thermal history of the sample.

The following investigations are based on TDBS and dielectric spectroscopy (DES). As usual DES has been performed in the plate condenser geometry at different temperatures in the range of 173 K– 393 K. The dielectric spectra are obtained for a frequency range between $f = 10^2$ Hz and

10^5 Hz and were fitted using Havriliak-Negami

$$\hat{\varepsilon}(\omega) = \varepsilon' - i\,\varepsilon'' = -i\,\frac{\sigma_{DC}}{\omega\,\varepsilon_0}$$

$$+ \sum_{k=\alpha}^{\gamma} \left[\frac{\Delta\varepsilon_k}{(1 + (i\,\omega\,\tau_k)^{\nu_k})^{\mu_k}} + \varepsilon_{\infty k} \right] \tag{3.50}$$

as a model function which in addition to the usual dipole relaxation processes (α, β, γ) of glass-forming materials includes a dc conductivity term. The quantities $\Delta\varepsilon_k$ designate the relaxation strength and the τ_k represent the related relaxation times. The $\varepsilon_{\infty k}$ are the frequency-clamped dielectric constants related to the different relaxation processes k.

Coming back to the time domain method of BS we need to describe the relaxing part of the frequency response in an analytic way. On approaching the TGT the relaxing part Δf^r needs increasingly time to evolve (Figs. 3.41, 3.42). The temporal evolution of the hypersonic α-relaxation process can be described by a Kohlrausch-Williams-Watts (KWW) law with a stretched exponential [102]:

$$f(T) = f^{\infty} - \left(f^{\infty} - f^{\text{inst}}\right) \cdot e^{\left\{-\left(\frac{t}{\tau_\alpha}\right)^{\beta}\right\}} \tag{3.51}$$

where f^{inst} is the instantaneous frequency response, f^{∞} is the relaxed value and β measures the distribution of the α-relaxation time ($\beta = 1$ corresponds to monodisperse processes).

The average relaxation time is derived from

$$\langle \tau_\alpha \rangle = \frac{\tau_\alpha}{\beta} \cdot \Gamma\left(\frac{1}{\beta}\right) \tag{3.52}$$

where the Gamma-function yields the time-average of the distribution.

Every fit yields an average relaxation time and a limiting value for the relaxed sound frequency at the related temperature jump. The relaxed sound frequency data are shown in Fig. 3.40. Since the exact location of the intrinsic glass transition temperature, T_{gs}, was not known prior to the TDBS experiments, it was not possible to perform a temperature jump ending precisely on T_{gs}. In reality a temperature jump was performed which started slightly above T_{gs} ($T_{gs} + \varepsilon$) and which ended slightly below T_{gs} ($T_{gs} - \varepsilon$). Astonishingly, at ($T_{gs} - \varepsilon$), that means already within the glassy state, a final sound frequency relaxation is observed (Fig. 3.43). At still lower temperatures further relaxations are not found. The asymptotic sound frequency value does definitely no more meet the linear extrapolation of the related sound frequency of the liquid state. This result was confirmed by estimating the potential relaxation frequency of the next temperature step within the glassy state. For this purpose the relaxation frequencies have been plotted in an activation diagram (Fig. 3.44). According to this estimation the relaxation frequency should have been observable but was not observed.

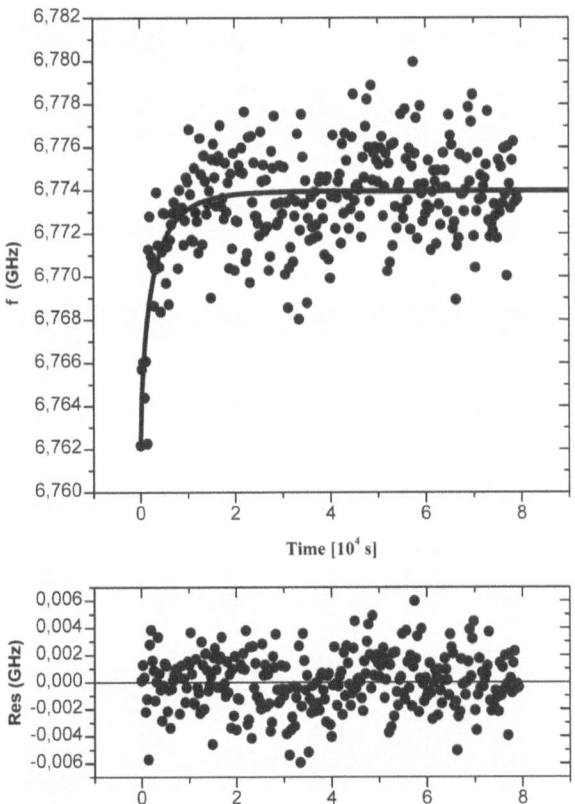

Fig. 3.41. Sound frequency response f of PVAc after a temperature step from 307.7 K to 305.7 K. The full line represents a KWW-fit which residuals are shown below

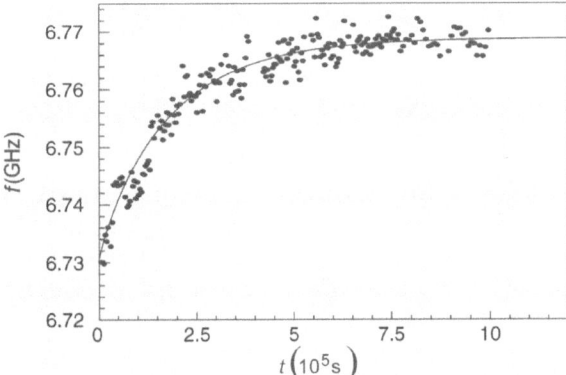

Fig. 3.42. Sound frequency response f of PVAc after a temperature step from 303.3 K to 300.4 K. The full line represents the KWW-fit

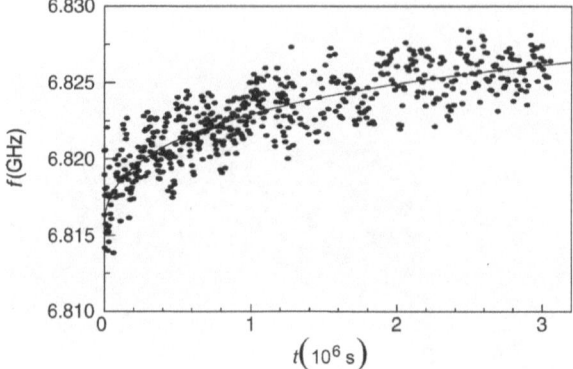

Fig. 3.43. Sound frequency response f of PVAc after a temperature step from 297.5 K to 294.6 K. The full line represents the KWW-fit

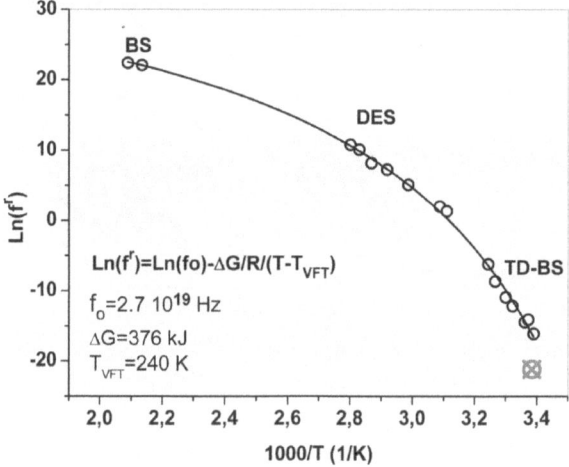

Fig. 3.44. Activation-plot (α-relaxation frequencies vs. inverse temperature $1000/T$) for PVAc. The frequencies have been determined by BS, dielectric spectroscopy (DES) and time domain (TD) BS

According to these TDBS-data even in the limit of extremely slow cooling there appears a kink in the sound frequency – temperature curve at a temperature T_{gs} (s.a. Fig. 3.40) which is close but slightly below the T_g measured under usual conditions realized in Brillouin experiments. Indeed, the measured T_{gs} is located well above T_0 predicted by the VFT representation.

The activation plot of Fig. 3.44 contains, beside dielectric data, Brillouin data measured in the frequency- as well as in the time domain. It seems that these data roughly follow a VFT law; however, a close statistical inspection contradicts this interpretation. Especially the low frequency relaxation frequencies behave Arrhenius- instead of VFT-like. The relaxation

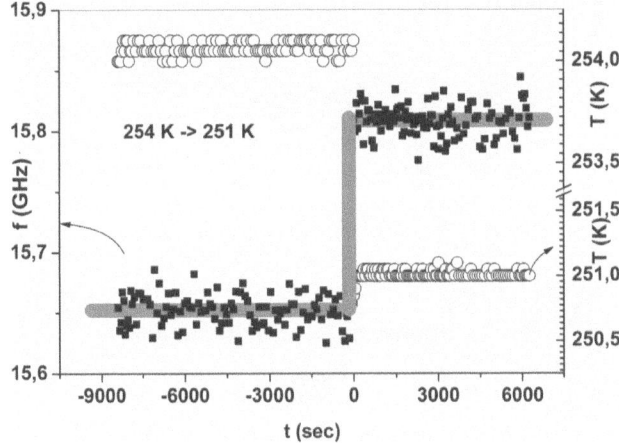

Fig. 3.45. Temperature step from 254 K to 251 K (right ordinate) at the reference time for the new temperature step $t = 0$ and the related response of sound frequency of DGEBA. The *horizontal grey lines* are least squares fits

frequency at the lowest temperature doesn't fit at all to the VFT-law. The relaxation frequency of this data point in the glassy state is two orders of magnitude lower than that of the preceding temperature.

The above discussed investigations on the TGT of atactic PVAc suggest the existence of an intrinsic glass transition at a well defined temperature T_{gs}. This intrinsic glass transition is evidenced by a strong anomaly of the α-relaxation frequency at T_{gs} and by a well defined kink in the sound frequency-temperature diagram.

Recent investigations of glass forming *diglycidyl ether of bisphenol A* (DGEBA) confirm for the primary glass relaxations [122, 123, 130] a strong deviation from the Vogel-Fulcher-Tamman (VFT)-behaviour. In the case of DGEBA the thermal glass transition point is evidenced by a kink in the sound frequency of the longitudinal polarized hypersonic mode and a sudden disappearance of the α-relaxation time flattening out as a function of decreasing temperature.

The experimental procedure of TDBS shown schematically in Fig. 3.40 was also used for DGEBA. As for PVAc the temperature-time scenario was started well above the freezing process e.g. at $T \geq T_{gs} + 10$ K and the phonon spectra were accumulated for a sufficiently long time t in order to be sure that the sound velocity (frequency) v (f) has relaxed to its equilibrium value v^∞ $(\propto f^\infty)$ of the liquid state.

Figure 3.45 demonstrates that a "temperature step" of $\Delta T = -3$ K at 254 K produces no relaxations of the sound velocity response at all. Consequently, the sound velocity response is as fast as the temperature step experimentally realized. Approaching T_{gs} and making a temperature step from 249 K to 247 K (Fig. 3.46) the sound velocity response is no more instan-

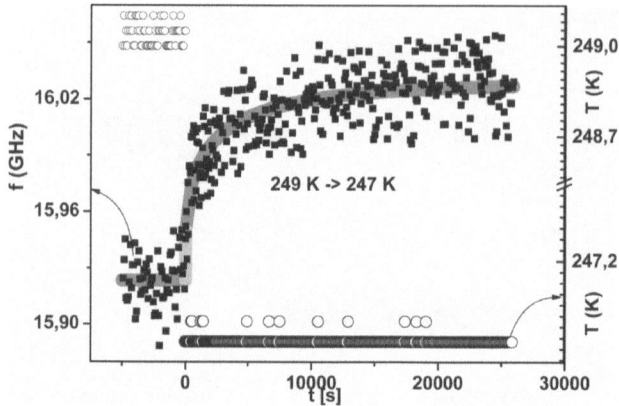

Fig. 3.46. Temperature step from 249 K to 247 K (*right scale*) and related sound frequency response of DGEBA (*left scale*). (For further explanations, see text)

taneous but shows a time lag. As for PVAc it was found that close to T_{gs}, the sound frequency response can be divided into two parts: an instantaneous response f^{inst} which responds as fast as the temperature equilibrates and a relaxing part that can be described by the Kohlrausch-Williams-Watts law Eq. (3.51) [101].

The temperature step from 243 K to 241 K (Fig. 3.47) yields a decreased total sound velocity response but no more hint for any relaxation process. From the relaxation times measured at the foregoing temperature steps a hypothetical relaxation time has been estimated for the step from 243 K to 241 K. This estimated sound velocity versus time relaxation curve is also shown in

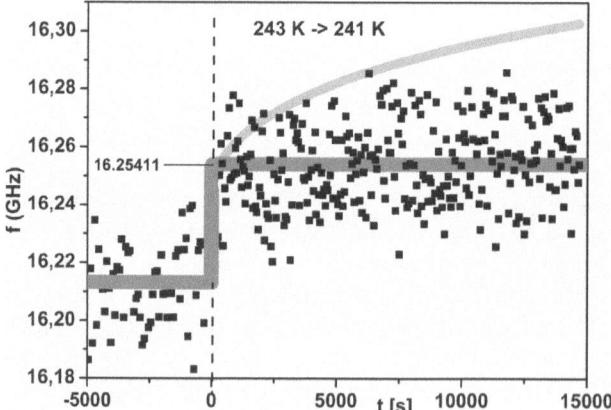

Fig. 3.47. Sound frequency response of DGEBA after a temperature step from 243 K to 241 K. The *horizontal dark gray lines* is a linear least-squares fit. The *light gray curve* gives an extrapolation to fluid conditions. (See text for further explanations)

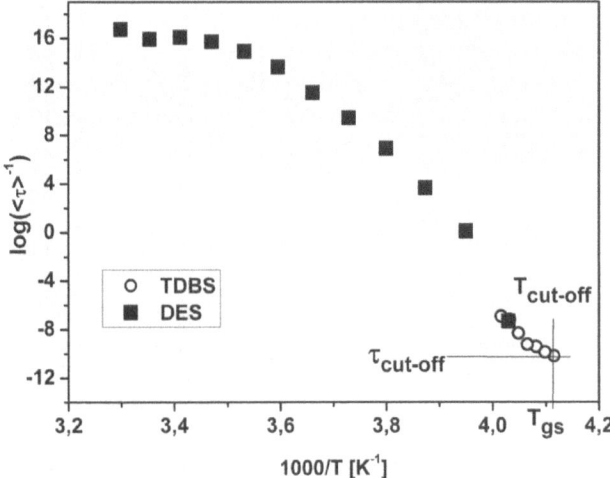

Fig. 3.48. Activation-plot of the inverse averaged α-relaxation times of DGEBA as measured by DES and TDBS

Fig. 3.47. From this estimate, it can be concluded that at least this kind of relaxation should have been observable if it existed. This result implies that under the given conditions in DGEBA below 243 K, no α-relaxation process exists anymore.

The values calculated there for $\langle \tau_\alpha \rangle$ are presented for all measured temperature steps in Fig. 3.48. The step from 245 K to 243 K is the last temperature step still showing a relaxing sound-frequency response as a consequence of the step-wise temperature excitation. Accordingly, the subsequent temperature step from 243 K to 241 K shows only an instantaneous sound velocity response of the glassy state (Fig. 3.36). It is therefore natural to identify $T_{gs} = 243$ K as the intrinsic TGT (see Subsect. 3.5).

Astonishingly, the sets of relaxation times as derived from the DES and TDBS agree quite well with each other, within the margin of error. Taking into account different thermodynamic boundary conditions, irreversible thermodynamics would in principle have predicted different relaxation times for different external variables involved in the relaxation process. As expected at higher temperatures, the α-relaxation behaviour as a function of $1/T$ is compatible with the expected VFT-relation. However, on approaching T_{gs} the activation plot flattens out (Fig. 3.48) and shows a cut-off at 243 K indicating a sudden disappearance of the α-relaxation process. At the same temperature the $f(T)$-curve shows a kink which, according to the genesis of the sound-frequency data is neither due to the cross-over of the experimental time scale with the α-process nor obscured by it.

The main conclusion of the current chapter is that the main glass relaxation process, the α-relaxation process, does not follow the VFT-law through the glass transition down to a VFT-temperature T_0. TDBS investigations on three different glass forming liquids show that the α-relaxation process exhibits a cut-off well above the VFT-temperature T_0. This means, that the "cross-over" argument does not hold for these three materials and that the glass transition temperatures obtained from the TDBS-procedure are intrinsic freezing temperatures specific for each of the three materials.

3.5 Static Properties at the Thermal Glass Transition

Taking into account that the TGT in canonical glasses affects predominantly their mechanical properties, one could argue that these properties are suitable probes in order to elucidate the mechanism of the quasi-static glass transition and eventually of the ideal glass transition proposed above [131,132]. In this context **Brillouin spectroscopy** (BS) is an experimental method of particular interest because BS measures mechanical properties of the glass former in a completely non-destructive manner and without mechanical contact with the sample [7,8,133–136]. Moreover, even some ten degrees above T_g, BS measures the instantaneous elastic response in the "slow motion regime" in the sense that $\tau_\alpha \omega \gg 1$ holds (τ_α structural relaxation time, ω sound frequency). The latter property provides the possibility to determine, in addition to the hypersonic properties, the refractive indices from pure Brillouin spectroscopic data [137]. Unfortunately, the quasi-static GT is usually only hinted by an undramatic, more or less sharp, kink in the sound frequency vs. temperature curve, which yields no hint for a non-ergodic instability. In addition, this kink depends to a certain extent on the thermal history of the sample, see Sect. 3.1. A main task for the experimentalist is therefore to find experimental scenarios which give answers to the question whether there may exist an intrinsic glass transition and a glass transition temperature which is experimentally accessible and at which some static phenomenological properties change in a well-defined manner.

Static phenomenological properties should be measured either with a static measurement technique in internal thermodynamic equilibrium or with dynamical techniques in frequency-clamped equilibrium. In terms of irreversible thermodynamics, internal equilibrium means that the relevant internal variables ξ_k can be completely expressed by the external thermodynamic variables $\xi_k = \xi(p, T, \vec{E}, \ldots)$. Dynamic measurements of phenomenological properties in internal equilibrium imply

$$\omega \cdot \tau_k \ll 1 \quad \forall \quad k \tag{3.53}$$

where the τ_k are the structural relaxation times connected to the internal variables ξ_k and where $\omega = 2 \cdot \pi \cdot f$ is the probe frequency of the dynamic experiment [139]. In contrast, clamped equilibrium means that for the probe frequency involved all relevant internal degrees of freedom ξ_k are frequency-clamped, i.e. $\xi_k(\omega) = $ const for $k = 1, 2, 3 \ldots$. In relation to the relevant relaxation times τ_k and the probe frequency ω the condition of clamped internal variables then means that the system is excited in the so-called "slow motion regime"

$$\omega \cdot \tau_k \gg 1 \quad \forall \quad k \tag{3.54}$$

The measurements of static phenomenological properties like specific volume (mass density) or the properties of the static specific heat capacity are experimentally extremely difficult to realize, especially in the vicinity of the thermal glass transition. Rehage et al. [131], Schwarzl et al. [132] and Kovacs et al. [140, 141] e.g. have demonstrated the experimental difficulties which have to be overcome in order to determine the specific volume across the transition from the liquid to the glassy ("solid") state. One of the problems concerns sticking of the sample to the container walls and accompanying internal stresses. Similar difficulties occur in the course of adiabatic calorimetry measurements [142]. On the other hand this latter technique is the only one which gives reliable information about the static specific heat capacity in the vicinity of the TGT.

Dynamical measurements of clamped phenomenological susceptibilities like dielectric constants measured at optical frequencies, $\varepsilon_{\mathrm{opt}} = n^2$ (n refractive index), or clamped elastic constants, c_{ii}^∞, measured at hypersonic frequencies in the vicinity of the glass transition yield equilibrium (frequency-clamped) properties provided the temperature changes necessary in order to reach the thermal glass transition (TGT) are sufficiently small and are performed sufficiently slowly (see Sect. 3.4). Figure 3.49 shows the typical behaviour of the longitudinal and of the shear stiffness of a bisphenol A (I1) [143, 144]. The experimental fact that the shear stiffness c_{44}^∞ is observed at Brillouin frequencies above T_g proves that both moduli c_{44}^∞ and c_{11}^∞ are measured in the "slow motion" regime. They are therefore high frequency-

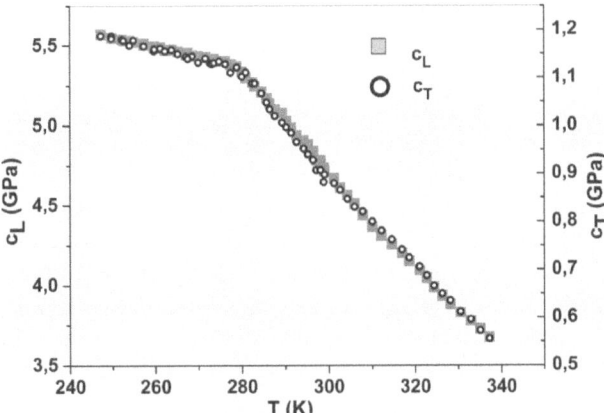

Fig. 3.49. Sound velocity of the longitudinal (*full symbols*) and shear (*open symbols*) mode of I1

clamped equilibrium properties with respect to the longitudinal and shear deformation. Both curves show a kink-like temperature dependence at the operative glass transition temperature T_g. The question whether the operative kink temperature T_g can be shifted to lower temperatures for infinitely slow cooling decides about the existence of an intrinsic glass transition at a well-defined temperature. The fact that both elastic stiffness curves could be brought to coincidence using an appropriate scaling of the coordinate axes signifies that I1 follows the generalized Cauchy-relation in the whole temperature range, which is an additional proof for the absence of hypersonic relaxations.

The method of time domain Brillouin spectroscopy (TDBS) introduced in Sect. 3.4 is one of the key experimental methods to study the static aspects of the thermal glass transition. TDBS enables us to investigate the elastic constants with respect to the mechanical deformation in the slow motion regime. In the course of Brillouin investigations of the TGT the only remaining external variable is then the temperature T. Indeed a sudden temperature change ΔT, necessary in order to approach the TGT at T_g, can initiate a relaxation of the sound velocity (sound frequency) with a relaxation time τ_α which increases on approaching T_g from above. Thus TDBS is able to probe mechanical relaxations with respect to temperature changes ΔT necessary to investigate the TGT. Particularly, TDBS is able to clarify in which temperature regime above and below T_g thermo-acoustic relaxations are active and especially whether they do persist below T_g. The first systematic investigations were reported for polyvinylacetate (PVAc). The investigations on PVAc and other glass-forming materials were performed persuing the following scheme: At some ten Kelvin above T_g hypersonic frequency measurements were started in the slow motion regime $\omega \cdot \tau_\alpha \gg 1$. In order to approach the glassy state small temperature steps $-\Delta T$ (see Fig. 3.50) were performed and the isothermal sound velocity (-frequency) at fixed wave vector was simultaneously recorded until saturation of the velocity/frequency response. Only the fully relaxed sound velocity data were then used for the sound velocity/temperature plot. Figure 3.50 shows the results for PVAc. The most important result of these TDBS investigations is that there exists a well-defined temperature T_g where every α-relaxation disappears, or at least spontaneously increases its value by several orders of magnitude. It is worth noting that T_g is well above the hypothetical VFT-temperature T_0.

In order to prove that the hypersonic measurements were really performed in the slow-motion regime $\omega \cdot \tau_\alpha \gg 1$ the opto-acoustic dispersion function (D-function, Eq. (3.55), s.a. Sect. 3.2) was measured with BS and compared with high performance refractive index measurements. If hypersonic relaxations are present the D-function shows a convex deviation from the refractive index curve. The maximal deviation occurs at that temperature where $\omega \cdot \tau_\alpha = 1$.

$$D(T) = \sqrt{\frac{\left(\frac{f^{90R}}{f^{90A}}\right)^2 + 1}{2}} \tag{3.55}$$

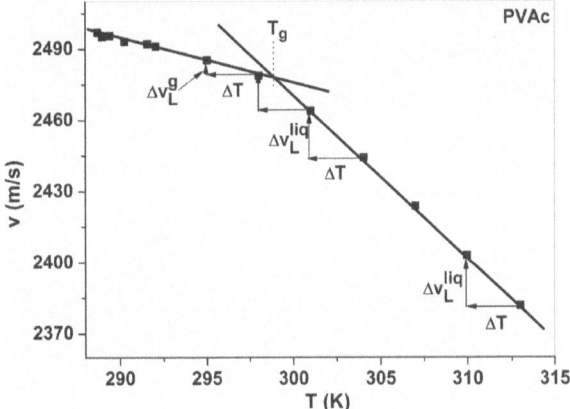

Fig. 3.50. Sound velocity of the longitudinally polarized sound mode of PVAc measured with TDBS. ΔT temperature jump, $\Delta v_L^{\text{lic, }g}$ relaxed sound velocity response in the liquid (liq) and the glassy (g) state

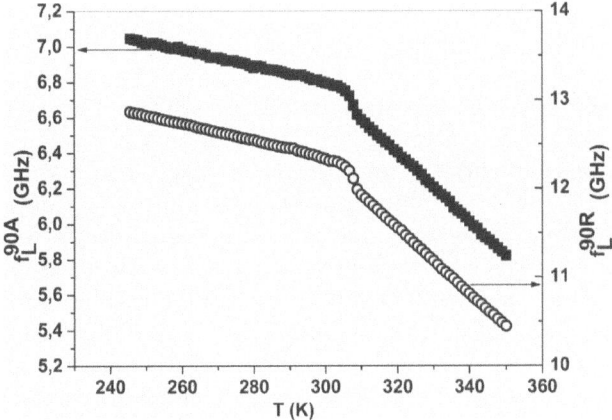

Fig. 3.51. Temperature dependence of the hypersonic frequencies of PVAc measured on heating after extremely slow cooling (see Fig. 3.50) using the 90A- and 90R-scattering geometry

The D-function was derived from simultaneously determined sound frequencies measured in the 90A- and in the 90R-scattering geometry (Sect. 3.2) and Fig. 3.51.

Since the sound frequencies presented in Fig. 3.51 were measured on heating after an extremely slow cooling of the sample into the glassy state in the course of the TDBS experiment mentioned above, both sound frequency curves show an overshoot (bumper) in the temperature regime of the TGT. This metastability occurs because the material was cooled much more slowly

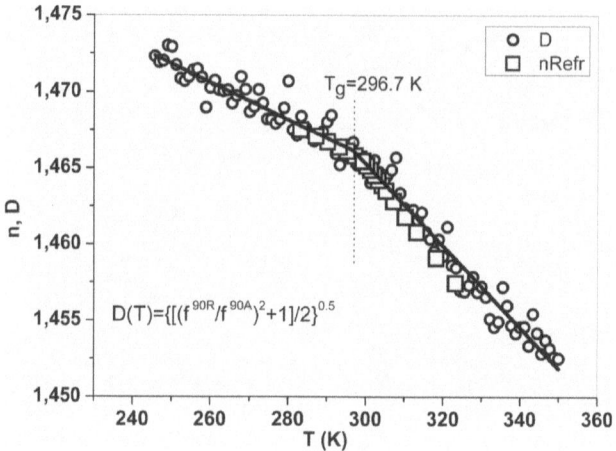

Fig. 3.52. Temperature dependence of the opto-acoustic dispersion function (D-function) and the refractive index function (n-function) of PVAc

to the glassy state (over months) than heated up during the BS measurement (5 days) (see Fig. 3.4).

The refractive indices were measured on step-like cooling over two months using an Abbé refractometer (Fig. 3.52). Within the margin of error the D-function coincides with the refractive index function $n(T)$ yielding $D(T) = n(T)$ (Fig. 3.52). Since the D-function is derived from two independently but simultaneously measured sound frequency curves the larger data scatter of this curve in comparison to the $n(T)$-curve is comprehensible. In the margin of error the glass transition temperatures measured with both techniques are identical. In consequence in the temperature range of this investigation any hypersonic relaxations or optical relaxations are either absent or act the same way on the two phonons measured in the 90A- and 90R-scattering geometry.

It is extremely important to note that the two "bumpers" at T_g shown in Fig. 3.51 are absent in the D-function (Fig. 3.52). These bumpers are of course signs for metastability or even for instability and depend on time and temperature. The reason for the disappearance seems to be a common factor which guides the evolution of $f_L^{90A}(T)$ as well as of $f_L^{90R}(T)$.

Taking into account that the D- and the n-function are in very good agreement (Fig. 3.52), the disappearance of the overshoot within the D-function is not expected to be an artifact of the data-analysis.

Keeping in mind this result, the effect of q-vector dependent relaxations on the temperature dependence of the D-function has to be investigated. Figure 3.53 shows the temperature dependence of the D-function of PVAc up to 530 K. According to Eq. (3.55), $D(T)$ coincides with $n(T)$ outside of relaxation regimes and deviates in a convex manner from the n-function. This is exactly what happens for PVAc above 350 K. The maximum convex deviation

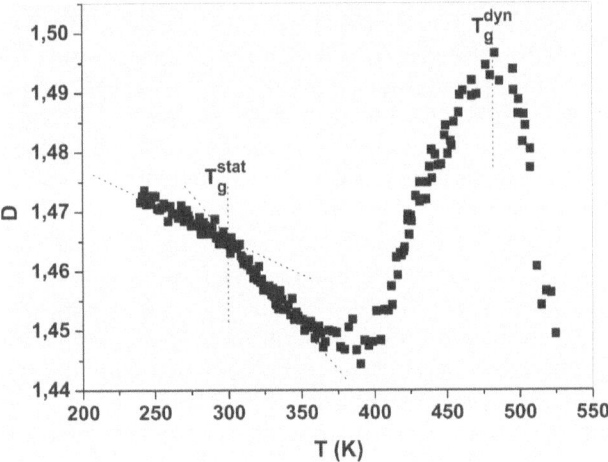

Fig. 3.53. Opto-acoustic D-function of PVAc measured over a wide temperature range including the range of the dynamical glass transition at hypersonic frequencies

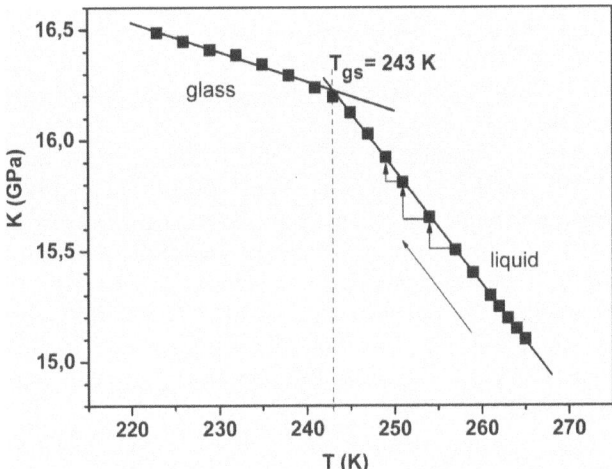

Fig. 3.54. Relaxed compressional modulus as a function of temperature for DGEBA. The *straight lines* are fit curves. The *arrow* indicates the cooling regime

from the n-function defines the dynamic glass transition temperature at the Brillouin frequencies involved in measurement.

In order to elucidate the nature of the TGT the temperature behaviour of the D-function within the glassy state ($T < T_g$) deserves a final discussion: (i) As the $D(T)$- and the $n(T)$-functions coincide totally within the glassy state, $(D(T) - n(T)) = 0$, there is no evidence for acoustically relevant relaxations in the GHz-frequency regime. (ii) The $D(T)$-representation suppresses slowly varying effects of metastability present in the constituting functions, i.e. the related phonon frequencies. In other words, the $D(T)$-function acts as a filter for changes in acoustic properties due to metastability and displays only equilibrium properties. It is therefore concluded that the $D(T)$-function displays equilibrium properties for the glassy state of PVAc and that in particular the kink at T_g is not due to a cross-over between intrinsic relaxation times and the experimental time scale. Accordingly, the glass branches of the sound frequency (velocity) curves are not metastable.

In literature the TGT is often brought in close relation to the so-called α-relaxation process. The α-relaxation frequency f^α seems to slow down, according to the VFT-law if the temperature of the glass-forming sample approaches T_g and seems to go to zero if the temperature approaches the so-called VFT-temperature T_o (see Sect. 3.1). Hence it appears that the α-relaxation process is in some way the leading process for the glass transition. This holds true for the dynamical glass transition well above the static glass transition when the probe frequency crosses the relaxation frequency. The situation close to the TGT is much more complicated.

At first, the true behaviour of $f^\alpha(T)$ in the vicinity of T_g is an open question and depends evidently on the experimental probe. Moreover, the predicted VFT-behaviour is somewhat similar to the behaviour of a soft mode frequency at a structural phase transition. Keeping in mind that soft-mode frequencies never really go to zero it is thus interesting to know what happens with the α-relaxation frequency close to T_g.

Secondly, if the α-relaxation follows a VFT-behaviour on one hand and if the TGT is investigated with a high frequency probe on the other hand, there should be no probe frequency which could cross the α-relaxation frequency. Consequently the only possibility would be that the waiting time between subsequent temperature-changes is too small compared to the thermoelastic relaxation time. Only the latter event could eventually provide a kinetic influence on the glass transition. So the question arises again what happens if this latter kinetic effect is experimentally avoided and what happens in this case with α-relaxation processes.

Having established for DGEBA simultaneously the kink-like anomaly in the curve of the frequency clamped sound velocity and the cut-off of the α-relaxation process (Sect. 3.4) [145] we conclude that PVAc is not an exception but that also for this material there exists an intrinsic glass transition approximately 20 K above the VFT-temperature T_o. Taking into account that the average relaxation time for the temperature jump to $T_{gs} = 243$ K amounts to 5.4 hours and that sound frequency changes in the range of some per mille have to be resolved unambiguously during hours and / or days after a single temperature jump, it is obvious, that the experimental proof of a sample-inherent glass transition process is experimentally not an easy task.

In order to elucidate in addition to the extremely low frequency relaxation behaviour the behaviour at hypersonic frequencies we have simultaneously investigated the sound velocity v_L and the related hypersonic attenuation Γ_L of DGEBA over a wide temperature range (Fig. 3.55).

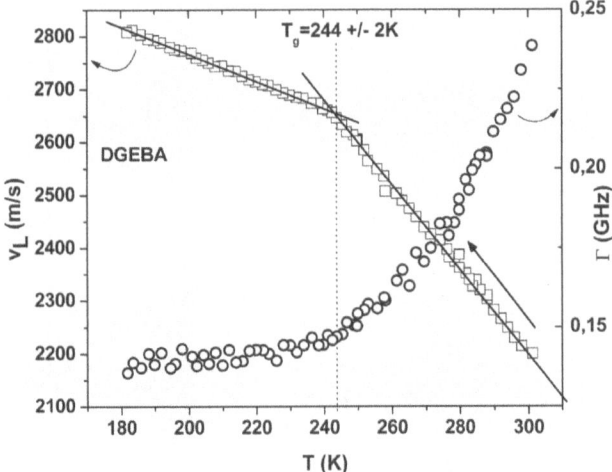

Fig. 3.55. Temperature dependence of the longitudinal sound velocity v_L and the hypersonic attenuation Γ_L of DGEBA

Astonishingly, and in contrast to the attenuation behaviour of many other glass forming liquids, the hypersonic attenuation maximum comes down directly at the TGT. In other words, for DGEBA until the TGT there exist hypersonic relaxations, which slow down only at T_g. For this material it is therefore not true that all α-relaxation times are extremely large close to the glass transition. From the TDBS- and the dielectric measurements it is clear that there exist α-relaxations which slow down to 5.4 h, but at the same time the hypersonic attenuation behaviour clearly demonstrates the existence of pronounced hypersonic relaxations which only stop at the TGT. So the question arises, why these high frequency relaxations are stopped by the TGT if the TGT itself is believed to be caused by these relaxations?

Taking into account the importance of this observation for the interpretation of the TGT it is interesting to look for further examples, which show a similar relaxation behaviour. A very impressive example for the evolution of the hypersonic dynamics around the TGT was published recently for an epoxy (EPON) [146, 147].

Figure 3.56 shows the hypersonic velocity and the opto-acoustic dispersion function of EPON, both as a function of temperature. The hypersonic velocity behaves as expected in showing a rather sharp kink at the TGT. Really astonishing is the temperature dependence of D-function. Within the glassy state $D(T)$ behaves as expected like $n(T)$. Precisely at T_g the D-function changes the slope, indicating the onset of hypersonic relaxations. Again this result is in contradiction to the general statement, that close to T_g all relaxations connected to the glass transition should be of very low frequency. Rather, the observed hypersonic relaxation process dies out at the TGT. Whether the existence of this hypersonic relaxation process is without any doubt, its

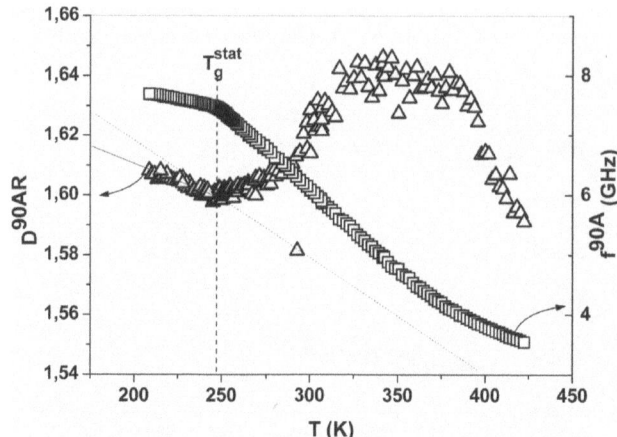

Fig. 3.56. Hypersonic velocity and opto-acoustic dispersion function (D-function) for EPON, see text for details

origin is not yet clear. From the fact that the D-function emerges from two phonon-frequencies related to two acoustic wave vectors of different magnitudes it follows that the convex deviation from the n-function (Fig. 3.56) is due to a relaxation process. The distribution of the relaxation process seems to be either double-valued or alternatively extremely wide.

The complex dendrimer molecule G1 shown in Fig. 3.57 is a model molecule for a glass-forming material, which cannot crystallize [148]. Consequently, the glass transition of the material G1 cannot be masked by a recrystallization process. Whereas the longitudinal sound velocity behaves as expected as a function of temperature (Fig. 3.58), the temperature dependence of the hypersonic attenuation behaves in a strange manner. On approaching the TGT from above the hypersonic attenuation Γ increases permanently until T_g and remains constant within the glassy state.

The hypersonic attenuation Γ is rather small in the whole temperature regime but shows a minimum at $T \sim 215\,\mathrm{K}$. The attenuation then increases with decreasing temperature towards the TGT. Assuming that the Brillouin measurements are performed in the slow-motion regime this increase can hardly be explained. Another cause for this attenuation could be acoustic scattering on internal stresses, which are built up if the temperature approaches the TGT. Once the dendrimer is frozen, the elastic scattering remains constant. But even if this explanation is not true it is clear that the lack of certain dynamics below the TGT is responsible for the constancy of the sound attenuation.

Having elucidated the relation between the static and the dynamic properties at the TGT a reinvestigation of the relation between kinetic and static properties is reasonable. Taking the temperature and time dependence of the high-frequency longitudinal elastic modulus c_L^∞ as a reference, Fig. 3.59

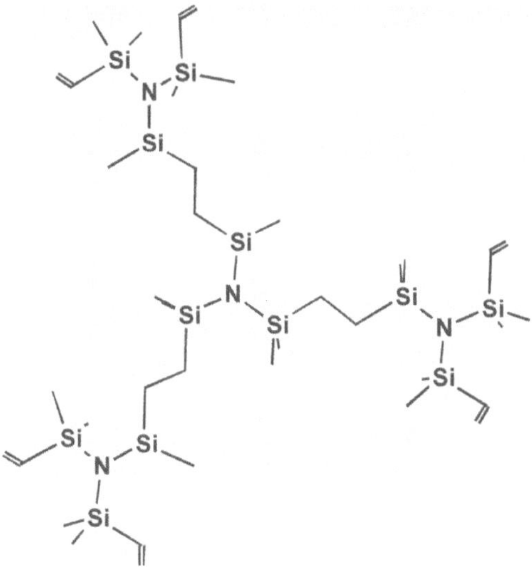

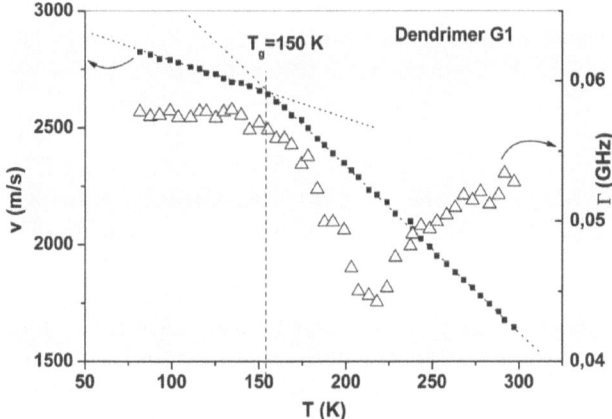

Fig. 3.57. Two-dimensional projections of the structure of the Dendrimer G1. The carbon atoms and the hydrogen atoms have been omitted

Fig. 3.58. Temperature dependence of the sound velocity v and the hypersonic attenuation Γ for G1

displays schematically the expectation for this temperature-time dependence from the kinetic point of view on the left side and the experimental observation on the right side. The equilibrium liquid branch of c_L^∞ $(T \geq T_g)$ has a unique tangent

$$m_\ell = \frac{\partial}{\partial T} c_L^\infty \ (T \geq T_g) = \text{const} .$$

At sufficiently high temperatures the glass-forming material is in the liquid state. As it is well known [149] the packing of the liquid state is almost solid-

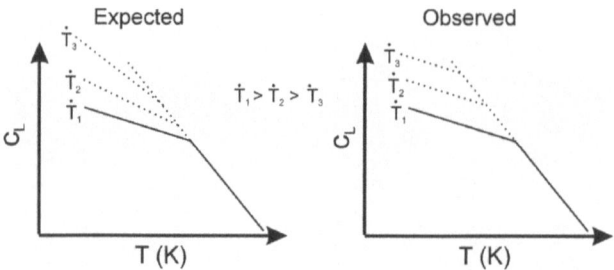

Fig. 3.59. *Left*: expected behaviour of the temperature dependence of the elastic modulus c_L around the TGT due to different cooling rates. *Right*: actually observed behaviour of this temperature dependence around the TGT

like, the same holds true for the local order. However, there remains some free volume, which gives space for some random flight of the molecules. Since at the TGT the specific volume behaves continuously (see Sect. 3.3) the liquid and the glassy state have the same specific volume at T_g.

Assuming a rather high reference temperature T and a probe frequency f which is sufficiently high providing that still a high frequency clamped elastic modulus is measured, the related $c_L^\infty(T \geq T_g)$-value is located on the equilibrium liquid branch. At slow cooling the measured c_L^∞-data remain on this liquid branch. At a sufficiently large cooling rate $\dot{T}_1$ the c_L^∞-curve bends away from the equilibrium liquid curve (Fig. 3.59). It should be stressed that this bending occurs despite the fact that the molecular dynamics related to the formation of the glassy state is already frequency-clamped. Consequently, what is observed is the temperature-rate dependence of a frequency-clamped quantity $c_L^\infty(\dot{T})$. The question arises then about the tangent of the c_L^∞-curve during fast cooling and its isothermal recovery after quenching. According to all existing experimental data, the change of the tangent $\partial c_L^\infty(T)/\partial T$ behaves continuously as a function of time and of temperature. In the temperature interval between the start of the quenching process and the TGT the absolute value $c_L^\infty(T, \dot{T})$ and the tangent $\partial c_L^\infty(T)/\partial T$ converges versus the related equilibrium values of the liquid state. The relaxation time of this isothermal recovery process increases strongly on approaching the TGT (see Sect. 3.4). The question arises, following a kinetic view of the glass transition, what happens below T_g with $c_L^\infty(T \leq T_g, \dot{T})$ and $\partial c_L^\infty(T \leq T_g)/\partial T$? Intuitively, a smaller cooling rate $\dot{T}$ should bring these values closer to the expected equilibrium data which are located on the extrapolated liquid branch (Fig. 3.59, left).

However, that's not what is observed! A schematic drawing of what actually is observed is given on the right side of Fig. 3.59. Away from the immediate departure from the equilibrium curve, the $c_L^\infty(T < T_g, \dot{T})$-data are located on a straight line with a tangent $m_g = \partial c_L^\infty(T)/\partial T = \text{const}$ which is independent of the cooling rate $-\dot{T}$. It is worth noting that the y-axis intercept depends on $-\dot{T}$ and decreases with the increase of the cooling rate. In the other limit, given by $\dot{T} \to -0$, the kinetic view demands again that the $c_L^\infty(T, \dot{T} \to -0)$-values are located on the extrapolated liquid branch. The latter argument demands a discontinuous change from m_g to m_ℓ for $T \to T_g$ as well as for $T \to 0$. The independence of m_g from the temperature rate $-\dot{T}$ for $T < T_g$ demands a special structure of the glassy state: Independently from the cooling rate there is formed a glassy state which is randomly closed packed (rcp) where the rcp-state is rigid ($c_{44} > 0$). This rcp-state includes voids (free volume) which are surrounded by the rcp's. The temperature gradient of this state is determined by the temperature gradient of the rcp-matrix which has to be m_g. Consequently, there exists a spontaneous transition from the liquid state to a well defined matrix state with a temperature gradient m_g and additional free volume which does not contribute to the thermal expansion of the glassy material but which depends on the cooling rate. This transition from the liquid to the glassy matrix state takes place at the TGT. Therefore, with respect to the free volume the glassy state depends on the cooling rate $-\dot{T}$. With respect to the matrix the glassy state is unique. For very low cooling rates the free volume goes to zero and there remains a well defined glassy state.

The freezing process of glass forming liquids in porous glasses shows a further aspect of the static and dynamic behaviour of the acoustic properties at the TGT. Dibutylphtalate (DBP) is known to be a good glass-forming liquid with a freezing point at about 175 K. Porous glasses with average pore diameters of 20 nm and 2.5 nm were filled with DBP. The glass transition behaviour was investigated with temperature modulated differential scanning calorimetry (TMDSC) and with Brillouin spectroscopy (Fig. 3.60).

First of all there seems to be no T_g-shift between DBP-filled porous glasses of 2.5 nm and 20 nm. This result is astonishing for DBP in nanopores of 2.5 nm

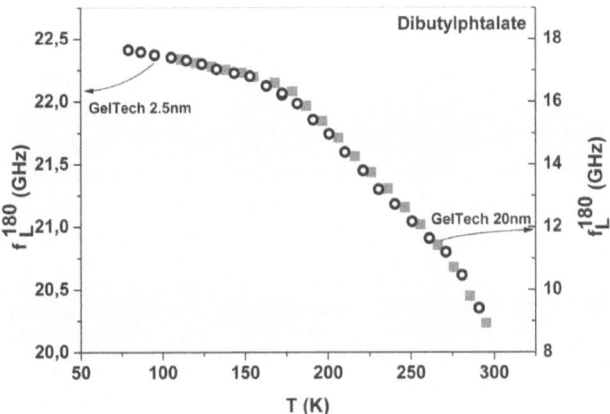

Fig. 3.60. Temperature dependence of the hypersonic frequency of DBP in porous glasses with pore diameters of 20 nm and 2.5 nm

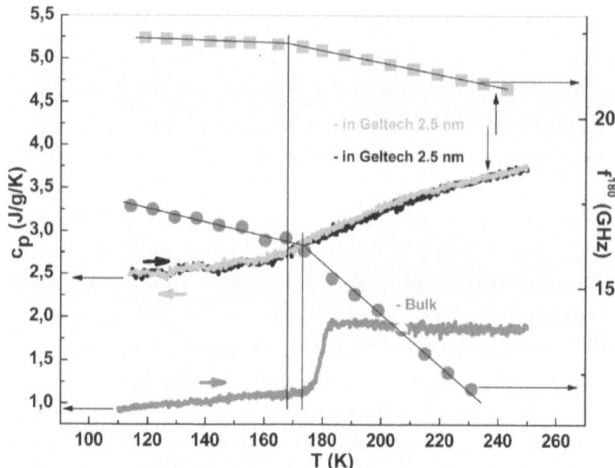

Fig. 3.61. Temperature dependence of the specific heat capacity and of the hypersonic velocity of DBP in the bulk and in a porous glass of a pore diameter of 2.5 nm

because there seem to exist mainly DPB molecules which are in direct contact to the wall of the pores. These "DBP-wall molecules" are usually believed to be immobile. Consequently, from a dynamic point of view they should not undergo a glass transition, but they obviously do. Brillouin data as well as specific heat capacity data clearly show a TGT for pore diameters of 2.5 nm. For pore diameters as large as 20 nm there are of course other than "DBP-wall molecules". For this reason a change of the glass transition behaviour is expected between 2.5 nm- and 20 nm diameter pores.

Figure 3.61 shows a comparison of specific heat capacity data and Brillouin data between the bulk DBP and DBP in glass pores of 2.5 nm diameter. For these two extreme situations a clear T_g-shift of 5 K and a change in the phenomenological behaviour of the c_p' – curves are observed, whereas the sound frequencies show for both cases the usual kink-like behaviour. Since the amount of DBP per unit-volume in the pore-system is much smaller than in the bulk material the kink in the pore system is much weaker than in the bulk system. The anomaly of the c_p' – curve for DBP in 2.5 nm pores can be interpreted as a kink whereas for the bulk DBP a step rather than a kink is observed.

This step-like behaviour (Fig. 3.61) is usually interpreted as the transition from the fast motion- to the slow motion regime, i.e. the inflection point of the c_p' - curve appears at $2 \cdot \pi \cdot f \cdot \tau \cong 1$. Consequently, steps within the c_p' – curve should be accompanied by a maximal loss, i.e. a $c_p''(T)$– maximum. Figure 3.62 shows that the expected $c_p''(T)$ – maximum is indeed present but does not scale at all with the height of the jump of the c_p' – curve. The jump amounts to about 1 J/g/K whereas the peak height of the $c_p''(T)$– maximum amounts to

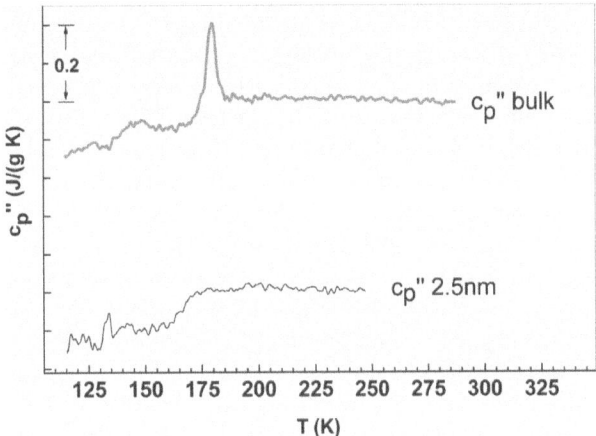

Fig. 3.62. Temperature dependence of the imaginary part c_p'' for bulk material of DPB and for DBP in a porous glass with pore diameters of $2.5\,\mathrm{nm}$

$0.2\,\mathrm{J/g/K}$. On the other hand the width of the peak is rather small so that a broad distribution of relaxation times can be excluded. Consequently, it seems likely that the measured c_p' – curve is composed by a static c_p-background and an additional dynamic relaxation anomaly (probe frequency $f = 16.7\,\mathrm{mHz}$).

Comparing the c_p' – and $c_p''(T)$ – data measured for the bulk with those for the nano-pores of $2.5\,\mathrm{nm}$ an astonishing result is obtained (Fig. 3.62). The relaxation maximum in $c_p''(T)$ for DBP in $2.5\,\mathrm{nm}$ pores has completely disappeared. This result would be in complete agreement with the idea mentioned above that in $2.5\,\mathrm{nm}$ pores all molecules are pinned wall-molecules. This idea seems to be in contradiction with the existence of a TGT. If something freezes below T_g it should be allowed to move to some extent above T_g. Together these arguments imply that the expected relaxation process lies outside the temperature interval of investigation. If this latter argument holds true, the related TGT has nothing to do with a cross-over between the intrinsic relaxation time and probing time! Consequently the c_p-curve measured for DBP in $2.5\,\mathrm{nm}$ pores represents static specific heat capacity data. Another consequence concerns the general role of the α-relaxation process: the α-process is not a prerequisite for the thermal glass transition! It is worth noting that similar behaviours were found for glycerol, I1 and salol in porous glasses of different pore diameters [150].

3.6 The Role of Non-Linear Elastic Behaviour at the Thermal and Chemical Glass Transition

In Sect. 3.1 it was shown that orientational glasses, i.e. single crystals with frozen orientational disorder, show the same anomalies within their phenom-

enological properties at their thermal glass transition as structural glass formers do. Above the glass transition of these crystals with orientational disorder
the material is dynamically disordered with respect to the molecular orientation [151–155]. However, the positional order is given, at least on average.
At the glass transition of these dynamically disordered crystals, the disorder
of the molecular orientation freezes but the positional order is maintained.
The specific volume, the refractive index and the high-frequency clamped
elastic constants show a kink-like behaviour at the glass transition temperature T_g (Figs. 3.6, 3.7). The habitual concept of free volume, which goes to
a minimum at T_g, in connection with a kind of molecular percolation, becomes meaningless for these orientational glasses. The concept of a molecular
interaction potential which changes spontaneously at T_g seems to be more
useful [156–162]. Therefore it makes sense to reinvestigate the glass transition in structural glass formers under the aspect of a spontaneously changing
molecular interaction potential.

In spin glasses the non-linear behaviour of the order parameter plays a
significant role [cf. 163]. Whereas the linear susceptibilities just show a kink,
the nonlinear susceptibilities have been found to diverge in the vicinity of
the freezing temperature T_g. This latter fact yields an additional argument to
study non-linear elastic properties around the TGT of structural glasses.

Usually non-linear elastic properties are not easy to measure. This holds
especially true for the glass transition zone of structural glasses in which the
material transforms from the liquid to the "solid" state. A quantity which
reflects non-linear elastic properties on one hand [cf. 157] and which can be
determined exclusively from Brillouin spectroscopic data [164, 165] on the
other hand are the acoustic **mode-Grüneisen parameters** (MGP) [157].
To our knowledge there exist no theoretical predictions for the temperature
behaviour of the MGP's at the glass transition of structural glasses. In an
isotropic solid the MGP relates the sound frequency of a $p-$ polarized sound
mode of a given wave vector q to its mass density ρ [10].

$$\gamma(\boldsymbol{p}, \boldsymbol{q}) = \frac{\partial Ln(f(\boldsymbol{p}, \boldsymbol{q}))}{\partial Ln(\rho)} \tag{3.56}$$

Since for isotropic materials there exist only longitudinally and transversely
polarized modes ($\boldsymbol{p} = L, T$) we use in the following the notation L and T. It
is expected that the concept of the MGP can be extended to acoustic waves
propagating in liquids, and that within the solid or glassy state, the MGP [166]
varies only slightly with temperature. γ_L of a liquid still reflects anharmonic
properties and structural changes of the liquid state [59, 156–158, 160, 166, 167].

There exist only a few Brillouin investigations on the temperature dependence of the MGP at the TGT [156, 164, 165]. Brody et al. [156] investigated
polystyrene by Brillouin spectroscopy and reported at T_g a small step of the
MGP defined by Eq. (3.56). However, because of their rather large margin of
experimental errors they did not arrive at a definitive statement concerning
the step-like behaviour of γ_L. In Ref. [164, 165] the observed discontinuities

of $\gamma_L(T)$ at the glass-transition temperatures T_g of polymethylmethacrylate, polystyrene and polyvinylacetate were related to a change of the molecular interaction forces. The discontinuity of γ_L of polystyrene was found to be 4.3 which is by a factor of 3.6 larger than that reported in Ref. [156]. This illustrates some of the difficulties to determine reliable data for MGP's.

In order to determine a MGP, given by Eq. (3.56), one has to study the relative change of the mode frequency $f_{L,T}$ as a function of a relative change of mass density ρ. Usually this is realized by changing the pressure p or the temperature T of the sample yielding

$$\gamma_{L,T}(x) = \frac{\rho(x)}{f_{L,T}(x)} \frac{\partial f_{L,T}(x)}{\partial \rho(x)} = \frac{\delta_{L,T}(x)}{\alpha(x)} \tag{3.57}$$

with

$$\delta_{L,T}(x) = \frac{1}{f_{L,T}(x)} \frac{\partial f_{L,T}(x)}{\partial(x)} \tag{3.58}$$

and

$$\alpha(x) = \frac{1}{\rho(x)} \frac{\partial \rho(x)}{\partial(x)} \tag{3.59}$$

with x being the pressure p or the temperature T. If $x = T$, then α is the usual thermal volume expansion coefficient. $\alpha(x)$ is a generalized volume expansion coefficient and $\partial_{L,T}(x)$ a generalized frequency coefficient.

It is important to remember that BS probes the acoustic modes at GHz-frequencies. In consequence, BS measures around the TGT predominantly frequency clamped acoustic properties, i.e. the glass forming material under investigation behaves even in the liquid state solid-like with respect to the probe frequency. Modern BS is able to measure acoustic mode frequencies with a relative accuracy of better than 0.1%. This accuracy is sufficient for the determination of MGP's.

More difficult is the determination of sufficiently accurate mass density data. In the context of density measurements around the glass transition the technical problems are significantly increased due to the fact that the sample passes from the liquid to the solid state or vice versa. Of course measurement techniques for these two states need to differ from each other. This problem of accuracy becomes reinforced through the sticking of the sample on the container wall and the accompanying internal stresses. A serious additional problem of accuracy occurs if mode frequencies and mass density data are not measured under precisely the same external conditions. In case that the mode frequencies and mass densities are measured on different samples at different choices of temperatures or pressures an interpolation or curve fitting is needed in order to create values for $f_{L,T}$ and ρ at the same temperatures or pressures. Since the behaviour of MGP's is studied in the vicinity of the TGT, different cooling conditions of the samples in use may destroy the validity of the calculated result. The same holds true for inhomogeneous samples (inhomogeneities may arise from internal stresses). Then the physical information

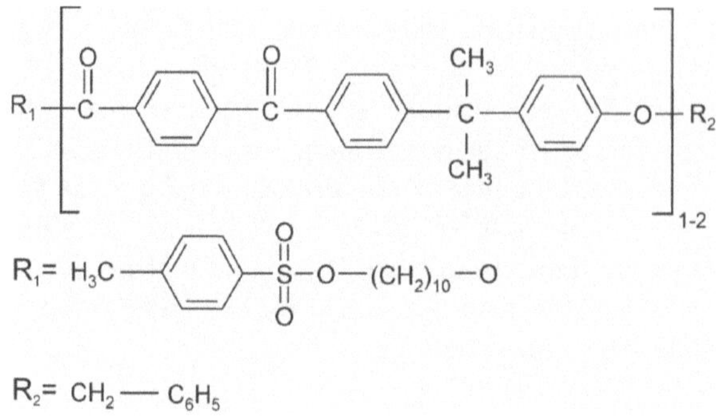

Fig. 3.63. Chemical formula of the oligoarylate mixture I2

about the mode frequency and the mass density is obtained from different sample volumes.

A suitable way to overcome most of the problems mentioned above is to obtain the information about the phonon frequencies and the mass density at the same time and temperature or pressure from almost the same sample volume. This is possible using high-performance Brillouin spectroscopy (BS) in the 90A- and the 90R- or 180-scattering geometry. The combination of these scattering techniques provides simultaneously the desired frequencies of the acoustic modes and refractive index data as a function of temperature T [18, 19].

The Lorentz-Lorenz relation

$$\frac{n(T)^2 - 1}{n(T)^2 + 2} = r \cdot \rho(T) \tag{3.60}$$

yields the necessary relation between the refractive index n and the mass density ρ. Taking the specific refractivity r as a constant, the generalized expansion coefficient $\alpha(T)$ can be calculated from

$$\alpha(T) = \frac{6 \cdot n(T)}{n(T)^2 - 1)(n(T)^2 + 2)} \cdot \frac{\partial n(T)}{\partial T} \tag{3.61}$$

Calculating the frequency coefficient $\delta_{L,T}(T)$ directly from the measured sound frequencies and using the data given by Eq. (3.54) for the volume expansion coefficient $\alpha(T)$, the MGP's $\gamma_{L,T}(T)$ can be determined according to Eq. (3.57). Whereas the MGP's in the glassy state are clearly connected to anharmonic behaviour, the interpretation in the regime of the clamped fluid is more difficult because of the possible influence of entropic degrees of freedom.

The following analysis of MGP's was made for an oligoarylate mixture which will be called I2 (Fig. 3.63) for convenience. The Brillouin data were measured on slow cooling and the sound frequencies $f_{L,T}^{90A}$ and f_L^{90R} were determined simultaneously. Figures 3.64 to 3.66 show the raw data.

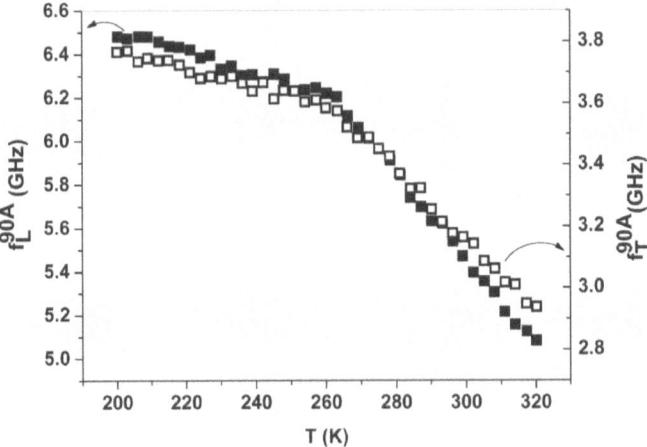

Fig. 3.64. Temperature dependence of the longitudinal and transverse 90A-mode of the oligoarylate mixture I2

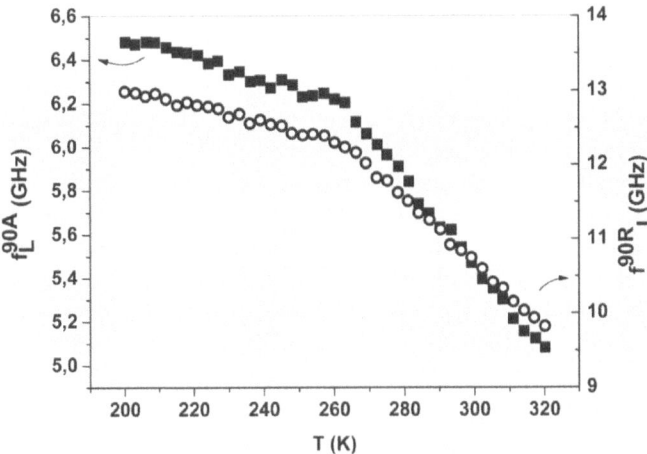

Fig. 3.65. Temperature dependence of the longitudinal 90A- and 90R-modes of the oligoarylate mixture I2

The quasi-static TGT is at about 262 K. Up to 320 K the shear phonon could be detected without major difficulties. The opto-acoustic dispersion function (D-function) could be derived from the 90A- and the 90R-scattering geometry (Figs. 3.65, 3.66). According to Fig. 3.66 the D-data have been interpreted as data for the refractive index. This is reasonable since the refractive index measured with an Abbé refractometer (crossed circle, nD) is in accordance with the D, n-value obtained from BS.

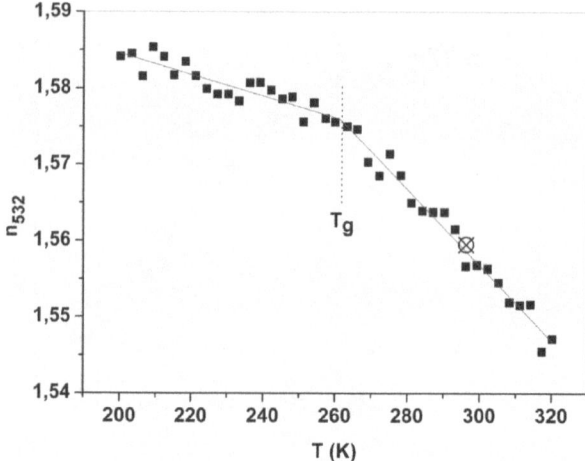

Fig. 3.66. Temperature dependence of the refractive index n of I2 as measured by the D-function. The crossed circle represents a value measured by Abbé-refractometry

As for other oligomeric glass formers, the square of the acoustic mode frequencies fulfill a generalized Cauchy relation (Fig. 3.67) with a slope of almost 3 as usual (see Sect. 3.3).

The calculation of the longitudinal and transverse mode-Grüneisen parameters poses some numerical problems. Due to the scatter of the measured sound frequency and the refractive index data numerical differentiation gave no reliable results for the frequency coefficient δ and the volume expansion coefficient α. In order to obtain reliable results the measured data were first interpolated and then smoothed (moving average) to a sufficient degree. Figure 3.68 shows the sound frequency data treated in this way.

The immediate vicinity of the TGT has been removed from the numerical treatment because of a possible misleading interpretation of the kink feature. Figure 3.69 displays calculated data for the refractive index and the volume expansion coefficient using the Lorentz-Lorenz relation Eq. (3.53). The data of the immediate vicinity of the TGT have been again removed from the data. The refractive indices obtained from the D-function correspond to an optical wavelength of 532 nm and not to nD.

Finally, Fig. 3.70 displays the temperature dependence of the longitudinal and transverse acoustic MGP's $\gamma_L(T)$ and $\gamma_T(T)$. In the margin of error both MGP's are identical and behave jump-like at the TGT. The magnitude of the jumps $\Delta\gamma_L$ and $\Delta\gamma_T$ of the mode Grüneisen parameters are of the order of 20%. The tangent of the MGP's is larger in the fluid phase which may be due to the aforementioned entropic influences in the fluid phase.

The question arises about the significance of these results for the nature of the TGT of structural glass formers. If we interpret the TGT as a purely ki-

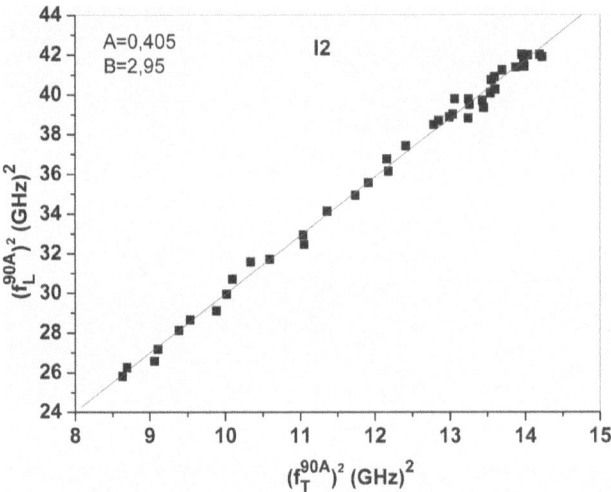

Fig. 3.67. Generalized Cauchy relation of I2

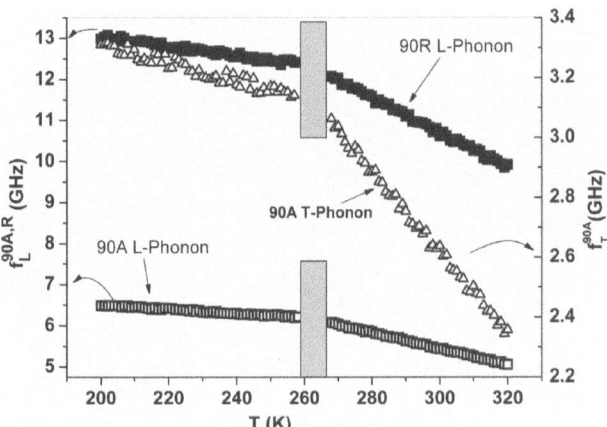

Fig. 3.68. Temperature dependence of the longitudinal 90A and 90R phonon frequencies and the transverse 90A phonon frequency

netic phenomenon in the sense that the freezing of the sample can be avoided on an infinitely long experimental time scale, the anomalies $\Delta\gamma_{L,T}$ in the TGT zone would have to disappear. This disappearance has to be discontinuous because on finite experimental time scales $\Delta\gamma_L$ seems to be a conserved quantity, as it shows no explicit dependence on the time scale with the exception of very fast cooling, where $\Delta\gamma_L$ even increases. Consequently, it is hard to believe that $\Delta\gamma_L$ disappears on any experimental time scale.

The jump-like change of the MGP's at the TGT gives a hint on what happens physically at the TGT. In a first approach, the nonlinear elastic be-

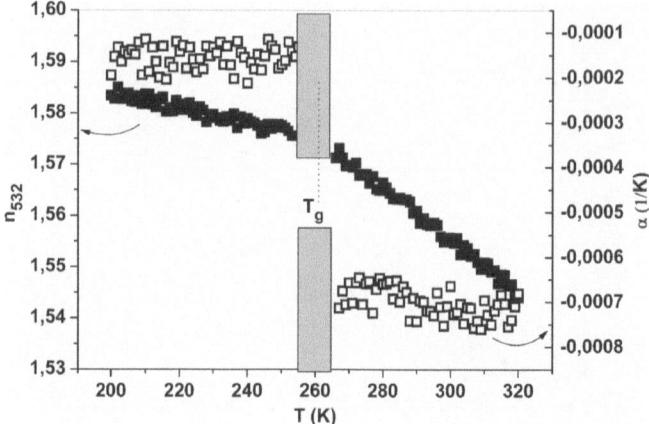

Fig. 3.69. Temperature dependence of the thermal expansion coefficient α and of the refractive index n of I2

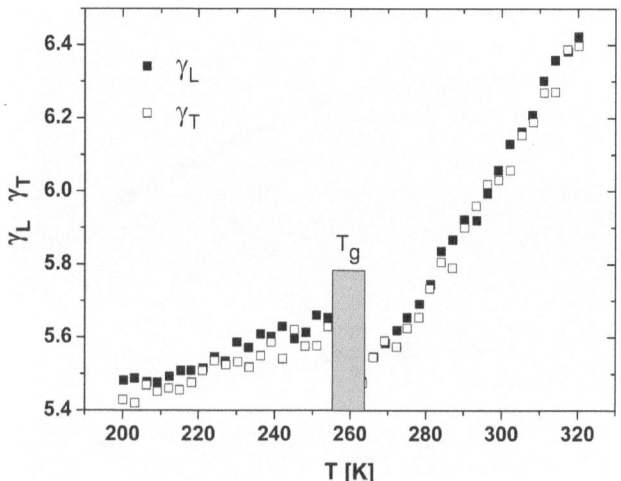

Fig. 3.70. Temperature dependence of the longitudinal and transverse MGP's γ_L and γ_t of I2

haviour of the isotropic state is described by six elastic constants of third order [17, 166]: $c_{111}, c_{112}, c_{123}, c_{144}, c_{155}, c_{456}$ together with three isotropy conditions:

$$c_{112} = c_{123} + 2 \cdot c_{144} \tag{3.62a}$$

$$c_{155} = c_{144} + 2 \cdot c_{456} \tag{3.62b}$$

$$c_{111} = c_{123} + 6 \cdot c_{144} + 8 \cdot c_{456} \tag{3.62c}$$

Thus only three independent third order elastic constants remain.

These third-order elastic constants together with the second-order elastic constants can be used to calculate the two acoustic MGP's γ_L and γ_T [16,20]. The reversal is not true:

$$\gamma_L = \left(\frac{K}{2c_{11}}\right) - \frac{(c_{111} + 2 \cdot c_{112})}{6 \cdot c_{11}} - \frac{1}{3} \tag{3.63a}$$

$$\gamma_T = \left(\frac{K}{2 \cdot c_{44}}\right) - \frac{(c_{144} + 2 \cdot c_{155})}{6 \cdot c_{44}} - \frac{1}{3} \tag{3.63b}$$

with K being the compressional modulus.

Recalling that the second-order elastic constants (c_{11}, c_{44} and K behave continuously at the TGT and assuming the temperature difference ε to be infinitesimally small, at least one of the three third-order elastic constants behaves discontinuously

$$\gamma_L^{T_g+\varepsilon} - \gamma_L^{T_g-\varepsilon} = \frac{1}{6 \cdot c_{11}(T_g)}[(c_{111}(T_g + \varepsilon) - c_{111}(T_g - \varepsilon))$$
$$+ 2(c_{112}(T_g + \varepsilon) - c_{112}(T_g - \varepsilon))] \tag{3.64a}$$

and

$$\gamma_T^{T_g+\varepsilon} - \gamma_T^{T_g-\varepsilon} = \frac{1}{6 \cdot c_{44}(T_g)}[(c_{144}(T_g + \varepsilon) - c_{144}(T_g - \varepsilon))$$
$$+ 2(c_{166}(T_g + \varepsilon) - c_{166}(T_g - \varepsilon))] \tag{3.64b}$$

It seems likely that the third-order elastic constants show a jump-like behaviour at the TGT. This latter interpretation suggests the idea that the TGT is accompanied by a jump-like change of the molecular interaction potential.

A structural picture of this jump-like change of the molecular interaction potential has been discussed for the TGT of polyvinylacetate (PVAc) in terms of a percolation of clusters of minimum free volume [171, 172] in the sense of Ref. [162, 173]. Within this hypothesis the jump-like behaviour of $\gamma_L^{90A}(T)$ at T_g is interpreted as a consequence of the spontaneous structural but not symmetry-breaking changes at T_g and that it is therefore unavoidable.

In order to elucidate the nature of the glass transition and in particular to discriminate between the kinetic and the phase transition aspects of the TGT, one of these two aspects should affect the experimental data very little and might even be completely removed from it. As it was shown in the foregoing sections, this can be done by an appropriate choice of experimental conditions. Extremely slow cooling experiments combined with TDBS decreased or even eliminated the influence of kinetics at the TGT. The inspection of the generalized Cauchy relation in combination with fast quenching emphasized the kinetic aspect of the TGT and its connection to ageing phenomena. The study of the non-linear elastic behaviour via the study of thermal MGP's again provided further evidence in favour of our phase transition view of the TGT.

A further approach to the understanding of the glass transition is to leave the thermal version of this phenomenon in order to become completely rid of the kinetics due to the cooling or heating procedures of the sample. Indeed, freezing can proceed in different ways although the thermally induced solidification discussed above is the most frequently investigated process. As discussed above the TGT always suffers from the kinetic restrictions imposed by the strong increase of the structural relaxation time τ_α above but close to the TGT.

The "cross-over" between the intrinsic structural relaxation time scale τ_α and the kinetic time scale (cooling rate, probe frequency, etc.) does not exist in an isothermal polymerizing experiment, where the macromolecular structure vitrifies in the course of the chemical reaction (any textbook of polymer chemistry or physics [174–176]). The related **chemical glass transition** (CGT) can be performed without any change of external variables while the internal variables equilibrate due to the chemical reaction.

The structural relaxation time τ_α increases in the course of the chemical reaction, as new intermolecular bonds hinder the translational motions. In order to visualize such a CGT, an experimental probe is needed that will neither significantly disturb this equilibration process nor is influenced by dynamic effects inside the sample. Provided the probe works in the limit of linear response, the experimental time constants $\tau_{\exp}$ (inverse probe frequencies) have to be chosen in a way, that the relevant internal thermodynamic variables do either move freely, i.e. $\tau_\alpha/\tau_{\exp} \ll 1$, or behave as clamped quantities, $\tau_\alpha/\tau_{\exp} \gg 1$.

The isothermal, isobaric curing of an epoxy resin can be considered as a model for the CGT. The curing process starts from a two-component oligomer. Once started, the polymerization, driven by the related chemical activities, runs under appropriate conditions to the final brittle (glassy) polymer network [174–176]. As a particular feature, the CGT is not dominated by changes in molecular conformations and packing but by changes of the molecular structure due to the chemical conversion. Physical properties like mass density ρ, chemical turnover u, refractive index n, frequency-clamped elastic constants [15] c_{11}^∞ and c_{44}^∞, etc. behave completely continuously in the course of the polymerization process. Different to the TGT, the CGT does not appear as a distinct anomaly in phenomenological properties (with the exception of the specific heat capacity c_p) measured in linear response.

In the following we will study the curing process of our model substance epoxy. The starting ingredients of this epoxy are diglycidylether of bisphenol A (DGEBA, 100 mass parts) and diethylenetriamine (DETA, 14 mass parts). In the curing experiment, temperature ($T \approx 296\,\mathrm{K}$ and pressure (ambient) are set. The gross epoxy group consumption u serves as the leading internal variable.

A suitable probe for the intrinsic glass transition which is little sensitive to kinetic influences is provided by the acoustic MGP discussed before in this section. Discontinuous changes of the MGP's which are indicative for spontaneous changes of the molecular interaction potential are only expected at phase transitions (e.g. [158]). As a matter of fact, the strength of the discontinuity is more or less independent of the cooling or heating rate although these rates can slightly modify the temperature position of the discontinuity. Having in mind this discontinuous behaviour of the longitudinal acoustic MGP at the TGT the question appears whether or not acoustic MGP's could provide a clearer picture of the location of the CGT on the scale of the chemical turnover and about the nature of this reaction-driven freezing process.

Accordingly, the role of MGP as a sensitive probe for the glass transition during chemical freezing is the matter of debate.

The basic relation for an acoustic mode-Grüneisen parameter as a function of chemical turnover appears at a first sight strange:

$$\gamma^{p,q}\left(f^{p,q}\left(u\right),\rho\left(u\right)\right)=\frac{\frac{1}{f^{p,q}(u)}\frac{df^{p,q}(u)}{du}}{\frac{1}{\rho(u)}\frac{d\rho(u)}{du}}=\frac{\delta^u}{\alpha^u} \tag{3.65}$$

In Eq. (3.65, p $(=L,T)$ denotes the polarization of the sound mode, q is the wave vector. α^u represents a generalized volume expansion coefficient, and δ^u is a generalized gradient of sound velocity.

As the chemical reaction changes continuously the system itself, the usual concept of MGP as a measure for anharmonicity is not applicable in this case. Even if the potential was harmonic but dependent on u, the MGP would be changed due to the reaction. Therefore to differentiate between the classic MGP and the one defined for a chemical reaction, the latter one is denoted as CMGP. The physical meaning of the CMGP is somewhat clouded.

As for the TGT the CMGP's are best measured with Brillouin spectroscopy or more precisely with Time Domain Brillouin spectroscopy (TDBS) (see Sect. 3.4) using the so-called 90A-scattering geometry [163, 165]. This scattering technique meets exactly the measuring condition for the sound frequency, which is needed to calculate the related CMGP's [156, 158, 159, 165, 166]: a constant phonon wave vector in the course of a changing phonon frequency f. A laser wavelength $\lambda_{\mathrm{opt}} = 532\,\mathrm{nm}$ yields an acoustic wave vector:

$$q^{90A}=\{(4\cdot\pi\cdot\sin\left(\pi/4\right))/532\,\mathrm{nm}\} \tag{3.66}$$

The hypersonic velocities are then obtained from the usual dispersion relation

$$v_{L,T}^{90A}\left(t\right)=\frac{2\cdot\pi\cdot f_{L,T}^{90A}\left(t\right)}{q^{90A}} \tag{3.67}$$

The determination of CMGP's at the CGT has to be done at constant temperature. As discussed above, the determination of CMGP's needs the knowledge of the mass density measured under the same conditions under which the related phonon frequencies were determined. Again, the mass density can best be obtained via the optical refractive index with help of Eq. (3.53). At ambient temperature a very precise way to measure refractive index data is Abbé refractometry. With this technique the temporal evolution of the refractive index of the model epoxy can be determined at the same wavelength, $\lambda_{opt} = 532$ nm as used for Brillouin spectroscopy. The refractive index $n = \sqrt{\varepsilon^\infty(\lambda_{opt})}$ is directly related to the dielectric constant measured at optical frequencies ($f_{opt} \approx 5 \cdot 10^{14}$ Hz).

Figure 3.71 gives an overview over the temporal evolution of the refractive index n, of the longitudinal sound velocity v_L and of the specific heat capacity c_p for a curing process at 296 K. It is obvious from Fig. 3.71 that the apparent anomalies in the three susceptibilities occur at different times and therefore at different degrees of chemical turnover. Consequently there is no unique indication for the location of the CGT.

Whereas for the TGT the natural driving parameter is the temperature T, the natural parameter for the CGT is the time t. Of course, the time t is not a property of the material under study. A more physical driving parameter would be the amount of chemical conversion (turnover) u (s.a. Fig. 3.72). Attenuated total reflection infrared spectroscopy (IR-ATR) yields the chemical turnover u as a function of curing time t. With the intensity $I(1510 \text{ cm}^{-1})$ of the phenylene band as internal standard, the intensity $I(915 \text{ cm}^{-1})$ of the epoxy band is normalized to: $I_{EP}^{norm}(t) = I_{915}(t)/I_{1510}(t)$. Then, the IR-spectroscopic degree of epoxy group consumption u is calculated from $u(t) = [1 - \frac{I_{EP}^{norm}(t)}{I_{EP}(0)}]$.

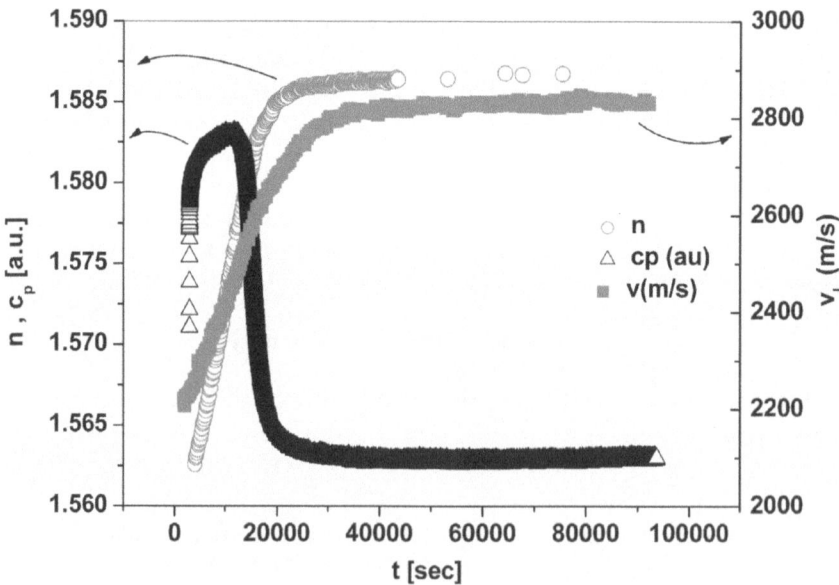

Fig. 3.71. Temporal development of the refractive index n, the specific heat capacity c_p and the longitudinal sound velocity v_L during the cross-linking of a DGEBA/DETA mixture of 100/14 at $T = 296$ K

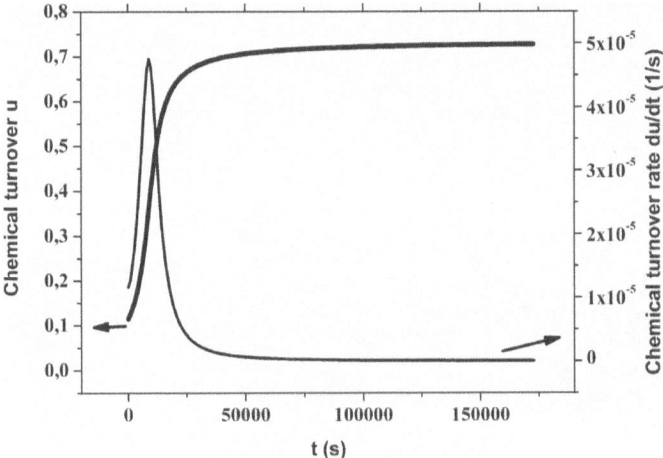

Fig. 3.72. Chemical turnover u and chemical turnover rate $\frac{du}{dt}$ as a function of curing time t for the cross-linking of DGEBA/DETA 100/14

Figure 3.72 shows for the model epoxy mentioned above how the chemical turnover u and its rate $\partial_t(u)$ depend on the curing time t. $u = u(t)$ behaves completely smooth and $\partial_t(u)$ shows a peak in the short time regime at about $t = 9000\,\text{s}$. It is evident, that even after differentiation of $u = u(t)$ no discontinuity emerges from the measured data.

As mentioned above, CMGP's are derived from static or frequency-clamped acoustic properties. Therefore it has to be checked whether the sound modes are measured in a relaxation-free time regime.

Figure 3.73 shows the evolution of the structural relaxation time τ_α as measured by dielectric spectroscopy (DES) within the time interval between roughly $t_{\text{cure}} = 0\,\text{s}$ and $t_{\text{cure}} = 25 \cdot 10^3\,\text{s}$. The relaxation time increases within this time interval from about $\tau_\alpha = 10^{-9}\,\text{s}$ to $\tau_\alpha = 10\,\text{s}$. Following usual conventions the DES data can be used to define an operative glass transition time: $t_g^{\text{op}} = 2.5 \cdot 10^3\,\text{s}$.

According to Fig. 3.73 the CGT is accompanied by a strong increase of τ_α. In the time interval $I_{\text{soft}} = [0, t_g^{\text{op}}]$ the hypersonic frequency f_L^{90A} increases from about 5.5 GHz to 7.5 GHz (Fig. 3.73). As a result, BS as a probe for hypersonic frequencies f offers the possibility to measure clamped mechanical properties in almost the full interval: $2 \cdot \pi \cdot f \cdot \tau_\alpha \gg 1$. As a result, neglecting curing times up to 5000 s the related elastic constants c_{11}^∞ and c_{44}^∞ are frequency-clamped quantities.

Figure 3.74 gives the frequency clamped sound velocities $v_L^{90A}(t)$ and $v_t^{90A}(t)$ and the refractive index $n = n(t)$ as measured with an Abbé refractometer. All Brillouin data which eventually do not represent the frequency-clamped state have been suppressed. The longitudinal and the transverse

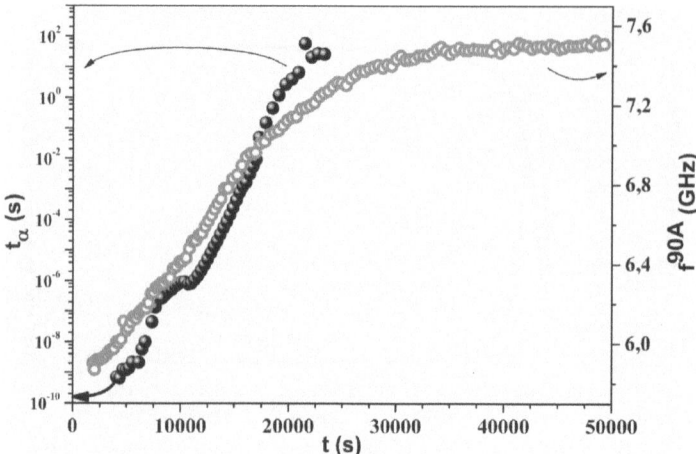

Fig. 3.73. Sound frequency f_L^{90A} and structural α-relaxation time as a function of curing time t for DGEBA/DETA 100/14

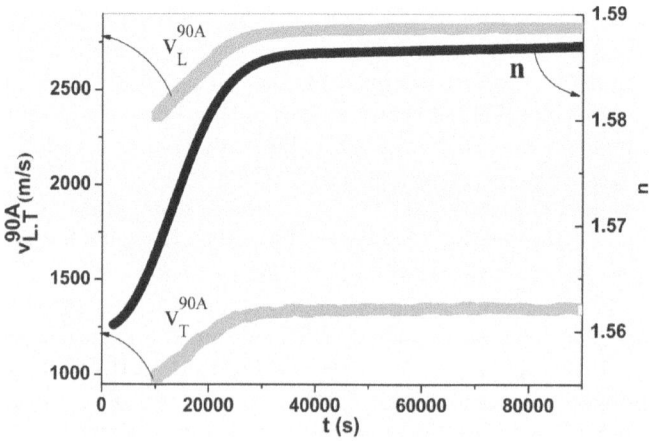

Fig. 3.74. Sound velocity (L-mode: f_L^{90A}; T-mode: f_L^{90A} and the refractive index, n_D^{296K}, as a function of curing time t

sound velocities $v_L^{90A}(t)$ and $v_t^{90A}(t)$ behave strongly non-linearly but both show a very similar increase with the curing time.

We have recently shown that the changes of c_{11} and c_{44} in the course of the thermal glass transition usually follow a generalized Cauchy condition [177–179]

$$c_{11}(x) = A + 3 \cdot c_{44}(x) \qquad (3.68)$$

with

$$\partial_x c_{11} = 3 \cdot \partial_x c_{44} \qquad (3.69)$$

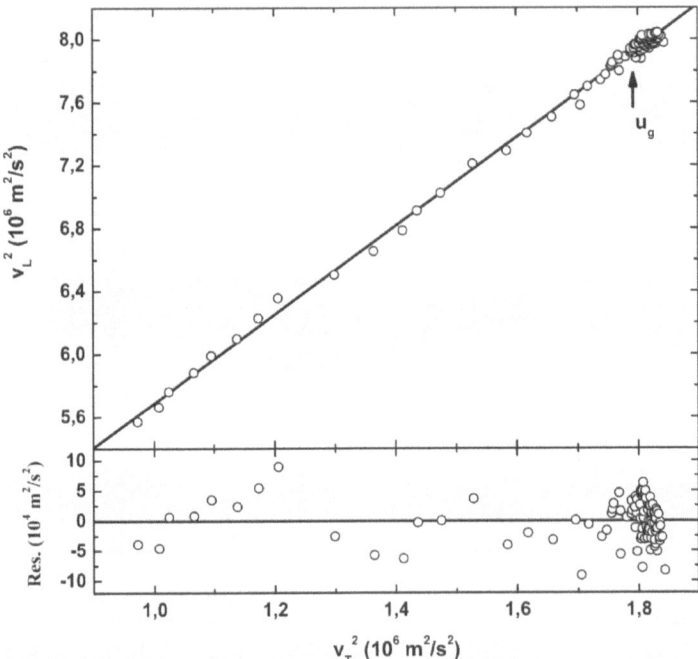

Fig. 3.75. Linear fit of the squared hypersound velocities $(v_L^{90A})^2$ vs. $(v_L^{90A})^2$. The residuals of the fit show the validity of the generalized Cauchy condition $(v_L^{90A})^2 = \tilde{A} + 3 \cdot (v_L^{90A})^2$

(x = time, temperature, turnover). The same relation is found for our model epoxy during the chemical reaction even across the chemical glass transition.

Figure 3.75 shows the result using the sound velocity representation [178, 179]

$$\left(v_L^{90A}\right)^2 = \tilde{A} + B \cdot \left(v_T^{90A}\right)^2 \tag{3.70}$$

with $\tilde{A} = A/\rho = 2.3 \cdot 10^6$ m^{2}s^2 and $B = 3$. This Cauchy relation holds true throughout the whole curing process.

Taking into account that any curing process reflects a succession of non-equilibrium transitions it is really surprising that the generalized Cauchy relation holds true.

This result is even more astonishing if we take into account results on thermally quenched canonical glass formers mentioned in Sect. 3.3: the quenching process has destroyed the Cauchy relation yielding a higher slope of the $c_{11} = c_{11}(c_{44})$-curve. If even the non-equilibrium curing process does not violate the Cauchy relation, one can conclude that the totally hidden CGT represents at least not a strong non-equilibrium process in the sense that the longitudinal and the shear elastic constant get out of equilibrium with respect to each other.

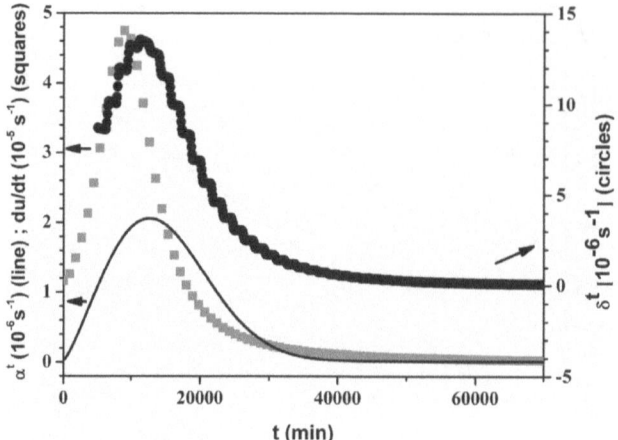

Fig. 3.76. Temporal volume expansion coefficient α^t, temporal sound frequency gradient δ^t of the L-mode and the chemical turnover rate $\frac{du}{dt}$ as a function of curing time t

It remains the question whether the acoustic CMGP's do respond more sensitively to the CGT than the Cauchy relation does. The acoustic CMGP's γ are related to the sound mode frequencies f and to the mass density ρ, which in turn are functions of u. Since the external parameters like temperature and pressure are kept fixed during the measurement, all time scales relevant for the chemical freezing process are controlled internally by the chemical reaction. Hence, the cross-over problem as observed at the TGT is avoided, the CGT cannot be obscured by kinetic effects.

For the evaluation of the CMGP the quantities f, ρ and u are independently measured as a function of curing time t at 296 K but in order to calculate the related

$$\gamma_{L,T}^{90A} = \gamma_{L,T}^{90A} = \frac{\delta_{L,T}^{90A}(u)}{\alpha(u)} \tag{3.71}$$

the δ and α-values have to be provided for the same $u(t)$. Again interpolation between measured data and the application of moving averages solve the numerical problems. Figures 3.75 and 3.76 depict the temporal sound propagation coefficient

$$\left(\delta_{L,T}^{90A}\right)^t = \frac{1}{f_{L,T}^{90A}} \frac{\partial f_{L,T}^{90A}}{\partial t} \tag{3.72}$$

as calculated from $f_{L,T}^{90A}(t)$.

Since the determination of the MGP's needs the generalized volume expansion coefficient α^u this quantity has to be derived from the time dependence of the refractive index $n(t)$ and of the chemical turnover $u(t)$. In a first step α^t is calculated

$$\alpha^t = \frac{1}{\rho} \cdot \frac{\partial \rho}{\partial t} = \frac{6n(t)}{n^4(t) + n^2(t) - 2} \cdot \frac{\partial n}{\partial t} \tag{3.73}$$

Figure 3.76 shows the temporal evolution of this quantity. The refractive index $n = n_D^{296K}$ was measured with an Abbé refractometer at a temperature of 296 K (black curve in Fig. 3.74). The $n(t)$-data were measured on the same epoxy batch which was splitted for the different experiments after preparation. With the available $\partial_t u$-data (Fig. 3.72), the generalized expansion coefficient α^t can be transformed into

$$\alpha^u = \frac{1}{\rho} \cdot \frac{\partial \rho}{\partial u} = \frac{\alpha^t}{du/dt} \tag{3.74}$$

Having calculated all ingredients, we are able to determine the CMGPs from Eq. (3.70). Figure 3.78 provides the results for γ_L^{90A} and γ_T^{90A} of these calculations as a function of u. The chemical turnover u_g corresponding to the peak position of $\gamma_L^{90A}(u_g)$ and $\gamma_T^{90A}(u_g)$ is interpreted as the degree of chemical conversion for which the ideal glass transition takes place as explained in the following.

Firstly, the peaks of the MGP's occur by far later than those of δ^u and α^u (Figs. 3.76, 3.77). Secondly $\frac{\delta_{L,T}^u}{\alpha^u}$ remains constant in the region of the strongest variations of these two quantities whereas the peaks of $\gamma_{L,T}^{90A}$ appear far in the almost flat wings of δ^u and α^u for $u > 0.6$ (Fig. 3.79). Consequently, the appearance of the $\gamma_{L,T}$-peaks results from the different levelling of δ^u and α^u at high degrees of curing: on approaching the CGT the volume expansion coefficient α slows faster down than the frequency expansion coefficient δ does and finally levels. This means, that the frequency change per change of density increases on approaching the CGT. If at still higher degrees of chemical

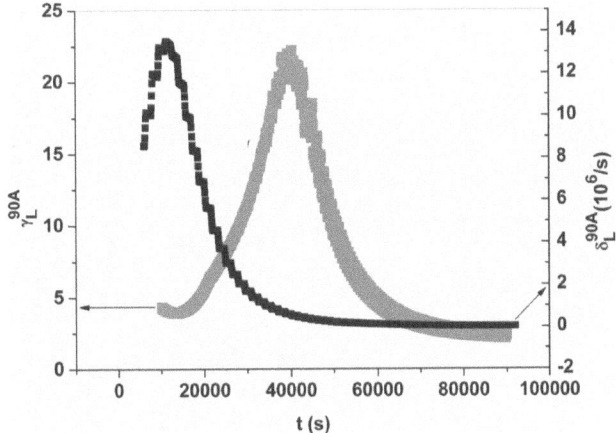

Fig. 3.77. Temporal longitudinal frequency coefficient δ^t and temporal MGP $(\gamma_L^{90A})^t$ of the L-mode as a function of curing time t

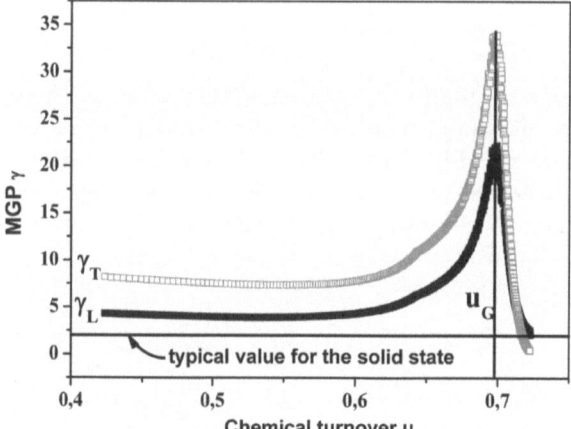

Fig. 3.78. Grüneisen parameters γ_L and γ_T as a function of the degree of chemical conversion $u \cdot u_g$ gives the "critical" degree of chemical conversion at which the ideal glass transition is assumed to take place

turnover the frequency coefficient also starts to level, the CMGP slows down again.

For the epoxy samples of 14 mass percent DETA the DETA concentration is sufficiently high (over-stoichiometric) that, at least in principal, the chemical reaction could be completed. It is therefore interesting to note that the chemical turnover u for the sample of 14 mass percent DETA does not reach the value of 100% but stops at 70% if the curing process takes place

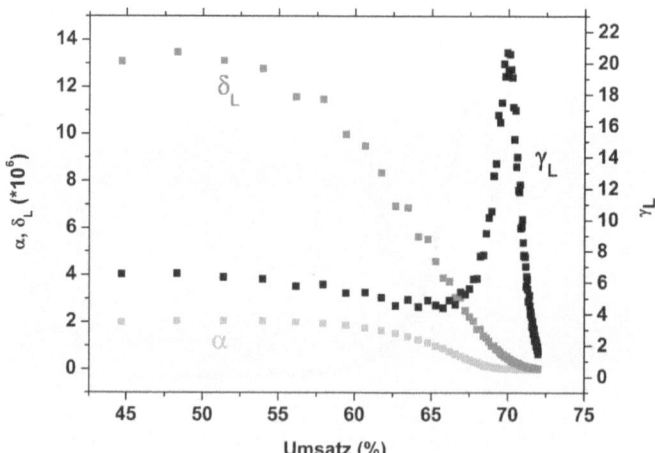

Fig. 3.79. Sound frequency coefficient δ, expansion coefficient α and longitudinal mode-Grüneisen parameter γ_L spectroscopic chemical conversion u of DGEBA/DETA 100/14

at $T_{cure} = 296$ K. It seems that the CGT stops or at least significantly slows down the curing process. This explains the anomaly of the volume- and the frequency coefficients around the critical concentration u_g (Fig. 3.79). It is therefore expected that the CGT's of reactive epoxy systems depend on the concentration of DETA and shift to shorter times with increasing concentration of the latter. At a sufficiently low amount of DETA the formation of network knots is limited so strongly that no more chemical freezing takes place during polymerization at ambient temperature. The evolution of the CMGP as a sensitive indicator for the CGT should therefore depend on the DETA concentration. Figure 3.80 confirms this interpretation. The sample of 18 mass percent DETA shows its CGT after about 300 min of curing whereas the CGT of the sample of 14 mass percent DETA freezes only after about 500 min. The sample with only 6 mass percent of DETA does not freeze at all at ambient temperature. Of course, these results do not give information about a possible shift of the critical turnover with the concentration of DETA. Recent measurements on polyurethanes show in principal the same behaviour at the CGT [180]. In Fig. 3.80 another feature of the CMGP's at the CGT becomes evident which concerns the long time behaviour. As a matter of fact, after a sufficiently long curing time there should be neither a change of the mass density nor of the phonon frequencies, all quantities should saturate. Consequently, the CMGP's should become constant but that information is by definition inaccessible. An approximative information about the limiting values for the CMGP's can be calculated from δ and α-values reliably larger than zero.

In conclusion, the reactive system DGEBA/DETA (100/14) transforms during the crosslinking process from a two-component liquid via a percolated

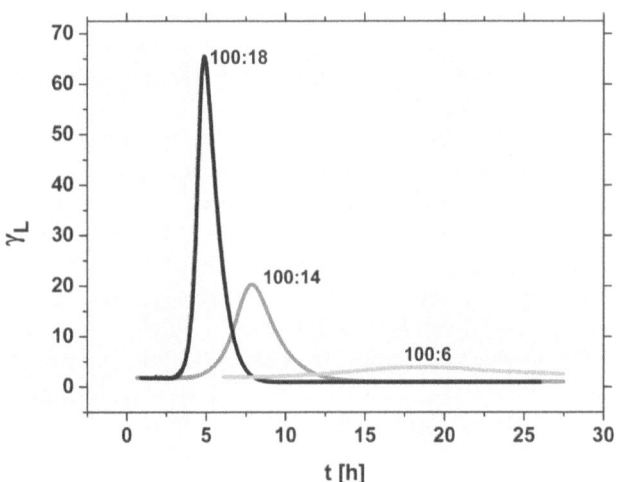

Fig. 3.80. Temporal evolution of the longitudinal mode-Grüneisen parameters for the model epoxy system DGEBA/DETA 100/18, 100/14 and 100/6

epoxy network into a chemically induced glass. The most pronounced change of behaviour is the given by the maximum of the CMGP, so that this maximum has to be identified with the chemical glass transition. As kinetic influences are excluded an intrinsic glass transition seems to exist for this model epoxy at the degree of conversion, u_g ~0.70. In the phase of low chemical conversion, i.e. in the "liquid phase" with $u \ll u_g$, the Grüneisen parameters γ_L and γ_T exceed those of the glassy state ($u > u_g$) respectively. Although a percolation transition exists below u_g, this transition seems not or only weakly coupled to the measured and calculated phenomenological parameters. At u_g, the longitudinal as well as the transverse acoustic mode-Grüneisen parameters go through a rather sharp maximum.

It should be stressed that for $u = u_g$ the TGT of the material coincides with the curing temperature $T_g(u_g) = T_{\text{cure}}$. Therefore it exists an inherent correlation between the CGT and the TGT. But in as much as the material can continue to cure in the chemical glassy state the related thermal glass transition temperature will exceed the curing temperature.

To this end it is interesting to clarify the different significances of the MGP's for the thermal and the chemical freezing process. For the TGT it should be kept in mind, that constant MGP's indicate, that the related molecular interaction potential is independent of temperature and a discontinuous change of the MGP's at a definite temperature T_g indicates a discontinuous structural change within the material of interest. This point of view implies that in the liquid and the glassy state of a given canonical glass former respectively two different molecular interaction potentials are responsible for the temperature dependence of phenomenological properties and that these properties at a given temperature correspond to a related thermally excited state within the appropriate potential. In reality the MGP's are not completely independent of temperature. This holds especially true for the liquid state and indicates the appearances of slight but continuous changes of the local structure with changing temperature and/or the influence of entropic degrees of freedom.

During chemical freezing the situation is completely different. With increasing curing time the initially two-component liquid transforms continuously to a molecular network. This formation of a molecular network corresponds to a continuous change of structure and therefore should yield time dependent MGP's. This means that the changes of the CMGP's observed in the course of curing may reflect structural changes rather than pure changes of anharmonicity. This is understandable since all external variables remain constant during the curing process. It is interesting to note that in the early curing regime until the CGT, that is the regime were the formation of network knots is fast, the two acoustic CMGP's $\gamma_{L,T}^{90A}(u)$ remain almost independent of u (Fig. 3.78). Obviously, this behaviour corresponds to a roughly linear dependence of the elastic modulus on mass density, as $\delta = \gamma \cdot \alpha$. Indeed, over a wide range of ρ-values the elastic modulus c_{11} behaves linearly as a function of ρ. Only in the glass transition region a strong increase of c_{11} followed by

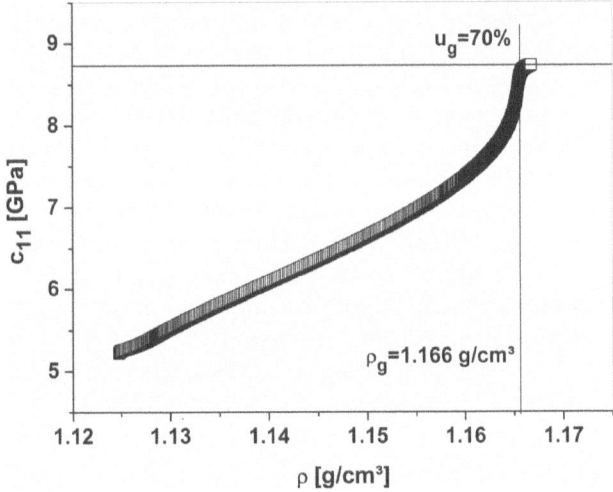

Fig. 3.81. Difference in the sum of third-order elastic constants c_{111} and c_{112} between glassy and fluid phase vs. chemical conversion u for DGEBA/DETA 100/14

a levelling of this quantity appears (Fig. 3.81). In other words, the structural changes accompanying the curing process do not affect the relative change of the phonon frequencies induced by the changes of mass density created by the same chemical process. This implies, that all changes in the frequency coefficients $\delta_{L,T}^{90A}(u)$ and the volume expansion coefficient $\alpha(u)$ can be described by the same factor $c(u)$ with

$$\delta_{L,T}^{90A}(u) = \delta_{L,T}^{90A}(u_0) \cdot c_\delta(u) \qquad (3.75a)$$
$$\alpha(u) = \alpha(u_0) \cdot c_\alpha(u) \qquad (3.75b)$$

with $c_{\alpha,\delta}(u_0) = 1$.

Dividing Eq. (3.77a) by Eq. (3.77b) yields

$$\frac{c_\delta(u)}{c_\alpha(u)} = \text{const} = \frac{c_\delta(u_0)}{c_\alpha(u_0)} = 1 \qquad (3.76)$$

Of course this identity is only approximately fulfilled. Within the peak region representing the CGT the order relation $c_\delta(u_o) \geq c_\alpha(u_o)$ holds true.

It therefore has to be concluded, that from the mechanical point of view the chemical freezing process is much more dominant and effective than the driving polymerization and percolation process. Even in a molecular plastic crystal the same kind of discontinuity was found (see Sect. 3.1, DFTCE).

In the vicinity of the CGT the α-relaxation process is slow and the continuation of further curing is in concurrence to the freezing process. At the beginning of the CGT, due to the improved molecular packing, first the changes of the density are slowed down but still some curing can happen in the almost fixed molecular skeleton. The additional bonds, included in the spatially fixed molecular structure, further stabilize the existing skeleton and thus increase the phonon frequency due to an improved elastic stiffness. This additional stabilization of the given molecular skeleton causes the anomaly of the CMGP. After the CGT the system behaves like a solid, so that further bonds can influence the elastic behaviour only slightly. The glassy state hinders the translational diffusion of oligomers and cross-linked clusters but does not fully stop it, so that further reactions are still possible. These further reactions deprive the system of this potentially swelling molecular groups, so that the density is increased, which leads to a slowing down of the CMGP's.

It is self explaining that mechanical properties which describe mechanical stiffnesses are suitable probes for glass transitions. The stiffening at the transition from the liquid to the glassy state is at least partly caused by an improved molecular packing. But that is only part of the truth. As has been deduced in this chapter from the evolutions of the MGP's around the glass transitions, there exists an additional contribution to elastic stiffness caused by stiffened spring constants not related to density changes. The chemical glass transition demonstrates this effect drastically, because additional chemical bondings can at least in principal increase the mechanical stiffness without changing the mass density. The same observation of an excess stiffness as a function of density holds true at the thermal glass transition, although additional bondings don't play any role. This excess stiffness occurring at glass transitions seems to be one of the central features of glass transitions.

3.7 Conclusion

There is no doubt that kinetic phenomena accompany the thermal as well as the chemical glass transition. The leading role of the molecular kinetics for the thermal and the chemical glass transition is not confirmed. Beyond the influence of molecular kinetics there are sufficient proves for the existence of an intrinsic glass transition which nature has to be elucidated furthermore.

The most striking evidences for the existence of such an intrinsic feature are as follows:

The investigation of the generalized Cauchy relation on differently quenched glass forming liquids has shown that there exist two different glassy states: one which follows the generalized Cauchy relation and one which violates it. Significant quenching creates a metastable glass and thus violates the generalized Cauchy relation. Under the latter conditions aging can be observed. However, this aging process does not bring the material to the liquid phase as predicted by the kinetic view but to a stable reference glassy state.

Time Domain Brillouin Spectroscopy shows for different glass forming liquids a definite cut-off for the low-frequency relaxation processes often accompanying the thermal glass transition. As a consequence the move away from local equilibrium of the reference state due to a cross-over between the experimental time scale and the the α-relaxation time can be avoided and a "equilibrium glassy state" is observed.

The inspection of the opto-acoustic dispersion function (D-function) proves that in several glass forming liquids relaxation processes within the GHz-range are present which are eliminated only by the quasi-static glass transition. In contrast to the kinetic point of view, close to the quasi-static glass transition there are still very mobile relaxation processes which are cut off by the quasi-static freezing process.

Glass-forming liquids filled-in in nano-porous glasses prove that the quasi-static glass transition can occur without being accompanied by the α-relaxation process. As a consequence, the glass transition takes place without any cross-over of the experimental time with α-relaxation time.

The quasi-static glass transition in polymer liquid crystals has necessarily to appear in order to avoid unrealistic elastic constants (supplement to the so-called "Kauzmann Paradoxon").

Structural glass formers and orientational glass formers sometimes show identical anomalies at the quasi-static glass transition. The glass transition in orientational glasses doesn't need the ingredient of cooperative rearrangement units.

The analysis of mode-Grüneisen parameters shows that the quasi-static thermal glass transition is accompanied by a discontinuity of the physical property. This result implies a jump-like change of the molecular interaction potential at the glass transition indicating a spontaneous change of structure. This discontinuity creates jump-like changes of the involved third-order elastic constants. The latter observation is a clear hint for a phase transition.

Reactive polymers show a chemically induced freezing process called "chemical glass transition" during the curing process. This type of transition shows again an anomalous behaviour of the mechanical properties during freezing, indicating the existence of an intrinsic transition. The curing process is a completely free running process without any influence of the experimentalist on the ongoing freezing process.

It is therefore time to revise the actual view of the glass transition.

References

1. K. L. Ngai, "Universal Patterns of Relaxations in Complex Correlated Systems", in "Disorder Effects on Relaxation Processes", Richert/Blumen, Springer-Verlag, Berlin, Heidelberg (1994)
2. E. Donth, "Relaxation and Thermodynamics in Polymers, Glass transition", Akademie-Verlag, Berlin (1992)
3. W. Götze, L. Sjögren, Rep. Prog. Phys. **55**, 241–376 (1992)
4. J. H. Gibbs, E. A. DiMarzio, J. Chem. Phys. **28**, 373 (1958)
5. G. Gee, Contemp. Phys. **11**, 313 (1970)

6. J. J. Prejean, in "Dynamics of Disordered Materials", Ed. by D. Richter, A. J. Dianoux, W. Petry and J. Teixeira, Proceedings in Physics **37**, 242, Springer (1989)
7. J. Jäckle, Rep. Progr. Phys. **49**, 171 (1986)
8. J. Jäckle, J. Phys. Cond. Matter, **1**, 267 (1989)
9. J. K. Krüger, R. Roberts, H.-G. Unruh, K.-P. Frühauf, J. Helwig, H. E. Müser, Progr. in Coll. & Polym. Sci. **71**, 77 (1985)
10. K.-P. Frühauf, J. Helwig, H. E. Müser, J. K. Krüger, R. Roberts, Colloid & Polymer Science **266**, 814 (1988)
11. M. H. Cohen, D. Turnbull, J. Chem. Phys. **31**, 1164 (1959)
12. M. H. Cohen, D. Turnbull, J. Chem. Phys. **34**, 120 (1961)
13. K.-P. Bohn, J. K. Krüger, in "Structure and Properties of Glassy Polymers", ACS Symposium Series 710, ed. by M. R. Tant and A. J. Hill (1998)
14. W. Götze, in "Liquids, Freezing and Glass Transition", edited by J. P. Hansen, D. Levesque and J. Zinn-Justin, Elsevier, Amsterdam (1989)
15. J. K. Krüger, Th. Britz, J. Baller, W. Possart, H. Neurohr, Phys. Rev. Letter **89**, 285701 (2002)
16. S. Brawer, "Relaxation in Viscous Liquids and Glasses", American Ceramic Society (1985)
17. M. D. Ediger, Annu. Rev. Phys. Chem. **51**, 99–128 (1999)
18. S. R. Elliott, "Physics Of Amorphous Materials", Longman (1990)
19. C. A. Angell, K. L. Ngai, G. B. McKenna, P. F. McMillan and S. W. Martin J. Appl. P., **88**, 3113 (2000)
20. I. Gutzow and J. Schmelzer, "The Vitreous State" Berlin: Springer (1995)
21. P. G. Debenedetti "Metastable liquids – concepts and principles", Princeton University Press (1996)
22. D. Richter, A. J. Dianoux, W. Petry and J. Teixeira (eds.) "Dynamics of Disordered Materials", Springer-Verlag (1989)
23. A. J. Kovacs, Journal of Polymer Science **30**, 131 (1958)
24. G. Parisi, F. Zamponi, J. Chem. Phys. **123**, 144501 (2005)
25. T. M. Truskett, V. Ganesan, J. Chem. Phys. **119**(4), 1897 (2003)
26. G. Tarjus, D. Kivelson, in "Jamming and Rheology: Constrained Dynamics on Microscopic and Macroscopic scales", S. Edwards, A. Liu, S. Nagel Eds., Taylor and Francis, London (2001)
27. D. Chowdhury "Spin glasses and other frustrated systems" World scientific (1986)
28. J. K. Krüger, R. Jiménez, K.-P. Bohn, J. Petersson, J. Albers, K. Klöpperpieper, E. Sauerland, H. E. Müser, Phys. Rev. B Cond. Matter, **42**, 8537 (1990)
29. R. Jiménez, B. Jiménez, J. K. Krüger, J. Schreiber, F. Sayetat, F. Mauvy, Ferroelectrics **157**, 141 (1994)
30. Fischer & Hertz, "Spin glasses" Cambridge university press (1991)
31. K. Binder, A. P. Young, Rev. Mod. Phys. **58**, 801 (1986)
32. K. Knorr, A. Loidl, Phys. Rev. B **31**, 5387 (1985)
33. S. K. Satija, C. H. Wang, Solid State Comun. **28**, 617 (1978)
34. S. K. Satija, C. H. Wang, J. Chem. Phys., **69**, 1101 (1978)
35. C. H. Wang, S. K. Satija, Chem. Phys. Lett., **87**, 330 (1982)
36. R. Böhmer, A. Loidl, Phys. Rev. B **42**, 1439–1443 (1990)
37. R. Jiménez, K. P. Bohn, J. K. Krüger, Eur. Phys. J. B, **13**, 643 (2000)

38. J. Hessinger, K. Knorr, Phys. Rev. Letter, **63**, 2749 (1989)
39. K. H. Michel, Phys Rev B, **35**, 1405 (1987)
40. J. M. Rowe et al., Phys. Rev. Letter **53**, 1158 (1973)
41. K. Kishimoto, H. Suga, S. Seki, Bull. Chem. Soc. Japan **51**, 1691 (1978)
42. K. Kishimoto, H. Suga, S. Seki, Cond. Matter., 19 (1978)
43. J. K. Krüger, M. Prechtl, J. C. Wittmann, S. Meyer, P. Smith, J. F. Legrand, J. Pol. Sci. Part B **30**, 1173 (1992)
44. J. K. Krüger, M. Prechtl, J. C. Wittmann, S. Meyer, J. F. Legrand, G. Asseza, J. Pol. Sci. Part B, **31**, 505–512 (1993)
45. H. A. Lorentz, Wied. Ann. Phys., **9**, 641 (1880)
46. L. V. Lorenz, Wied. Ann. Phys., **11**, 70 (1880)
47. M. Goldstein, The Journal of Physical Chemistry, **51**(9), 3728 (1969)
48. M. Born, K. Huang, "Dynamical theory of crystal lattices", Clarendon Press, Oxford (1968)
49. J. K. Krüger, K. P. Bohn, R. Jimenez, Condensed Matter News **5**, 10 (1996)
50. J. K. Krüger, K. P. Bohn, R. Jimenez, J. Schreiber, Colloid & Polymer Science **274**, 490 (1996)
51. J. K. Krüger, K.-P. Bohn, J. Schreiber, Phys. Rev. B **54**, 15767 (1996)
52. J. W. Tucker, V. W. Rampton, "Microwave ultrasonics in solid state physics", North Holland Publishing Co. (1972)
53. P. G. de Gennes, "The physics of liquid crystals", Clarendon Press (1974)
54. S. Chandrasekhar "Liquid crystals", Cambridge monographs on physics (1977)
55. J. K. Krüger, C. Grammes, R. Jiménez, J. Schreiber, K.-P. Bohn, J. Baller, C. Fischer, D. Rogez, C. Schorr, P. Alnot, Phys. Rev. E, **51**(3), 2115 (1994)
56. C. Grammes, J. K. Krüger, K.-P. Bohn, J. Baller, C. Fischer, C. Schorr, D. Rogez, P. Alnot: Phys. Rev E, Vol. **51**(1), 430 (1995)
57. J. K. Krüger, L. Peetz, R. Siems, H.-G. Unruh, M. Eich, O. Herrmann-Schönherr, J. H. Wendorff,, Phys. Rev. A **37**, 2637 (1988)
58. J. K. Krüger, C. Grammes, J. H. Wendorff, in "Dynamics of Disordered Materials", Eds. D. Richter et al., Springer Proceedings in Phys. **37**, 216 (1989)
59. D. Forster "Hydrodynamic fluctuations, broken symmetry, and correlation functions", W. A. Benjamin, Inc. (1975)
60. J. Wong, C. A. Angell "Glass – Structure by spectroscopy" Marcel Dekker, Inc. (1976)
61. G. S. Fulcher, J. Am. Chem. Soc. **8**, 339, 789 (1925)
62. G. Tammann, "Der Glaszustand", Leopold Voss, Leipzig (1933)
63. B. J. Berne, R. Pecora, "Dynamical Light Scattering", John Wiley (1975)
64. B. Chu, "Laser Light Scattering", Academic Press (1974)
65. W. Hayes, R. Loudon, "Scattering of Light By Crystals", John Wiley (1978)
66. J. K. Krüger, in "Optical Techniques to Characterize Polymer Systems", edited by H. Bässler, Elsevier (1989)
67. L. D. Landau, E. M. Lifshitz, "Lehrbuch der Theoretischen Physik", Bd. VIII, "Elektrodynamik der Kontinua", Akademie Verlag Berlin (1966)
68. L. D. Landau, E. M. Lifshitz, "Lehrbuch der Theoretischen Physik", Bd. V, "Statistische Physik", Akademie Verlag Berlin (1966)
69. H.-G. Unruh, J. Krüger, E. Sailer; Ferroelectrics **20**, 3–10 (1978)
70. J. F. Nye, "Physical Properties of Crystals", Oxford Press (1972)
71. J. K. Krüger, A. Marx, L. Peetz, R. Roberts, H.-G. Unruh; Colloid & Polymer Sci., **264**, 403–414 (1986)

72. J. K. Krüger, J. Embs, J. Brierley, R. Jimenez; J. Phys. D – Appl. Phys. **31**, 1913 (1998)
73. C. A. Angell, K. L. Ngai, G. B. McKenna, P. F. McMillan S. W. Martin J. Appl. Phys. **88** 3113 (2000)
74. K. Binder, A. P. Young, Rev. Mod. Phys. **58**, 801 (1986)
75. K.-P. Bohn, J. K. Krüger, in "Structure and Properties of Glassy Polymers", ACS Symposium Series 710, ed. by M. R. Tant and A. J. Hill (1998)
76. M. H. Cohen, D. Turnbull, J. Chem. Phys. **31**, 1164 (1959)
77. M. H. Cohen, D. Turnbull, J. Chem. Phys. **34**, 120 (1961)
78. P. G. Debenedetti, "Metastable liquids – concepts and principles", Princeton University Press (1996)
79. E. Donth, Relaxation and Thermodynamics in Polymers, Glass transition, Akademie-Verlag, Berlin 1992
80. S. R. Elliott, "Physics of Amorphous Materials", Longman (1990)
81. Fischer, Hertz "Spin glasses" Cambridge University Press 1991
82. W. Götze, in "Liquids, Freezing and Glass Transition", edited by J. P. Hansen, D. Levesque and J. Zinn-Justin, Elsevier, Amsterdam, 1989
83. W. Götze, L. Sjögren, Rep. Prog. Phys., **55**, 241–376 (1992)
84. B. A. Auld, "Acoustic Fields and Waves in Solids", John Wiley (1973)
85. J. Jäckle, Models of the glass transition. Rep. Prog. Phys., **49**, 171 (1986)
86. A. J. Kovacs, Journal of Polymer Science **30**, 131 (1958)
87. J. K. Krüger, C. Grammes, J. H. Wendorff, in Dynamics of Disordered Materials, Eds. D. Richter et al., Springer Proceedings in Physics **37**, 216 (1989)
88. H. Baur, "Thermophysics of Polymers, I: Theory", Springer Verlag (1999)
89. K.-P. Frühauf, J. Helwig, H. E. Müser, J. K. Krüger, R. Roberts, Colloid & Polymer Science **266**, 814 (1988)
90. R. Zwanzig, R. D. Mountain, J. Chem. Phys. **43**, 4464 (1965)
91. J. K. Krüger, J. Baller, A. le Coutre, R. Peter, R. Bactavatchalou, J. Schreiber, Physical Review B, **66**, 012206-1/4 (2002)
92. J. F. Nye, "Physical Properties of Crystals", Oxford Press (1972)
93. M. Born, K. Huang, "Dynamical theory of crystal lattices", Clarendon Press, Oxford (1968)
94. G. Grimvall, "Thermophysical properties of materials in Selected topics in solid state physics", Vol. XVIII, ed. by E. P. Wohlfarth, North Holland (1986)
95. J. P. Boon, S. Yip, in "Molecular Hydrodynamics" Mc Graw Hill (1980)
96. J. K. Krüger, J. Baller, A. le Coutre, Th. Britz, R. Peter, R. Bactavatchalou, J. Schreiber, Phys. Rev. B **66**, 12206 (2001)
97. J. K. Krüger, Th. Britz, J. Baller, W. Possart, H. Neurohr, Phys. Rev. Letter **89**, 285701 (2002)
98. S. Brawer, Relaxation in Viscous Liquids and Glasses, American Ceramic Society, (1985)
99. M. D. Ediger, Annu. Rev. Phys. Chem. **51**, 99–128 (1999)
100. A. K. Doolittle, J. Appl. Phys., **22**, 1471 (1951)
101. J. Wong, C. A. Angel, "Glass" Marcel Dekker (1976)
102. W. Williams, D. C. Watts, Trans. Farad. Soc., **66**, 80 (1970)
103. E. Donth, Relaxation and Thermodynamics in Polymers, Glass transition, Akademie-Verlag, Berlin 1992
104. P. G. Debenedetti, "Metastable liquids – concepts and principles", Princeton University Press (1996)

105. W. Götze, in Liquids, Freezing and Glass Transition, edited by J. P. Hansen, D. Levesque and J. Zinn-Justin (Amsterdam)
106. W. Götze, L. Sjögren, Rep. Prog. Phys., **55**, 241–376 (1992)
107. G. S. Fulcher, J. Am. Chem. Soc. **8**, 339, 789 (1925)
108. W. Cochran, "The Dynamics of Atoms in Crystals" in the Structures and Properties of Solids 3, Arnold (1973)
109. J. J. Alkonis, A. J. Kovacs, in "Contemporary Topics in Sciences", Ed. M. Shen, Plenum Press, **3**, 257 (1979)
110. G. Rehage, J. Macromol Sci., Phys., **818**, 423 (1980)
111. F. R. Schwarzl, F. Zahradnik, Rheol. Acta, **19**, 137 (1980)
112. C. A. Angell, "Strong and Fragile Liquids", in "Relaxations in Complex Systems", K. Ngai, G. B. Wright eds., Springfield, VA (1985)
113. C. A. Angell, "Perspectives on the Glass Transition", J. Phys. Chem. Sol. **102**, 205 (1988)
114. The Journal of Chemical Physics, **118**, 1593–1595 (2003)
115. T. Geszti, J. Phys. C. Sol. State. Phys., **16**, 5805 (1983)
116. J. K. Krüger, in "Optical Techniques to Characterize Polymer Systems", edited by H. Bässler, Elsevier (1989)
117. K.-P. Bohn, J. K. Krüger, in "Structure and Properties of Glassy Polymers", ACS Symposium Series 710, ed. by M. R. Tant and A. J. Hill (1998)
118. M. H. Cohen, D. Turnbull, J. Chem. Phys. **31**, 1164 (1959)
119. M. H. Cohen, D. Turnbull, J. Chem. Phys. **34**, 120 (1961)
120. G. M. Bartenev, J. V: Zelenev, "Physik der Polymere", VEB Verlag für Grundstoffindustrie, Leipzig (1979)
121. J. K. Krüger, R. Roberts, H.-G. Unruh, K.-P. Frühauf, J. Helwig, H. Müser Progr. in Coll. & Polym. Sci. **71**, 77 (1985)
122. J. K. Krüger, J. Baller, A. le Coutre, R. Peter, R. Bactavatchalou, J. Schreiber, Physical Review B, **66**, 012206–1/4 (2002)
123. J. K. Krüger, T. Britz, J. Baller, W. Possart, H. Neurohr; Phys. Rev. Letter, **89**(28), 285701 (2002)
124. J. K. Krüger, K.-P. Bohn, R. Jimenez; Condensed Matter News **5**, 10 (1996)
125. J. K. Krüger, K.-P. Bohn, M. Pietralla, J. Schreiber; J. Phys.: Condensed Matter **8**, 10863 (1996)
126. J. K. Krüger, K.-P. Bohn, J. Schreiber; Phys. Rev. **B 54**, 15767 (1996)
127. J. K. Krüger, K.-P. Bohn, R. Jimenez, J. Schreiber; Coll. Polym. Sci. **274**, 490 (1996)
128. K.-P. Bohn, J. K. Krüger; American Chem. Soc., Washington, DC, Developed from a symposium sponsored by the Division of Polymeric Materials: Science and Engineering at the 213th National Meeting of the American Chemical Society, San Francisco, California, April 13–17 (1997)
129. J. K. Krüger, K.-P. Bohn, M. Matsukawa; Phase Transitions **65**, 279–289 (1998)
130. J. K. Krüger, T. Britz, A. le Coutre, J. Baller, W. Possart, P. Alnot, R. Sanctuary, New Journ. Phys, **5**, 80.1–80.11 (2003)
131. G. Rehage, J. Macromol Sci., Phys., **818**, 423 (1980)
132. F. R. Schwarzl, F. Zahradnik, Rheol. Acta, **19**, 137 (1980)
133. W. Hayes, R. Loudon, "Scattering of Light By Crystals", John Wiley (1978)
134. B. Chu, "Laser Light Scattering", Academic Press (1974)
135. B. J. Berne, R. Pecora, "Dynamical Light Scattering", John Wiley (1975)

136. J. K. Krüger, A. Marx, L. Peetz, R. Roberts, H.-G. Unruh, Colloid & Polymer Sci., **264**, 403–414 (1986)

137. J. K. Krüger, in "Optical Techniques to Characterize Polymer Systems", edited by H. Bässler, Elsevier (1989)

138. J. K. Krüger, J. Embs, J. Brierley, R. Jimenez; J. Phys. D – Appl. Phys. **31**, 1913 (1998)

139. H. Baur, "Thermophysics of Polymers, I: Theory", Springer Verlag (1999)

140. A. J. Kovacs, in "Structure and Mobility in Molecular and Atomic Glasses", Ed. By J. M. O'Reilly and M. Goldstein, Ann. N. Y. Acad. Sci. **371**, 38 (1981)

141. A. J. Kovacs, Fortschr. Hochpolym. Forschung, **3**, 394 (1963)

142. K.-P. Frühauf, J. Helwig, H. E. Müser, J. K. Krüger, R. Roberts, Colloid & Polymer Science **266**, 814 (1988)

143. H. Seliger, M. B. Bitar, H. Nguyen-Trong, A. Marx, R. Roberts, J. K. Krüger, H.-G. Unruh; Macromol. Chem. **185**, 1335–1360 (1984)

144. J. K. Krüger, A. Marx, R. Roberts, H.-G. Unruh, M. B. Bitar, H. Nguyen-Trong, H. Seliger; Maromol. Chem. **185**, 1469–1491 (1984)

145. J. K. Krüger, Th. Britz, J. Baller, W. Possart, H. Neurohr, Phys. Rev. Letter **89**, 285701 (2002)

146. M. Matsukawa, N. Ohtori, I. Nagai, K.-P. Bohn, J. K. Krüger, Jap. J. Appl. Phys. **36**, 2976 (1997)

147. J. K. Krüger, K.-P. Bohn, M. Matsukawa, Phase Transitions **65**, 279–289 (1998)

148. J. K. Krüger, M. Veith, R. Elsäßer, W. Manglkammer, A. le Coutre, J. Baller, M. Henkel, Ferroelectrics, **259**, 27–36 (2001)

149. Frenkel, "Kinetic Theory of Liquids", Dover Publications, New York, 1955

150. R. Holtwick, "Niedermolekulare Flüssigkeiten in nano-porigen Träger-materialien – zur Natur des Glasübergangs", Dissertation Universität des Saarlandes, Saarbrücken (1998)

151. K. Knorr, A. Loidl, Phys. Rev. B **31**, 5387 (1985)

152. J. Hessinger, K. Knorr, Phys. Rev. Letter, **63**, 2749 (1989)

153. R. Böhmer, A. Loidl, Phys. Rev. B **42**, 1439–1443 (1990)

154. J. K. Krüger, R. Jiménez, K.-P. Bohn, J. Petersson, J. Albers, K. Klöpperpieper, E. Sauerland, H. E. Müser, Phys. Rev. B Cond. Matter, **42**, 8537 (1990)

155. R. Jiménez, B. Jiménez, J. K. Krüger, J. Schreiber, F. Sayetat, F. Mauvy, Ferroelectrics **157**, 141 (1994)

156. E. M. Brody, C. J. Lubell, CH. L. Beatty, J. Pol. Sci., Pol. Phys. Ed., **13**, 295 (1975)

157. E. Grüneisen, Ann. d. Physik, **26**, 211 and 393 (1908)

158. W. Ludwig, "*Festkörperphysik*", Akademische Verlagsgesellschaft (1978)

159. G. Leibfried, W. Ludwig, Solid State Physics, 12 (1961)

160. T. H. K. Barron, J. G. Collins, G. K. White, Adv. Phys. **29**, 609 (1980)

161. K. Brugger, T. C. Fritz, Phys. Rev. **157**, 524 (1967)

162. M. H. Cohen, D. Turnbull, J. Chem. Phys. **31**, 1164 (1959)

163. J. J. Prejean, in "*Dynamics of Disordered Materials*", Ed. by D. Richter, A. J. Dianoux, W. Petry and J. Teixeira, Proceedings in Physics **37**, 242, Springer (1989)

164. J. K. Krüger, R. Roberts, H.-G. Unruh, K.-P. Frühauf, J. Helwig, H. E. Müser, Progress in Colloid & Polymer Science **71**, 77 (1985)

165. J. K. Krüger, in "*Optical Techniques to Characterize Polymer Systems*", Ed. by H. Bässler, Elsevier, Amsterdam, Oxford, New York 1989
166. G. Grimvall, "*Thermophysical Properties of Materials*" North-Holland (1986)
167. Frenkel, "Kinetic Theory of Liquids", Dover Publications, New York, 1955
168. H. A. Lorentz, Wied. Ann. Phys., **9**, 641 (1880)
169. L. V. Lorenz, Wied. Ann. Phys., **11**, 70 (1880)
170. J. K. Krüger, K. P. Bohn, R. Jimenez, Condensed Matter News **5**, 10 (1996)
171. J. K. Krüger, K. P. Bohn, R. Jimenez, J. Schreiber, Colloid & Polymer Science **274**, 490 (1996)
172. M. P. Stevens, "Polymer Chemistry an Introduction", Addison-Wesley Publishing Company (1975)
173. M. H. Cohen, D. Turnbull, J. Chem. Phys. **34**, 120 (1961)
174. J. R. Rabek, "Experimental Methods in Polymer Chemistry", John Wiley (1980)
175. B. Wunderlich, "Macromolecular Physics", Vol. **1–3**, Academic Press (1973)
176. R. Zwanzig, R. D. Mountain J. Chem. Phys **43**, 4464 (1965)
177. J. K. Krüger, J. Baller, A. le Coutre, R. Peter, R. Bactavatchalou, J. Schreiber, Physical Review B, **66**, 012206–1/4 (2002)
178. J. K. Krüger, T. Britz, A. le Coutre, J. Baller, W. Possart, P. Alnot, R. Sanctuary, New Journ. Phys, **5**, 80.1–80.11 (2003)
179. C. Vergnat, M. Philipp, Diplomarbeiten, Saarbrücken, 2006, to be published
180. C. Truesdell, "Mechanics of Solids", Vol. IV, Waves in Elastic and Viscoelastic Solids" Springer Verlag (1974)

4

Glassy Behaviours in A-Thermal Systems, the Case of Granular Media: A Tentative Review

O. Dauchot

SPEC, CEA-Saclay, L'Orme des merisiers, F-91191 Gif-sur-Yvette, France
olivier.dauchot@cea.fr

Abstract. Granular media, commonly referred to as a-thermal systems, obey a dissipative dynamics a priori very different from an Hamiltonian evolution. However everyday life and recent experiments suggest that a thermodynamical description of granular media might be feasible. Especially in the context of gentle compaction of grains, strong similarities with the behaviour of thermal glassy systems have been underlined. Given that granular media consist in a large number of grains, there is a strong motivation for providing a statistical ground to this hypothetic thermo-dynamical description. It has been argued by Edwards and collaborators that the dynamics is controlled by the mechanically stable – the so-called blocked – configurations and that all such configurations of a given volume are statistically equivalent. This immediately leads to the definition of a configurational entropy and the associated state variable, the "compactivity", the formal analogy of a temperature. First attempts to test this flat measure assumption have been conducted. However, clear evidence in real granular media is still lacking. In this lecture, we shall first discuss the meaning of thermal vs. a-thermal systems, second review old and new results revealing the strong similarities between granular media close to the jamming transition and super-cooled liquids close to the glass transition, and finally present and discuss Edwards proposal, together with recent experimental results on the volume statistics inside a granular packing.

4.1 Introduction

Granular media composed of large enough grains ($d \geq 250\,\mu\text{m}$) are often referred to as dissipative a-thermal systems. Indeed the energy necessary to move a grain is much larger than $k_B T$, and the interaction between the grains, whether it is friction or inelastic collisions, involves dissipation. For such systems, despite evidences of thermodynamical properties, such as experimentally reproducible relations between macroscopic quantities, a proper statistical approach remains to be constructed. Also, there are many similarities between thermal systems close to the glass transition and granular media close to the so-called jamming transition. These similarities have inspired a lot

O. Dauchot: *Glassy Behaviours in A-Thermal Systems, the Case of Granular Media: A Tentative Review*, Lect. Notes Phys. **716**, 161–206 (2007)
DOI 10.1007/3-540-69684-9_4

	Thermal systems	a-thermal systems
Stationary dynamics	Gibbs equilibrium	a-thermal stationary states
ageing dynamics	thermal glasses	a-thermal glasses

Fig. 4.1. Equilibrium vs. glassy behaviour of thermal vs. a-thermal systems. Temperature is well defined in the context of equilibrium. Although the present lecture concentrates on the glassy behaviour of granular media (*second line of second column*), we try in the first section to clarify the difference between thermal and a-thermal systems in the simpler context of stationary dynamics (*first line*)

of recent work towards a statistical description of granular media. However, it is important to note that there are a priori two different issues, one being the description of glassy systems (thermal or not) in the ageing regime, the other one being the identification of a precise prescription for the statistical description of a-thermal systems in general. Figure 4.1 summarizes the four corresponding situations which have to be considered.

In the present lecture, we first try to clarify what is meant – at least here – by a-thermal systems, and present a possible illustration in the context of stochastic dynamics. Then we review experimental results on dense granular media. Some results clearly deal with the glassy behaviour of these systems, others concentrate on the stationary or "super-cooled liquid" regime. In the following, we introduce the prescription proposed by Edwards as a ground for a statistical description of granular media. We discuss the various elements of this proposal, especially focusing on the conditions required to test them experimentally. Finally we present some experimental results on the statistical properties of a dense granular sample.

This lecture is the result of a research under progress. Many concepts remain to be clarified. Despite enormous effort in the recent years, many experimental results are still lacking and those existing may well find new interpretations in a close future due to the progresses on the theoretical side. The reader shall take it as it is: a number of thoughts which we hope will help and motivate him on his way towards the fascinating world of the so-called a-thermal systems.

4.2 Thermal vs. A-thermal Systems

4.2.1 Definitions and General Considerations

Let us first clarify what we mean by "dissipative a-thermal system". By thermal system one means a system which couples to the usual thermal environment: the individual components of the system exchange energy with the individual components of the surrounding. The molecules of a gas in a box kinetic energy with the molecules of the gas surrounding the box. Matter in general is thermal because the microscopic components of matter, the atoms, are of the same scale.

By a-thermal system one means a system whose individual components are of such a large scale compared to the components of the surrounding that the energy received from the thermal environment cannot make them move. One also calls such individual components non-Brownian particles. The thermal environment only contributes to thermalize the matter of which these components are made. Millimetric steel beads won't rearrange by thermal motion, but the steel itself is of course at the room temperature.

Now, a last concept to discuss is dissipation. Consider three scales, the thermodynamic scale, the particles scale and a cutoff scale below which the internal degrees of freedom of the particles are excluded from the description. In the case of a gas, one may choose to include in the description the electrons and their excitation levels, but not the nucleons. As long as the energy of interaction between the particles is low enough not to excite these sub-cutoff internal degrees of freedom, the dynamics is conservative. Dissipation occurs when there is a flux of energy from the scale of the particles to the scale of the excluded degrees of freedom.

Figure 4.2 illustrates the above concepts. In the case of usual thermal systems (Fig. 4.2a), the particles inside the system exchange energy with each other and with the particles outside of the system. The dynamics both inside and outside the system is conservative and the internal degrees of freedom are not excited. In the stationary state, the fluxes of energy are described by the usual equilibrium statistical physics, and lead to the equilibration of the well defined usual temperature. In the case of a a-thermal dissipative system surrounded by a usual thermal environment – for instance a granular system in a lab (Fig. 4.2b) – the fluxes of energy are different. The particles inside the system not only exchange energy during their interaction but also excite internal degrees of freedom excluded from the description – such as the phonons. These degrees of freedom, in turn, exchange energy with the thermal environment (Fig. 4.2c). However there is no transfer of energy from the thermal environment to the particles inside the system because of the scale gap. Such a system has to be forced to be maintained in a stationary state different from the rest.

With such images in mind, nothing prevents from imagining the situation where both the system and its surrounding are composed of large scale particles subject to a dissipative dynamics – for instance a small subsystem of a large granular system (Fig. 4.2d). In this case, one recovers a situation similar to that of the usual thermal systems, in the sense that particles inside the system exchange energy with the particles outside the system. However, the dynamics are not conservative anymore. Obtaining a stationary state requires to force both the system and its environment. Whether the fluxes of energy in such a stationary situation could be described by a generalized statistical

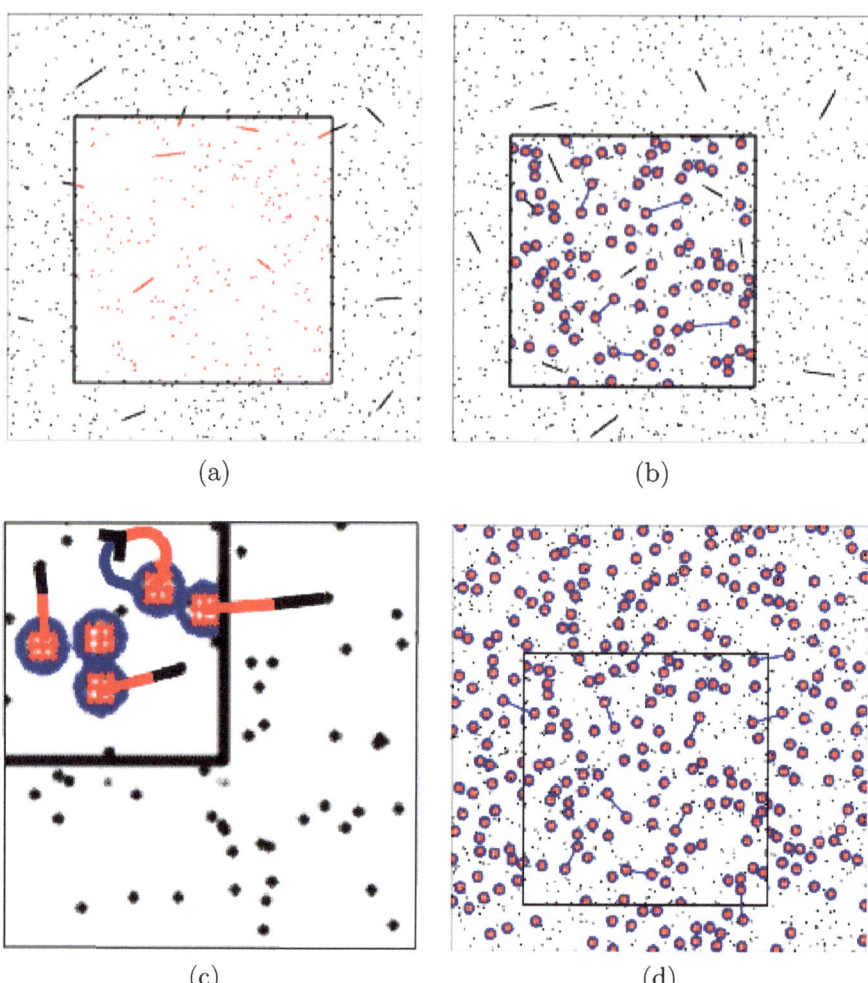

(a)   (b)

(c)   (d)

Fig. 4.2. Flux of energies in thermal conservative systems vs. a-thermal dissipative systems. The interactions among the various components are symbolized by straight segments. The dots are particles of gas inside the system in *red* and outside in *dark*. The *blue* circles are grains. The *red* grids inside the grains symbolize the internal degrees of freedom of the grains. (**a**): a gas embedded in a gas environment: the particles inside the system exchange energy with each other and with the surrounding particles. (**b**): a granular media embedded in a gas environment: grains interact with each other; the particle of gas also; the grains do not receive energy from the gas. But, as shown on (**c**): (zoom of (b)) the grains dissipate – flux of energy towards the internal degrees of freedom, as symbolized by the arrow loop–, which in turn exchange energy with the gas. (**d**): a subsystem of grains inside a larger system of grains. The gas is ignored in the description. The situation is very similar to that of (a) except for the dissipation which must be taken into account

physics, and thereby a generalized temperature, remains presently an open question of major importance.

4.2.2 Illustration in the Context of Stochastic Dynamics

In the following we shall describe a class of systems which may mimic the above situations in the context of a stochastic description – using the master equation formalism.

One crucial goal of a statistical approach for a-thermal dissipative systems would be to give a precise definition of thermodynamical intensive parameters and to predict their relationship with extensive macroscopic variables like energy or volume. Indeed many attempts have been made to define out of equilibrium temperature [1]. In the context of thermal glasses, which we shall focus on in the next sections, the notion of effective temperature has been defined recently as the inverse of the slope of the fluctuation-dissipation relation in the ageing regime [2]. This definition was inspired by the dynamical results obtained within a class of mean-field spin glass models [3]. A lot of numerical simulations [4–8] and experiments [9–14] have been conducted to test the validity of this definition. The situation we want to consider in this section is rather different from that of thermal glasses which are Hamiltonian systems following a non-stationary dynamics with very large relaxation times (first column – second line of Fig. 4.1). Here we want to focus on the stationary dynamics of a-thermal particles which follow a non-conservative dynamics (first line – second column of Fig. 4.1).

In order to find a stochastic model that describes in the best possible way a given complex Hamiltonian system, without knowing a priori the equilibrium distribution, one should at least preserve the symmetries of the original Hamiltonian system, which are the energy conservation and the time-reversal symmetry $t \rightarrow -t$. Energy conservation is easily implemented in the stochastic rules by allowing only transitions between states with the same energy. On the other side, the time-reversal symmetry in the Hamiltonian system can be interpreted in a stochastic language as the equality of two opposite transition rates between the micro-states α and β: $W(\beta|\alpha) = W(\alpha|\beta)$, a property called micro-canonical detailed balance or micro-reversibility. In the context of a dissipative dynamics, the energy is not conserved anymore and one expects the time-reversal symmetry and thereby micro-reversibility to break down. In the most general case, there are little chance to go any further in the description. Following [15, 16] we consider here a subclass of such systems for which we assume that the dynamics still conserves some other quantity – let us call it U.

The stochastic evolution is given by the master equation:

$$\frac{dP_\alpha(t)}{dt} = \sum_\beta W(\alpha|\beta)P_\beta(t) - W(\beta|\alpha)P_\alpha(t) \qquad (4.1)$$

where $P_\alpha(t)$ is the probability of the microstate α. In the hamiltonian case, the stationary regime is given by the uniform distribution, $P_\alpha = 1/\Omega(E)$, where $\Omega(E) = \sum_\alpha \delta(E_\alpha - E)$ is the number of states of energy E. When micro-reversibility is broken, the microcanonical stationary distribution is a priori not uniform anymore: $P_\alpha = f_\alpha/Z_\mu(U)$, where f_α is the statistical weight of the configuration α and $Z_\mu(U) = \sum_\alpha f_\alpha \delta(U_\alpha - U)$ can be called a microcanonical partition function. Indeed, the conservation of U was chosen so as to mimic a microcanonical situation. Yet, one sees that the absence of micro-reversibility already yields an important difference, the non uniformity of the stationary distribution.

In order to define a temperature in this context, one can try to follow a procedure similar to that of the equilibrium statistical physics. For an equilibrium system in the microcanonical ensemble, temperature is introduced in the following way. Considering a large system $\mathcal{S}$ with fixed energy, one introduces a partition into two subsystems $\mathcal{S}_1$ and $\mathcal{S}_2$, with energy E_ℓ ($\ell = 1, 2$). These two subsystems can mutually exchange energy; the only constraint is that the total energy E_{TOT} is fixed, and that the energy of interaction is small so that $E_{TOT} = E_1 + E_2$. The key quantity is then the number $\Omega_{\mathcal{S}_1}(E_\ell)$ of accessible states with energy E_ℓ in the subsystem $\mathcal{S}_\ell$. Assuming that the subsystems do not interact except by exchanging energy, the number of states of the system $\mathcal{S}$ compatible with the partition (E_1, E_2) of the energy is equal to $\Omega_{\mathcal{S}_1}(E_1)\Omega_{\mathcal{S}_2}(E_2)$. Since $E_1 + E_2$ is fixed, the most probable value E_1^* is found from the maximum, with respect to E_1, of $\Omega_{\mathcal{S}_1}(E_1)\Omega_{\mathcal{S}_2}(E_{TOT} - E_1)$. Maximizing this product with respect to E_1, one finds the usual result:

$$\frac{\partial \ln \Omega_{\mathcal{S}_1}}{\partial E_1}\bigg|_{E_1^*} = \frac{\partial \ln \Omega_{\mathcal{S}_2}}{\partial E_2}\bigg|_{E-E_1^*} \qquad (4.2)$$

Defining the microcanonical temperature T_ℓ of subsystem ℓ by the relation

$$\frac{1}{T_\ell} = \frac{\partial \ln \Omega_{N_\ell}}{\partial E_\ell}\bigg|_{E_\ell^*} \qquad (4.3)$$

one sees from Eq. (4.2) that $T_1 = T_2$, i.e. that the temperatures are equal in both subsystems (throughout the lecture, the Boltzmann constant k_B is set to unity). In addition, it can be shown that the common value T does not depend on the partition chosen; as a result, T is said to characterize the full system $\mathcal{S}$.

Very interestingly, this microcanonical definition of temperature can be generalized in a rather straightforward way to the more general case that we consider. Yet, it should be noticed first that microscopic configurations compatible with the given value of the conserved quantity U are not equiprobable, so that Ω_S is no longer relevant to the problem. But starting again from a partition into two subsystems $\mathcal{S}_\ell = \{\alpha_\ell\}(\ell = 1, 2)$ as above, one can determine the most probable value U_1^* from the maximum of the conditional probability $P(U_1|U_{TOT})$ that subsystem $\mathcal{S}_1$ has $U = U_1$ given that the total conserved quantity is U_{TOT}. The conditional distribution $P(U_1|U_{TOT})$ is given by:

$$P(U_1|U_{\text{TOT}}) = \sum_{\alpha \in \mathcal{S}} P_\alpha(U_{\text{TOT}}) \, \delta \left(U_{\alpha_1} - U_1 \right)$$

$$= \frac{1}{Z_\mu(U_{\text{TOT}})} \sum_{\alpha \in \mathcal{S}} f_\alpha \, \delta \left(U_\alpha - U_{\text{TOT}} \right) \delta \left(U_{\alpha_1} - U_1 \right)$$

$$= \frac{1}{Z_\mu(U_{\text{TOT}})} \sum_{\alpha \in \mathcal{S}} f_\alpha \, \delta \left(U_{\alpha_2} - U_2 \right) \delta \left(U_{\alpha_1} - U_1 \right) \qquad (4.4)$$

Now assuming – this is a major assumption – that the stationary distribution factorizes (i.e. $f_\alpha(U_1 + U_2) = f_{\alpha_1}(U_1) \, f_{\alpha_2}(U_2)$), one obtains that $P(U_1|U_{\text{TOT}})$ may be written in a compact form as:

$$P(U_1|U_{\text{TOT}}) = \frac{Z_\mu(U_1) \, Z_\mu(U_{\text{TOT}} - U_1)}{Z_\mu(U_{\text{TOT}})} \qquad (4.5)$$

This result generalizes in a nice way the equilibrium distribution. Indeed at equilibrium $P(E_1|E) = \Omega(E_1) \, \Omega(E - E_1)/\Omega(E)$ and $Z_\mu(U)$ turns precisely into $\Omega(E)$. Finally the most probable value U_1^* satisfies

$$\left. \frac{\partial \ln P(U_1|U_{\text{TOT}})}{\partial U_1} \right|_{U_1^*} = 0 \qquad (4.6)$$

which yields

$$\left. \frac{\partial \ln Z_\mu}{\partial U_1} \right|_{U_1^*} = \left. \frac{\partial \ln Z_\mu}{\partial U_2} \right|_{U_{\text{TOT}} - U_1^*} \qquad (4.7)$$

So in close analogy with the equilibrium approach, one can define an intensive parameter Y_ℓ for subsystem $\mathcal{S}_\ell$ through

$$\frac{1}{Y_\ell} = \left. \frac{\partial \ln Z_\mu}{\partial U_\ell} \right|_{U_\ell^*} \qquad (4.8)$$

Then Eq. (4.7) implies that $Y_1 = Y_2$. It can be shown [16] that Y can be computed from the global quantity $Z_\mu(U)$ instead of $Z_\mu(U_1)$ or $Z_\mu(U_2)$, and is thus independent of the partition chosen. This intensive parameter associated to the conservation of the global quantity U characterizes the statistical state of the whole system.

Up to now, we have considered only the "microcanonical" (in a generalized sense) distribution $P_\alpha(U)$. Yet, it would be interesting to introduce also the analogous of the canonical distribution. To do so, one must compute the distribution $P_{\text{can}}(\alpha)$ associated to a small (but still macroscopic) subsystem $\mathcal{S}_{\text{can}} = \{\alpha\}$ of a large isolated[1] system $\mathcal{S} = \{(\alpha, \alpha')\}$. The configurations corresponding to the reservoir $\{\alpha'\}$ have to be integrated out and one finds under the same assumption of factorizability $f_{\alpha,\alpha''} = f_\alpha f_{\alpha''}$ the following distribution:

---

[1] Isolated here means that U is conserved inside the large system

$$P_{\text{can}}(\alpha) = \sum_{\alpha''} \frac{1}{Z_\mu(U)} f_{(\alpha,\alpha')} \delta \left(U_\alpha + U_{\alpha'} - U \right) \qquad (4.9)$$

$$= \frac{1}{Z_\mu(U)} f_\alpha \sum_{\alpha'} f_{\alpha'} \delta \left(U_\alpha + U_{\alpha'} - U \right) \qquad (4.10)$$

The above summation is nothing but the microcanonical partition function of the reservoir $Z'_\mu(U - U_\alpha)$, which can be expanded to first order as:

$$\ln Z'_\mu \left(U - U_\alpha \right) = \ln Z'_\mu(U) - \frac{1}{Y} U_\alpha \qquad (4.11)$$

assuming that $U_\alpha \ll U$, which is true as long as $\mathcal{S}_{\text{can}}$ is much smaller than $\mathcal{S}$. The derivative of $\ln Z'_\mu(U)$ has been identified with $1/Y$ using Eq. (4.8). Introducing this last result into Eq. (4.9), one finally finds

$$P_{\text{can}}(\alpha) = \frac{1}{Z_{\text{can}}(Y)} f_\alpha \exp \left(-\frac{U_\alpha}{Y} \right) \qquad (4.12)$$

where $Z_{\text{can}}(Y) = Z'_\mu(U)/Z_\mu(U)$ – note that U is the conserved quantity of the global system which includes the reservoir and that Y is the associated intensive parameter imposed to the subsystem $\mathcal{S}_{\text{can}}$.

At this stage, it is worth making a break and to summarize the above results. Basically, it has been shown that for a stochastic dynamics, which does not conserve energy but conserves another extensive quantity, and which does either not satisfy micro-reversibility:

- one looses the property of uniformity for the probability distribution in the microcanonical ensemble;
- if the microcanonical distribution factorizes, one can still define an intensive parameter associated with the conserved quantity;
- this intensive parameter equilibrates between subsystems;
- one can compute a canonical distribution, which is different from but similar to the Gibbs distribution

The last point calls for a special remark. Because of the factorization of the distribution in the non-uniform measure and the Gibbs weight, the thermodynamical algebra remains valid. First, one can show that the dynamical entropy defined as

$$S_U(t) = - \sum_\alpha P_\alpha(t) \ln \frac{P_\alpha(t)\delta(U_\alpha - U)}{f_\alpha} \qquad (4.13)$$

is a non-decreasing function of time, which is maximal in the stationary state with the corresponding value $S(U)$ given by:

$$S(U) = - \sum_\alpha P_\alpha(U) \ln \frac{1}{Z_\mu(U)} = \ln Z_\mu(U) \qquad (4.14)$$

Second, it is straightforward that

$$\langle U \rangle = -\frac{\partial \ln Z_{\text{can}}}{\partial \gamma} \tag{4.15}$$

$$\langle U^n \rangle - \langle U \rangle^n = (-1)^n \frac{\partial^n \ln Z_{\text{can}}}{\partial \gamma^n} \quad \text{for} \quad n > 1, \tag{4.16}$$

where $\gamma \equiv Y^{-1}$. Finally a generalized free energy $F(Y)$ is also naturally introduced through

$$F(Y) = -Y \ln Z_{\text{can}} = \langle U \rangle - Y S \tag{4.17}$$

To conclude this part we shall say that the above formal analogy, which looks encouraging for further developments, has its drawback: given the very strong similarity at the thermodynamical level with usual thermal equilibrium, it will be experimentally difficult to distinguish between the two statistics. We shall come back to this point in Subsect. 4.4.2.

4.3 Glassy Behaviour of Granular Media

In this part of the lecture we shall review a selection of experimental results, which underline the similarity between granular media close to the jamming transition and super-cooled liquids close to the glass transition. It is assumed that the reader is familiar with the glass transition. He might otherwise refer to the other chapters of the present textbook.

4.3.1 Experimental Evidence of the Analogy at the Macroscopic Level

The first set of experimental results concentrates on evidences of the analogy at the macroscopic level. Generically, one considers a three dimensional sample of grains under compaction. We have tried to classify these results according to the following scheme:

- Relaxation towards a stationary state
- Fluctuations and critical slowing-down
- Ageing and Memory effects

Accordingly we shall browse across the results obtained by different groups to illustrate these behaviours. For simplicity we shall refer to these experiments by their localization. Yet, let us first present the various experimental set-up and protocols. Obviously, we can not provide here with all the details of these experiments, which can be found in the original papers.

Figure 4.3(a) displays the device used in Chicago by Knight et al. [17]. Monodisperse, 2 mm diameter glass beads are confined in a 1.88 cm diameter

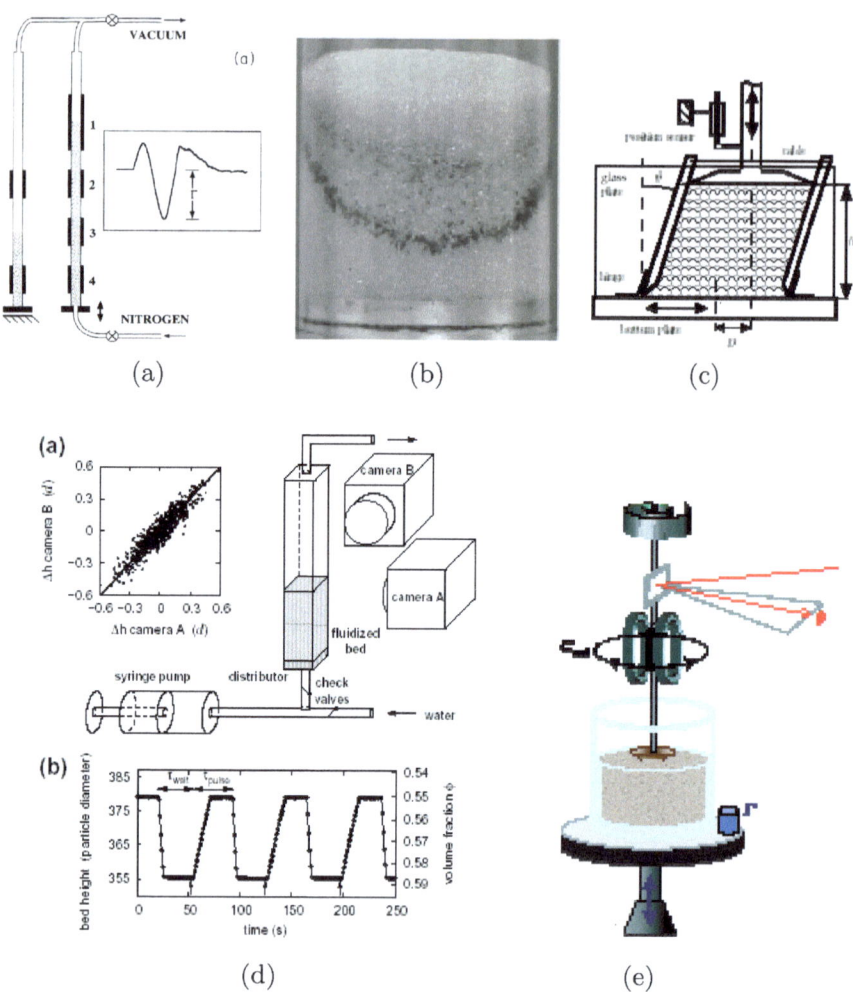

Fig. 4.3. Experimental devices. (**a**): Compaction under vibration in a large aspect ratio column. Knight et al. [17] (Chicago); (**b**): Compaction under vibration in a small aspect ratio cell. Philippe et al. [18,19] (Rennes); (**c**): Compaction under cyclic shear. Nicolas et al. [20] (Marseille); (**d**): Compaction in a fluidized bed. Schröter et al. [21] (Austin); (**e**): Vibration. D'Anna et al. [11,22] (lausanne)

1 m long Pyrex tube mounted on a vibration exciter. The beads are maintained under vacuum. They are prepared in a low density initial stage of packing fraction $\Phi_0 = 0.577$. No convection was observed. The vibration is composed of well separated taps of amplitude a. The acceleration profile of one tap is shown in the inset. $\Gamma = a\omega^2/g$ is the control parameter. The column density was measured with capacitors.

Figure 4.3(b) shows a picture of the set-up used in Rennes by Philippe and Bideau [18, 19]. A glass cylinder of diameter 10 cm, filled with 1 mm diameter glass beads up to 10 cm height, is shaken at regular intervals by an electromagnetic exciter delivering independent vertical taps of amplitude a. The experiments start from a reproducible loose packing $\Phi_0 = 0.583$. Boundary effects are restricted but convection is observed. $\Gamma = a\omega^2/g$ is again the control parameter. The average volume fraction in the bulk Φ is estimated by measuring the absorption of a γ-ray beam through the packing.

Figure 4.3(c) presents a different mode of compaction used in Marseille by Nicolas et al. [20]. A parallelipipedic box (10.5 cm high, 7.9 cm wide and 10.2 cm deep) full of 3 mm diameter glass beads is submitted to a horizontal shear through the periodic motion of two parallel walls. The granular packing is confined on the top by a rectangular plate mounted on a vertical rail. The volume fraction during the compaction process is recorded via the vertical position of the top plate. The mean initial volume fraction of the packing is $\Phi_0 = 0.592$. The lateral plates are oscillating quasi-statically between angles $\pm\theta$, θ being the control parameter. The volume fraction is recorded in the vertical position.

Figure 4.3(d) illustrates the set-up and protocol used in Austin by Schröter et al. [21]. In an original way, the compaction is conducted in a fluidized bed made of a square bore glass tube (24.1 mm × 24.1 mm) filled with about 3.6×10^6 glass beads of $250 \pm 13\,\mu$m diameter. The beads are fluidized with pulses of temperature-controlled de-ionised water. Flow pulses are generated by a computer-controlled syringe pump so that during a flow pulse the bed expands until its height reaches a stable value. After each flow pulse, the bed settles into a stable time-independent configuration, whose volume fraction is determined by measuring the bed height h with two CCD cameras at a 90° angle.

Finally Fig. 4.3(d) displays the apparatus used in Lausanne by D'Anna et al. [11, 22] for studying the jamming transition in weakly perturbed granular media. The granular material, glass beads of diameter $d = 1.1 \pm 0.05$ mm is contained in a metallic bucket of 150 mm height and 94 mm diameter, filled to a height of 130 mm. The system is subjected to taps, the control parameter being Γ, the peak acceleration of the container, normalized by the acceleration of gravity, g. The granular noise is measured with the help of a torsion oscillator, the rotating probe of which is immersed in the granular material.

Apart from these experiments, we shall also discuss the results obtained by Kabla and Debregeas [23] in Paris. In their experiment glass beads of diameter 45 μm, contained in a glass cell (30 mm, 10 mm, 2 mm), fully saturated with pure water, are very gently vibrated with a piezoelectric actuator on which the cell is rigidly mounted. The mean packing fraction is obtained by measuring the position of the upper surface of the pile with a CCD camera. One tap consists in a train of square wave vibrations. The microscopic dynamics induced by these gentle taps is probed by multi-speckle diffusive wave spectroscopy (MSDWS).

Relaxation Towards a Stationary State

The very first evidence of a "glassy" behaviour in dense granular media under compaction is the very slow relaxation towards a stationary state with a well defined volume fraction. Figure 4.4 presents the various compaction curves obtained in the experiments described above. Apart from the experiment in Austin (Fig. 4.4(c)), which is very specific and to which we shall come back in more details in Subsect. 4.4.2, the number of taps is always counted on a logarithmic scale. For both the experiments in Chicago (Fig. 4.4(a)) and Marseille (Fig. 4.4(b), it is not even clear that a stationary state is reached within the duration of the experiment. In the case of the experiment in Rennes (Fig. 4.4(d), a stationary state is obtained, but for large vibration amplitudes only.

To be more precise, various fits have been proposed to describe these experimental data. Both Chicago and Marseille experimental data are best fitted by the heuristic expression:

$$\frac{\Phi_\infty - \Phi(t)}{\Phi_\infty - \Phi(0)} = \frac{1}{1 + B \ln\left(1 + \frac{t}{\tau}\right)} \, , \tag{4.18}$$

whereas Rennes compaction curves are better described by a stretched exponential:

$$\frac{\Phi_\infty - \Phi(t)}{\Phi_\infty - \Phi(0)} = \exp\left[-\left(\frac{t}{\tau}\right)^\beta\right] \, , \tag{4.19}$$

where Φ_∞, B, τ and $\Phi(0)$ are free parameters depending only on Γ. The latter behaviour, introduced by Kohlrausch [24], Williams and Watts [25], often denoted the KWW law, is commonly observed in the relaxation of thermal glasses, the stretched exponential seemingly indicating the superposition of several relaxation times. Also, in the case of Rennes experiment, the relaxation time dependence is reminiscent of an Arrhenius law $\tau = \exp(\Gamma_0/\Gamma)$, for an activated process (Fig. 4.4f). There has been a lot of discussion about the validity of one or the other fit. As a matter of fact, both are plausible in the context of glassy dynamics. The Arrhenius dependence of the relaxation time is reminiscent of strong glasses, whereas one interprets the logarithm dependence as the signature of a fragile glass behaviour. Indeed, as emphasized by Boutreux and de Gennes [26], a Vogel-Fulcher dependence of the relaxation time $\tau = \exp(D\Gamma_0/(\Gamma - \Gamma_0))$ would lead to a logarithmic relaxation of the density.

Fluctuations of Density Around the Steady State

In statistical mechanics the study of fluctuations can be used to investigate the microscopic states that are accessible to a system maintained at a fixed temperature. In granular media, density fluctuations in the steady state are

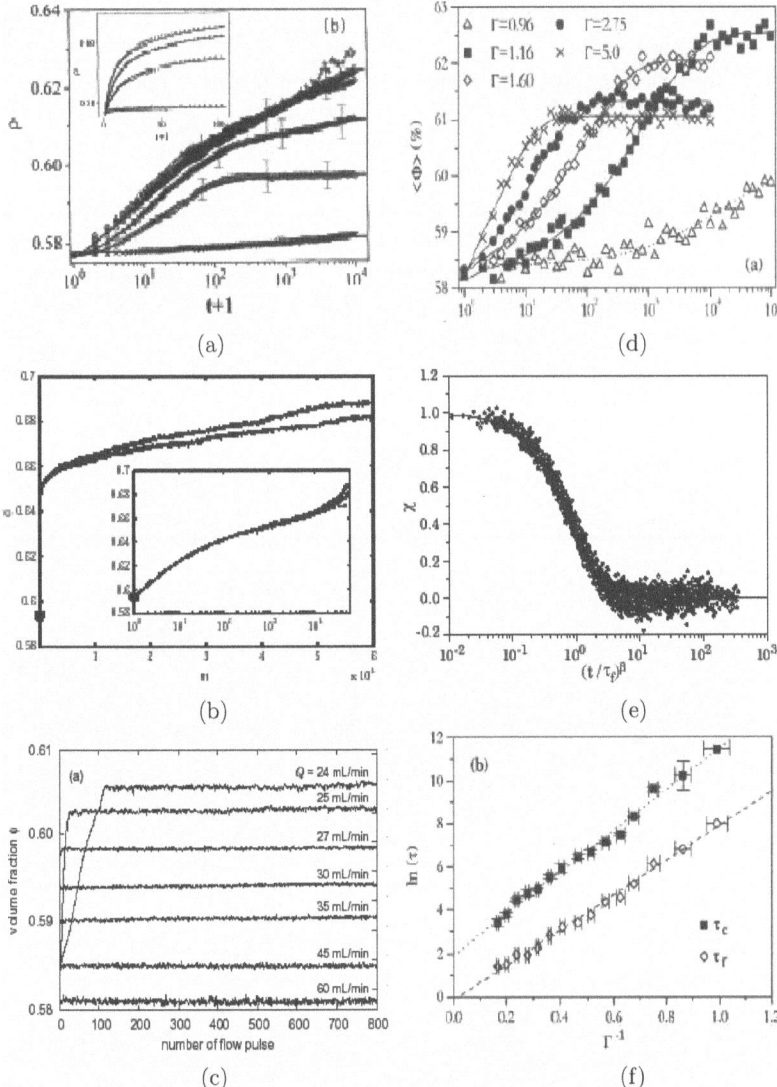

Fig. 4.4. Compaction experiments. (**a**): Chicago, packing density ρ as a function of the logarithm of the number of tap for various amplitude of vibration ranging from $\Gamma = 1.4$ to 5.4 (*inset* is the same plot in linear scale); (**b**): Marseille, compaction curves for $\theta = 5.4°$ for two different runs. Insert: semi-logarithmic scale; (**c**): Austin, the volume fraction of the sedimented bed for different flow rates Q; (**d**): Rennes, temporal evolution of the mean volume fraction for different tapping intensities ranging from $\Gamma = 0.96$ to 5.0; (**e**): Rennes, collapse of the compaction curves obtained with 15 values of Γ between 1.01 and 6.0. $\chi = (\Phi_{ss} - \Phi(t))/(\Phi_{ss} - \Phi(0))$ is plotted as a function of $u = (t/\tau)^{\beta}$. The *solid line* is the exponential function expected in the case of a stretched exponential law; (**f**): Rennes, two estimations of the relaxation time τ as functions of the inverse of the tapping intensity Γ

related to the different volume configurations accessible to the grains subject to an external vibration.

We shall come back in Subsect. 4.4.1 to the formal analogy proposed by Edwards to relate the role played by vibrations in a-thermal systems, such as granular media, and the role of temperature in thermal systems. For the moment let us come back to some experimental results obtained by Nowak et al. [27] in the Chicago experimental set up. We saw in the above section that for small values of Γ, it is difficult, if not experimentally impossible, to reach a steady-state by merely applying a sufficiently large number of taps of identical intensity. Nowak et al. showed that, in this case, it is possible to reach a steady state by "annealing" the system 4.5(a).

Experimentally, the value of Γ is slowly raised from 0 to a value beyond $\Gamma^* \simeq 3$, above which subsequent increases as well as decreases in Γ at a sufficiently slow rate $d\Gamma/dt$ lead to reversible, steady-state behaviour. If Γ is rapidly reduced to 0 then the system falls out of the steady state branch. Along

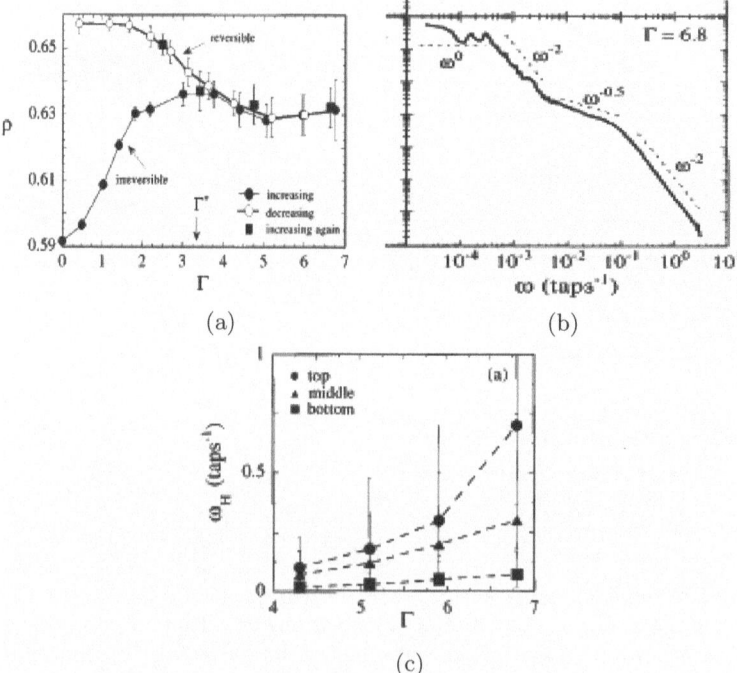

(a)

(b)

(c)

Fig. 4.5. Density fluctuations around the steady state in Chicago experiment (**a**): How to reach a reversible steady state branch; The sample is prepared in a low density initial configuration and then the acceleration amplitude is first slowly increased – solid symbols – and then decreased – open symbols. – The upper branch is reversible, see *square* symbols. (**b**): Power-spectrum of the density fluctuations; (**c**): Relaxation frequency as a function of the vibration amplitude

the reversible branch, the density is monotonically related to the acceleration. As Γ is increased both the magnitude of the fluctuations around the steady state and the amount of high-frequency noise increase. Figure 4.5(b) displays the power-spectrum of the density fluctuations $S(\omega)$, where the frequency ω is measured in units of inverse taps. Three characteristic regimes emerge: (i) a white noise regime, $S(\omega) \sim \omega^0$ below a low-frequency corner ω_L, (ii) an intermediate-frequency regime with nontrivial power-law behaviour, and (iii) a simple roll-off $S(\omega) \sim \omega^{-2}$ above a high frequency corner, ω_H. As shown on Fig. 4.5(c), both ω_L and ω_H increase as Γ is increased. Over the relatively small available range of Γ, the variation of ω_H is consistent with an activated process behaviour: $\omega_H = \omega_0 \exp(-\Gamma_0/\Gamma)$. Approximating to the first order in Γ the bi-univoque relation $\rho(\Gamma)$ characterizing the steady state branch, one sees that this mechanically activated law turns into a Vogel-Fulcher dependence in density, compatible with the observed logarithmic relaxation, as emphasized in the previous section. Note that according to this last remark, the distinction between strong and fragile glasses is not really relevant in the case of a transition controlled by density.

Towards the Jammed State

The above results were obtained for large enough external solicitations. We shall now turn to the behaviour of granular media when the external driving is reduced. Typically one expects a transition close to $\Gamma = 1$ since below this value, the grains are not allowed to lift off from the bottom of the container.

This is indeed the case as illustrated on Fig. 4.6. In the Chicago experiment (Fig. 4.6a), one sees that the densification after 10000 taps significantly increases for $\Gamma > 1.5$. Note that this is not a well defined threshold, since it depends on the number of taps, as well as on the details of the experiments. Figure 4.6(b) shows better evidence of the transition, where one clearly observes a sharp increase of the relaxation times when decreasing Γ below one. The slope variation in the log-lin plot, which indicates a jump in the 'energy barrier" of the mechanically activated process suggested by the Arrhenius laws, finds a natural interpretation in the difference of energy landscape seen by a grain, whether it lifts off or not!

Let us now turn to the Lausanne experiment by D'Anna et al. [22], where a critical slowing-down, qualitatively analogous to super-cooling towards the glass transition has been observed. The noise in Fig. 4.7(a) exhibits a $1/f^2$ spectrum, characteristic of a diffusive process, even for $\Gamma \ll 1$. This is already a clear indication that a weakly perturbed granular medium can display a diffusive behaviour well below the fluidization limit.

By the Wiener-Khintchine theorem, for a $1/f^2$ noise, the value of the noise at a given frequency is proportional to the diffusion coefficient. Hence, Fig. 4.7 displays the characteristic diffusion coefficient as a function of the vibration amplitude for very small amplitude. One observes at $\Gamma_f \simeq 1$ the signature of the vibration-induced fluidization. Second, the diffusion coefficient approaches

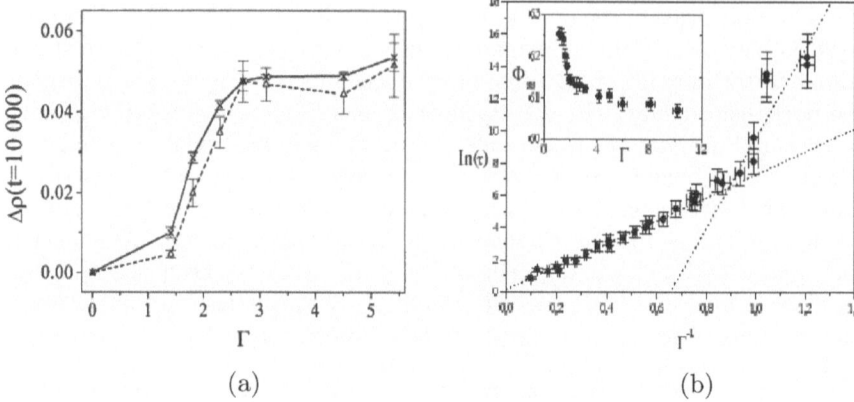

Fig. 4.6. The transition at weak amplitude of vibration (**a**): "Asymptotic" (after 10000 taps) density as a function of the vibration amplitude in Chicago experiment (the two curves correspond to two experimental determinations). (**b**): Arrhenius dependence of the relaxation time as a function of the vibration amplitude in Rennes experiment. *Inset*: variation of the final volume fraction in the cases where a steady state is actually reached

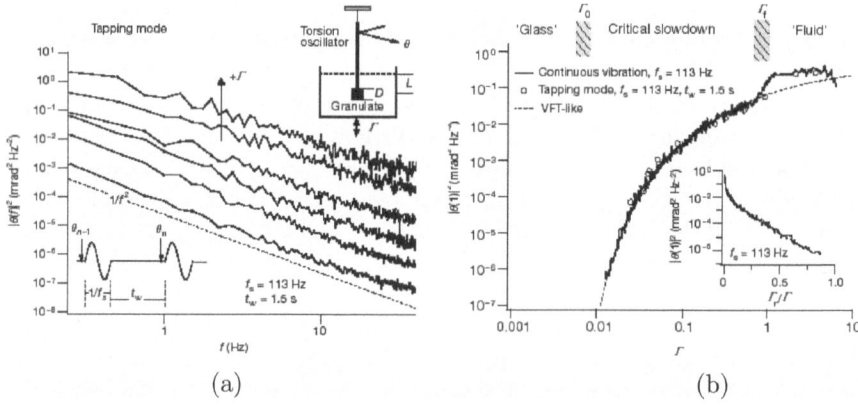

Fig. 4.7. Towards the jammed state in Lausanne experiment (**a**): Low-frequency power-spectrum of the torsion oscillator deflection for various intensities of the taps $\Gamma \in [0.025, 3.6]$ (**b**): Critical slowing-down. The power-spectrum level at 1 Hz, obtained from continuous vibration measurements. Some points (*circles*) are obtained from tapping spectra. The *dotted line* is obtained according to a Vogel-Fulcher fit. Γ_0 is the perturbation intensity where the configuration diffusivity, extrapolates to zero. Γ_f is the fluidization threshold. Inset, the same data as in the main panel in a semilogarithmic plot

zero critically, that is, the inverse noise level diverges. This critical approach to zero can be described by a modified Vogel-Fulcher form $A \exp[B(\Gamma - \Gamma_0)^p]$ with $\Gamma_0 = 0.005$ and $p = -0.4$.

All the above results clearly enforce the analogy between the granular behaviour and the physics of glass-forming liquids that super-cool.

Ageing and Memory Effects

Now that the analogy between thermal glasses and dense granular media has experimental grounds, it is tempting to look for specific behaviours of glasses such as ageing and memory effects in granular media close to the jamming transition.

Ageing was indeed experimentally observed by Kabla and Debregeas [23] in Paris using multi-speckle diffusive wave spectroscopy (MSDWS) to probe the micron-scale dynamics of a water saturated granular pile submitted to discrete gentle taps. The pile is first prepared in a reproducible way at low volume fraction, then submitted to high amplitude taps until it reaches a prescribed packing fraction. Only then the dynamics of contacts is probed by submitting the cell to very gentle taps. Figure 4.8(a) displays the compaction curves during the full procedure. One recognizes typical compaction curve during the first stage. In contrast, the low intensity vibrations do not induce significant further evolution of the packing fraction except for initially very loose packs. To quantify the internal dynamics, one measures the intensity correlation of speckle images – produced by the multiple scattering of photons through the sample – , taken between taps, as a function of the number of taps t that separate them. This function generally depends on the total number of small amplitude taps t_w that have been performed. Accordingly one computes

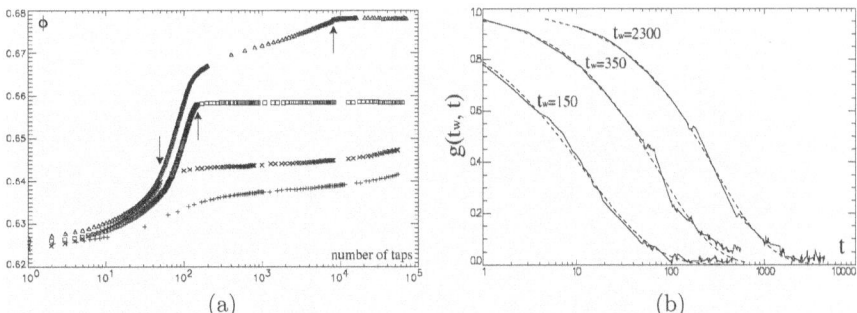

(a)   (b)

Fig. 4.8. Ageing is a gently vibrated granular media in Paris experiment (**a**): The packing fraction for four experimental runs. Each run consists of a first step in which high amplitude taps allow rapid compaction of the sample, followed by a sequence of gentle vibrations, during which the internal dynamics is probed. The *arrows* indicate the change in tapping intensity. (**b**): Two-time relaxation curves for different waiting time

the two-times correlation function $g(t_w, t)$:

$$g(t_w, t) = \frac{\langle I(t_w + t) I(t_w) \rangle_{spkl} - \langle I(t_w) \rangle^2_{spkl}}{\langle I(t_w)^2 \rangle_{spkl} - \langle I(t_w) \rangle^2_{spkl}}, \tag{4.20}$$

where I is the speckle intensity, $\langle \rangle_{spkl}$ denotes the average over several speckles.

Figure 4.8(b) shows three correlation functions obtained with the same sandpile at different values of t_w. These functions, well fitted by stretched exponentials, clearly demonstrate an increase of the relaxation time with t_w. This dynamical arrest is the signature of the ageing behaviour as exhibited in various glassy systems.

As for memory effects, they were observed both in Chicago by Josserand et al. [28] and in Marseille by Nicolas et al. [20]. In the case of Chicago, the granular sample is densified during a set of three experiments up to the same volume fraction Φ_0, but with three different accelerations Γ_0, Γ_1, and Γ_2. After Φ_0 is achieved at time t_0, the system is tapped with the same intensity Γ_0 for all three experiments. As seen in Fig. 4.9, the evolution for $t > t_0$ strongly depends on the history, which is the simplest form of memory effect. In the case of Marseille, a periodic shear with inclination angle θ_1 is first imposed to a random packing, and at a given time, the shear amplitude is suddenly changed to another value θ_2 and later switched back to θ_1. As can be seen, increasing the shear angle produces a rapid fall of volume fraction, followed by a slow and continuous increase. When shear angle is decreased back, a rapid increase of the packing fraction occurs, before recovering the slower one. This is another evidence of memory effect in the packing in the sense that points

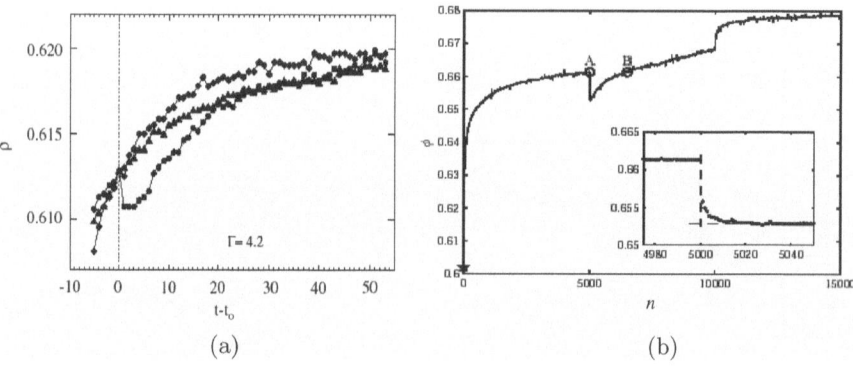

(a)   (b)

Fig. 4.9. Memory effect in granular media under compaction. (**a**): Chicago experiment, time evolution of packing fraction for a system which was compacted to $\rho_0 = 0.613$ at time t_0 using three different accelerations: $\Gamma_1 = 1.8(\bullet), \Gamma_0 = 4.2(\triangle)$, and $\Gamma_2 = 6.3(\diamond)$. After the density ρ_0 was achieved, the system was vibrated at acceleration Γ_0. (**b**): Marseille experiment, example of angle variation during the compaction process. The insert shows a close-up of the first jump

A and B in Figure correspond to packings having the same volume fraction, with different responses to the same shear amplitude.

Altogether, we have seen in this section that the jamming transition of granular media shares strong similarities – exceedingly slow relaxation, critical slowing-down, ageing, memory effects – with the glass transition of supercooled liquids. These similarities are not trivial given the very distinct microscopic processes underlying the dynamics in both systems: in glassy liquids, relaxation occurs by thermally activated rearrangements of the structure. In granular materials, the thermal environment is ineffective and relaxation results from the local yielding of contacts triggered by externally applied vibrations.

4.3.2 Recent Experimental Results at the Grain Scale

In this section, we shall report recent experimental results, which deal with the microscopic behaviour of granular materials under cyclic shear. The goal of these experiments is to find a microscopic ground for the analogy evidenced in the previous section. The first experiment was conducted in Marseille by Pouliquen et al. [29] in the device already presented. The second experiment was conducted in Saclay by Marty et al. [30] and Dauchot et al. [31] in a similar device, but significantly different in several aspects. We shall first summarize the results obtained in Marseille before describing in more details those, more recent and more complete, obtained in Saclay.

Fluctuating Motion During Compaction

In Marseilles experiment, the goal was to provide a link between the macroscopic dynamics and the microscopic structure of the packing during compaction by analyzing the individual motion of the grain. Accordingly the particles are tracked during compaction using an index matching method. The first experiments are performed at a constant shear angle. An example of the particle motion is presented in Fig. 4.10.1(b), where the plot represents the successive positions of the particles measured after each shear cycle. At first sight particles go down as expected for a macroscopic compaction – see the evolution of the volume fraction in Fig. 4.10.1(b) – . On top of this mean vertical displacement, one observes fluctuating motion characterized by ball-like regions as shown in the close-up of Fig. 4.10.1(c), revealing a caging process. The random motion of the particles is trapped for a while before escaping and being trapped again in another cage.

In order to further investigate the link with compaction, experiments are performed where the shear amplitude is discontinuously decreased. The corresponding volume fraction variation is plotted in Fig. 4.10.2(a). As expected from the results presented in the previous section, successive increasing steps in volume fraction are observed. The typical microscopic behaviour of a particle during this experiment is presented in Fig. 4.10.2(b). The volume explored

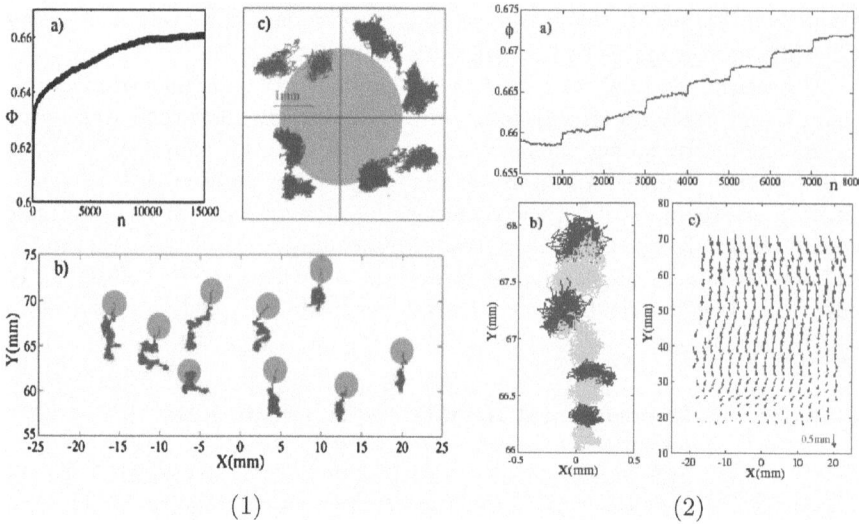

Fig. 4.10. Fluctuating particle motion in Marseille experiment. (1): Compaction for $\theta = 5.4°$; (**a**) Volume fraction as a function of the number of cycles. (**b**) Examples of trajectories during 15000 steps. The disks give the beads size and indicate the initial position of the tracers; (**c**) Examples of cages (trajectories plotted for time slots between 2500 and 5000 steps). (2): (**a**) Volume fraction as a function of cycles when θ varies stepwise (see text); (**b**) Corresponding trajectory of one particle. Changes in colour correspond to changes in θ; (**c**) Displacement field measured in the cell when θ changes from $10.4°$ to $1.4°$

by the particle during its random motion successively shrinks when the shear amplitude decreases because the mean particle displacement decreases. However, each time the shear angle changes, the other particles below the test particle experience the same decrease in their exploration volume. The result is a net downward motion observed when the angle changes. The observed volume fraction variation thus results from the change in the volume randomly explored by the particles. This becomes clear in Fig. 4.10.2(c) when looking at the displacement of all the particles during a sudden change of shear amplitude. In conclusion a simple scenario can be proposed for the compaction process and its memory effect. The slow dynamics of compaction observed in experiment at a constant amplitude is to be attributed to the changes of cages. These changes are irreversible and push the system towards more and more compact configurations. On the contrary, the rapid change of volume fraction observed when changing the amplitude is simply related to the change of the cage size, without important structural changes. This explains why this variation of volume is reversible and can be recovered by coming back to the previous amplitude of excitation. The existence of these two processes which affect differently the packing volume fraction explains that memory effects can be observed.

At this stage, it becomes obvious that a detailed statistical study of the particles displacements should bring a lot of information. What are the property of diffusion? The cage changes certainly involve complex cooperative processes. How are the correlations involved in such process? In Marseille experiment, the dynamics is not stationary and particles experience only a few cage changes before being trapped in their final location. Also, it was impossible to follow all particles in their 3D motion.

Cages and Diffusion Properties Without Compaction

Answering the above questions in a steady state situation, following all the grains, was the goal of the experimental set up built in Saclay. A prototype of the experimental set-up (Fig. 4.11a), allowed Marty and Dauchot to investigate experimentally the diffusion properties of a bi-dimensional bi disperse dry granular material under quasi-static cyclic shear. More specifically, they studied in detail the cage dynamics responsible for the sub-diffusion in the slow relaxation regime, and obtained the values of the relevant time and length scales. In a second version of the set-up (Fig. 4.11c), which allows to follow all the grains in a selected area of interest, measurements of multi-point correlation functions are produced. The intermediate scattering function and its self-part, displaying slower than exponential relaxation, suggest dynamic heterogeneity. Further analysis of four point correlation functions reveal that the grain relaxations are strongly correlated and spatially heterogeneous, especially at the time scale of the collective rearrangements. Finally, a dynamical correlation length is extracted from spatio-temporal pattern of mobility. The present section is devoted to the first set of results, the dynamical heterogeneities being described in the next section

The first experimental setup is as follows: a bi-dimensional, bi-disperse granular material, composed of about 6000 metallic cylinders of diameter 4 and 5 mm in equal proportions, is sheared quasi-statically in a horizontal deformable parallelogram of constant volume (volume fraction $\Phi \simeq 0.86$). The shear is periodic, with a shear amplitude $\theta_{\max} = 10°$. The authors follow

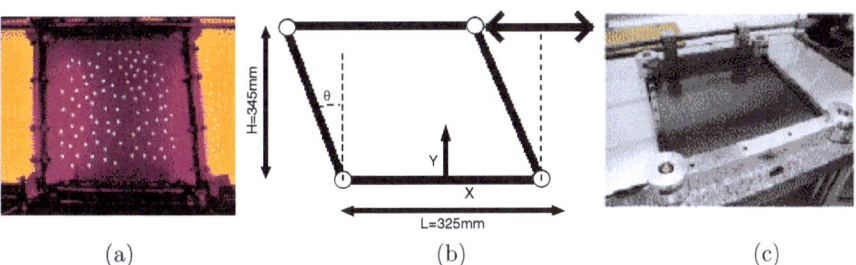

(a)   (b)   (c)

Fig. 4.11. Experimental set-up; (**a**) Prototype used for the measurement of the diffusion properties (**b**) Scheme of the shear cell; (**c**) Final set-up used for following all grains and measuring the spatio-temporal correlations

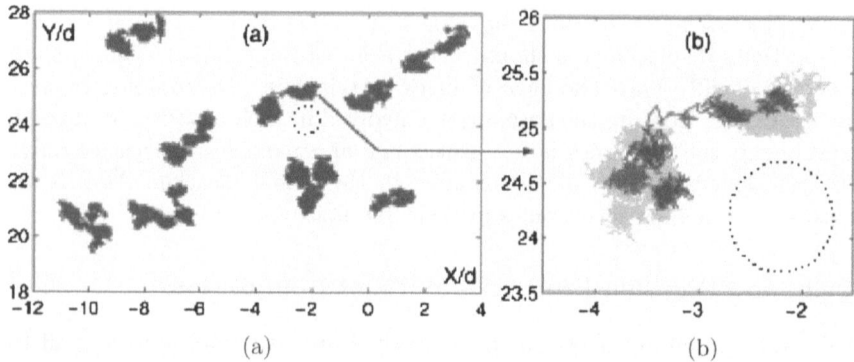

Fig. 4.12. Evidence of cages. (a) Some tracers trajectories. (b) *Gray*: a typical trajectory; *black*: 2000 consecutive steps of the same trajectory. The *circle* indicates the particle size

a sample of 500 of the grains with a CCD camera which takes a picture of the system each time it comes back to its initial position ($\theta = 0°$). The unit of time is then one cycle, a whole experiment lasting 10000 cycles. The unit of length is chosen to be the mean particle diameter d. The system is prepared by removing a fraction of the grains, shaking the remaining sample, putting back all the grains, and shearing the system during 10 to 20 cycles at high shear amplitude and rate. Figure 4.12 shows typical trajectories with well identified cages.

The probability distribution $P(\Delta X(\tau))$ of the displacements of one particle during a time step τ displayed on Fig. 4.13(a) for $\tau = 1, 10, 100, 1000$, exhibit fat tails compared to the Gaussian case, and thereby confirms the intermittent behaviour of the dynamics. The non-Gaussian parameter defined by $\alpha = (\langle \Delta X^4 \rangle / 3 \langle \Delta X^2 \rangle^2) - 1$ (inset of Figure) is indeed different from zero and is maximum, with a plateau, for $\tau \simeq 100$. For larger times, the distribution progressively recovers gaussianity. The root mean square displacement presents two regimes (Fig. 4.13b): at short times, the dynamics is sub-diffusive (logarithmic slope 1/4), while it becomes diffusive (logarithmic slope 1/2) at long times. These results confirm and precise the image of particles trapped in cages, where the cage size $r^* = 0.3d$ and the cage lifetime $t^* = 300$ are given by the the the crossover between the two regimes.

It is of interest to compare these results with those obtained by Weeks and Weitz [32] in a colloidal suspension of hard spheres, that is a *thermal* system. This system undergoes a glass transition for a packing fraction $\Phi_g = 0.58$. Typical trajectories shown on Fig. 4.14(left) and obtained via confocal microscopy for $\Phi = 0.52$ exhibit caged motion, with sudden cage rearrangements. The typical cage size is here also a fraction of the particle diameter. As shown on Fig. 4.14(right-a), the motion is diffusive at very short times, then becomes sub-diffusive at intermediate time scales, and finally recovers a diffusive behaviour at large time scales. The sole difference with the granular

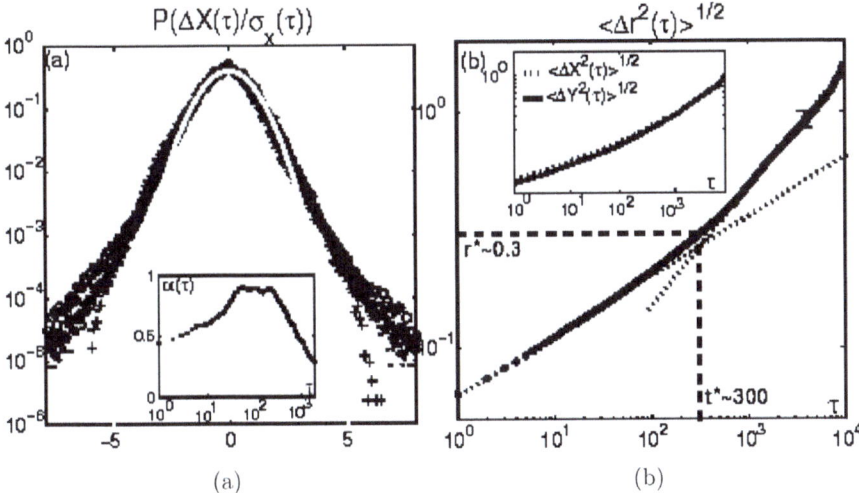

Fig. 4.13. Diffusion properties (**a**) pdf of the displacements $\Delta X(\tau)/\sigma_X$ for $\tau = 1(\bullet), 10(\star), 100(\circ), 1000(+)$; the *solid line* is the Gaussian distribution [*inset*: non-Gaussian parameter $\alpha(\tau)$]; (**b**) $\sigma(\tau) = \sqrt{\langle \Delta r^2(\tau) \rangle}$; *dotted lines* show the slopes $1/4$ and $1/2$; *dashed lines* indicate the position of the crossover (r^*, t^*) [*inset*: $\sigma_X(\tau)$ and $\sigma_Y(\tau)$; no anisotropy is observed]

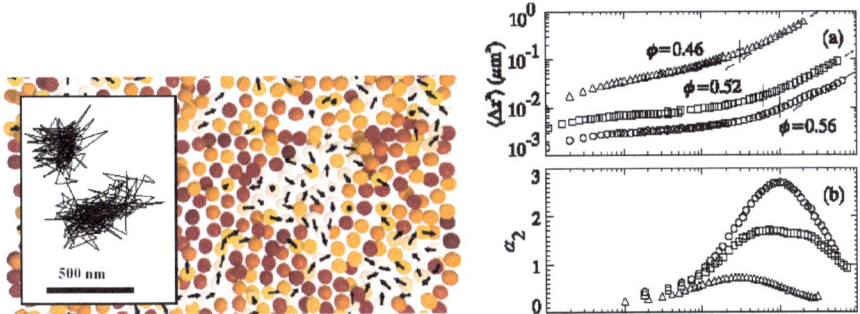

Fig. 4.14. Evidence of cages. *Left*: one layer of particles through a three-dimensional sample of the colloidal suspension, with *arrows* indicating the direction of motion for particles with displacements. *Lighter colours* indicate particles with larger displacements. *Inset*: 120 min. trajectory of one particle from this sample. *Right*: (**a**) Mean square displacements. (**b**) Non-Gaussian parameter

system is the diffusive motion at very short times, a signature of the thermal activation induced by the solvent of the colloidal suspension. Finally, the non-gaussian parameter (Fig. 4.14-right-b) also exhibits a peak which becomes sharper when Φ approaches Φ_g. The type of plateau that has been observed in Saclay typically occurs for $\Phi = 0.52$, that is at a relative distance to transition of 10%.

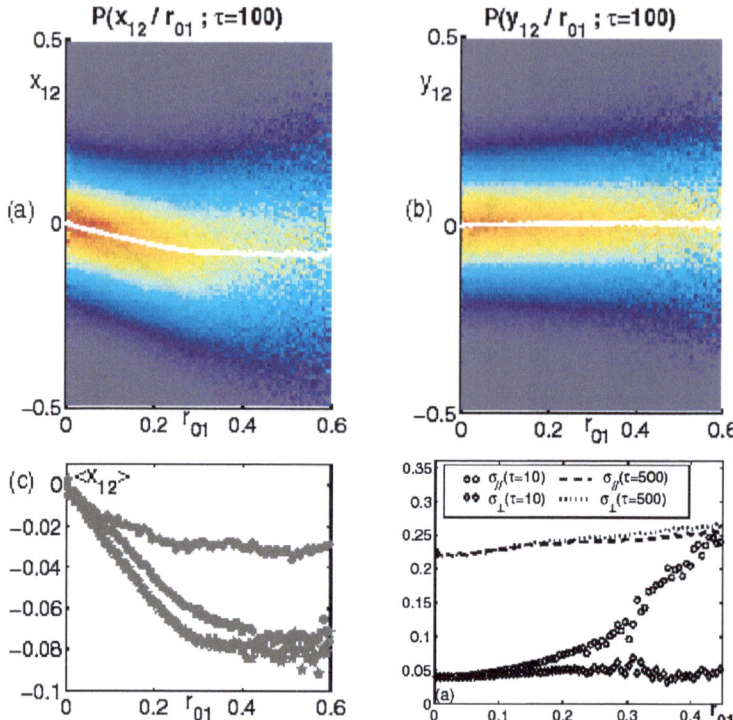

Fig. 4.15. Temporal correlations. *Top*: conditional probabilities (in colour scale) $P(x_{12}|r_{01})$ (*right*) and $P(y_{12}|r_{01})$ (*left*); the white traces are the mean values $\langle x_{12}\rangle$ and $\langle y_{12}\rangle$. *Bottom-left*: $< x_{12} >$ for different values of τ (from *bottom to top*: $\tau = 100; 300; 500$). *Bottom-right*: widths of the distribution of $x_{12}(\sigma_{//})$ and $y_{12}(\sigma_{\perp})$ versus r_{01} for $\tau = 10$ and $\tau = 500$

Let us now report the kind of analysis that can be conducted to better characterize the dynamics. A very convenient tool introduced by Doliwa and Heuer [33, 34], is the conditional probability $P(x_{12}|r_{01})$ (resp. $P(y_{12}|r_{01})$) of the projection x_{12} (resp. y_{12}) of the displacement during a given time interval τ along (resp. orthogonally to) the direction of the motion during a previous time interval of the same duration τ, conditioned by the length r_{01} of the motion during the previous time interval.

These quantities are displayed on Fig. 4.15. A first observation is that the mean value of y_{12} is zero, while the mean value of x_{12} is always negative. More precisely, for a given time interval τ, $\langle x_{12}\rangle$ decreases linearly with r_{01} for $r_{01} < r^*$, then saturates at a constant negative value. The decrease is stronger for small τ and disappears for $\tau > t^*$. On the contrary, the saturation always occurs at $r_{01} = r^*$, a strong indication that the dynamics is controlled by the cage size. Altogether for displacements smaller than r^*, the larger a step the more anti-correlated is the following step, which reflects a systematic back dragging effect experienced by the particle trapped in its cage. For displacements larger than r^*, a cage rearrangement has occurred, and so the anti-correlation does not increase any more. Yet, the constancy of $\langle x_{12} \rangle$ at this saturation value reveals some memory of the fact that part of the trajectory was made in a cage. At larger τ these effects become weaker, an indication that cages relax and adapt to the new positions of the enclosed particles. One can even go further in the interpretation of these distributions by extracting their widths $\sigma_{//}$ and $\sigma_\perp$. The increase of $\sigma_{//}$ with r_{01} reveals that large steps are more likely for particles which moved farther during the previous time interval. This is an indication of the existence of a population of fast particles, a typical feature of glass forming systems, as pointed out, for example, in [33, 35, 36]. Second, we see that for short time intervals, the increase of $\sigma_{//}$ is larger than the one of $\sigma_\perp$. This reflects some anisotropy in the motion, as observed in the string-like cooperation observed numerically by Donati et al. [37]. Both effects concern movements on short time scales, since they tend to disappear as the time interval τ is increased.

Let us now turn to the investigation of some spatial correlations. We choose to illustrate an other technique introduced by Hurley and Harrowell [35], based on relaxation times. For a particle i, the relaxation time $T_i(r)$ is defined as the time needed by the particle to reach a given distance r for the first time. The distribution of these relaxation times is shown in the inset of Fig. 4.16(a), for $r = r^*$.

Defining $T_{i,\ell}(r)$ as the mean relaxation time of the particles contained in a circle of radius ℓ centered on particle i, then the dependence on ℓ of the fluctuations of $T_{i,\ell}(r)$ should provide some information about the typical length L over which cooperative effects take place. A well normalized quantity

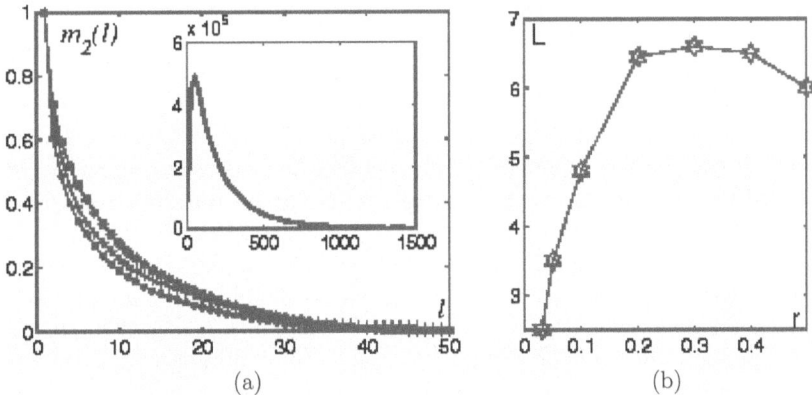

Fig. 4.16. Spatial heterogeneities (**a**) Second moment $m_2(\ell)$ of the relaxation time distribution, for different values of the cutoff distance $r[0.1(\bullet), 0.3(\star), 0.5(+)]$; the dependence of these curves on r is not monotonic (inset: relaxation time distribution for $r = 0.3$) (**b**) characteristic length L; it has a maximum L^* for $r = r^*$

to compute is $m2_\ell(r) = \langle (T_{i,\ell} - T_{i,avg})^2 \rangle / \langle (T_{i,1} - T_{i,avg})^2 \rangle$, where $T_{i,avg}$ is the mean relaxation time averaged over all particles. $m2_\ell(r)$ is plotted versus ℓ for different r on Fig. 4.16(a). $m2_\ell(r)$ naturally decreases with ℓ but is not monotonic with r. To quantify this, one can plot L (defined as the integral of $m2$ over ℓ) versus r and obtain the curve of Fig. 4.16(b). L reaches a maximum of 7 particle diameters for $r = r^*$ which means that cage rearrangements are phenomena which imply more cooperation than the dynamics at other scales and that about a hundred particles are involved in such rearrangements. One then sees that cage rearrangements are highly cooperative phenomena. This, added to the small value of r^* shows that the picture of a particle escaping from a static cage formed by its nearest neighbours is over simplified.

Apart from making precise the dynamics of the specific granular system presented here, this section also aimed at illustrating what can be done to characterize temporal and spatial correlations in systems in which one does not have access to the motion of all particles. In the same spirit, it is also possible to investigate some spatial correlations and discuss the existence of dynamical heterogeneities by considering multi-time correlation functions. However the analysis hardly leads to definitive conclusions and would lead us to discussions which are out of the realm of the present lecture. The reader who is interested can refer to the original work by Heuer et al. [38] and its application to the present system of interest by Marty et al. [30]. To obtain further evidences of the spatial correlations, and a better characterization of the dynamical heterogeneities, one can no longer avoid to follow all particles. The next section will present the kind of analysis conducted in Saclay in the case of granular media, taking benefit of the bi-dimensional geometry of the set-up.

Spatial Correlations and Dynamical Heterogeneities

The above Subsect. 4.3.2 provided a "microscopic" confirmation of the similarity between glass and jamming transitions. The typical trajectories of grains display the so-called cage effect and are remarkably similar to the ones observed in experiments on colloidal suspension [36] and in molecular dynamics simulations of glass-formers [39,40]. As for glass-formers, and contrary to standard critical slowing-down, this slow glassy dynamics does not seem related to a growing *static* local order. For glass-formers it has been shown numerically [35, 41–44] and experimentally [45] that instead the *dynamics* becomes strongly heterogeneous and *dynamical correlations* build up when approaching the glass transition. Recent theoretical works [46] and the end of the previous section suggest that this also happens close to the jamming transition.

The aim of the present section is to present the analysis of the slow dynamics close to jamming measuring multi-point correlation functions as it has been done for super-cooled liquids [39,40,47,48]. First, we shall focus on two point functions, in particular the intermediate scattering function and its self-part, whose slower than exponential relaxation suggests dynamical heterogeneity.

Then we shall turn to four point correlation functions. They have been introduced for glass-formers to properly measure dynamical correlations [47,48] and indeed reveal that the dynamics is strongly correlated and heterogeneous. Finally, we shall focus on spatio-temporal pattern of mobility, out of which we extract a direct measurement of a dynamical length-scale.

The second Saclay experimental setup 4.11(c) contains a bi-dimensional, bi-disperse granular material, composed of about 8,000 metallic cylinders of diameter 5 and 6 mm in equal proportions, which is again sheared quasi-statically in an horizontal deformable parallelogram. The shear is periodic, with an amplitude $\theta_{max} = \pm 5°$. The volume accessible to the grains is maintained constant and the the volume fraction is $\Phi = 0.84$. In this set up, it is possible to follow 2818 grains (located in the center of the device to avoid boundary effects) with a High Resolution Digital Camera which takes a picture each time the system is back to its initial position $\theta = 0$. These conditions are very similar to those of the prototype and by repeating the same analysis the cage radius is found to be $r^* = 0.2$ and the cage lifetime $t^* = 300$.

The intermediate scattering function and its self part are commonly used in the literature when describing the structure and the dynamics of a liquid or a glass. We still recall here some useful algebra which will allow us to introduce a more general quantity – the density overlap – and give us the opportunity to introduce our notations. The very first quantity of interest is the instantaneous density field.

$$\hat{\rho}(r,t) = \sum_i \delta(r - r_i(t)) \tag{4.21}$$

where $r_i(t)$ is the position of the ith particle at time t. One has that

$$\langle \hat{\rho}(r,t) \rangle = \bar{\rho} = \mathrm{cst} \quad \text{and} \quad \int dr\, \hat{\rho}(r,t) = N \quad \text{hence} \quad \bar{\rho} = \frac{N}{V}. \tag{4.22}$$

Here and in the following, the hatted quantities denote the non average observable. In the experiment the average $\langle \cdot \rangle$ means a time average over 300 steps separated by 10 cycles each, taking care that on such time scales the processes are stationary. One then introduces a generalized density correlation function by considering

$$W_a(t) = \langle \hat{W}_a(t) \rangle = \frac{1}{N} \int dr dr' \langle \delta\hat{\rho}(r,t) w_a(r - r') \delta\hat{\rho}(r',0) \rangle, \tag{4.23}$$

where $\delta\hat{\rho} = \hat{\rho} - \bar{\rho}$ and $w_a(r - r')$ is some kernel with a space scale a to be specified later. Replacing $\hat{\rho}$ by its definition (4.21), one obtains after a small calculation:

$$W_a(t) = \frac{1}{N} \int dr dr' \sum_{i,j} \langle \delta(r - r_j(t)) w_a(r - r') \delta(r' - r_i(0)) \rangle - \bar{\rho} \int dr\, w_a(r) \tag{4.24}$$

$$= \frac{1}{N} \left(\left\langle \sum_{i,j} w_a(r_j(t) - r_i(0)) \right\rangle \right) - N \langle w_a \rangle_V, \tag{4.25}$$

where $\langle . \rangle_V$ is the mean value of the kernel function over the sample volume. The self part of this correlation function is given by considering only one particle, hence the same formula replacing $r_j(t)$ by $r_i(t)$ and summing over one particle only. When considering the self part of such a correlation function, one obtains information about the single particle relaxation. When dealing with the non self correlation, one gains information about the structural relaxation. Using $\exp(ik.r)$ for $w_a(r)$, where the space scale a is given by $2\pi/k$, $W_a(t)$ is nothing but the intermediate scattering function $F(k,t)$. When $w_a(r)$ is some overlap function decreasing from one in $r = 0$ to zero for increasing r, $W_a(t)$ is called the density overlap correlation function, further noted $Q(a,t)$, as introduced by Franz and Parisi [48] and largely used by Donati et al. [47]. Practically, in the following computations $\delta(r)$ is approximated by a Gaussian of width 0.3.

The self part of the intermediate scattering function $F_s(k,t)$, where the subscript s here and in the following is for "self part", is plotted on the left of Fig. 4.17(a) as a function of time for different values of k ranging from 1 to 29. Contrary to glass-formers there is no visible plateau in this correlation function although from trajectories it was possible to identify a clear cage effect as seen in the previous section. A possible explanation is that the difference between the time-scales for the relaxation inside the cage and the escape from the cage is not large enough to give rise to a clear plateau. Except for very small k the decrease of $F_s(k,t)$ is slower than exponential in time. A good fit is provided by a stretched exponential: $\exp[-(t/\tau(k))^{\beta(k)}]$. We plot on the right of Fig. 4.17 $\tau(k)$ (top) and $\beta(k)$ (bottom) as a function of k. At small k the relaxation time scales as k^{-2} and the exponent $\beta(k)$ is one. As expected, the grains perform a Brownian motion on large length and time scales and therefore $F_s(k,t) \simeq \exp(-Dk^2t)$ for small k and large t [D is the self-diffusion coefficient of the grains]. Increasing k the stretched exponent decreases and is of the order of 0.7 for k of the order of 2π, corresponding to the inter-grain distance, and even lower for higher values of k. A very similar behaviour has been found in numerical simulations of glass-formers [39, 40]. Also the decrease of $\tau(k)$ steepens sharply for large k. This might be related to the sub-diffusive behaviour observed in the previous section: at short time the displacement distribution is roughly Gaussian with a variance varying as $t^{1/2}$ (not t like for standard diffusion). Hence, it would be natural that at large k the relaxation time went as k^{-4}. An overall very similar behaviour for the intermediate scattering function $F(k,t)$ (not plotted here) is obtained.

Dynamical heterogeneity is one of the possible explanation of the non-exponential relaxation of $F_s(k,t)$ (and of $F(k,t)$): the relaxation becomes slower than exponential because there is a wide spatial distribution of time-scales [45]. However this is not the only possible scenario [45,49]. In the following we want to go one step further and show direct "smoking gun" evidences of dynamical correlations. The proper way to unveil these correlations is through the fluctuations of the correlations [48]. The idea is that the temporal correlation is itself the order parameter of the transition. Accordingly, its fluctuations

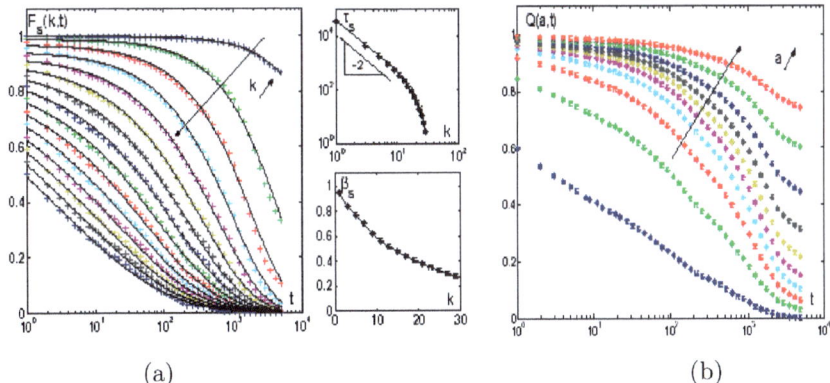

(a)   (b)

Fig. 4.17. Time correlations (**a**): $F_s(k, t)$ as a function of time for different values of the wave-vector $k = 1, 3, \ldots, 29$. The *black lines* represent fits of the form $\exp[-(t/\tau(k))^{\beta(k)}]$; on the *right*: $\tau(k)$ (*top*) and $\beta(k)$ (*bottom*) as a function of k. (**b**): $Q(a, t)$ as a function of time for $a = 0.05, 0.1, \ldots, 0.5$

should unveil correlations exactly as fluctuations of the magnetization unveil magnetic correlations close to a ferromagnetic transition. These fluctuations are characterized by four points correlation functions generically defined as:

$$\chi_4^W(t) = N \left\langle \left(\hat{W}_a(t) - \langle \hat{W}_a(t) \rangle \right)^2 \right\rangle \tag{4.26}$$

where W_a can be the intermediate scattering function, the density overlap, or their self part. It happens that the complex exponential kernel used to construct the intermediate scattering function -historically justified by the light scattering experiments- induces artificial fluctuations which prevent from properly computing the corresponding χ_4. From that point of view, the density overlap is much more convenient. Figure 4.17(b) displays $Q(a, t)$, where the overlap function $w_a(r)$ has been chosen as a non-normalized Gaussian: $w_a(r) = \exp(-r^2/2a^2)$. The evolution of $Q(a, t)$ is a measure of how long it takes for the systems to de-correlate from its density profile at time $t = 0$. One can verify that the behaviour of $Q(a, t)$ is very similar to that of $F_s(k, t)$, as for glass-formers [43, 48].

Figure 4.18(a) displays $\chi_4^{F_s}(t)$ for $k = 1, 3, \ldots, 29$. It has the form found for glass-formers [41–44, 50]: it is of the order of one at small and large times and displays a peak at a time somewhat larger than the cage lifetime. The largest $\chi_4^{F_s}(t)$ is obtained for $k = 9$ corresponding to a length of the order of the cage size. The behaviour at small and large times is in a sense expected since in these limits $\chi_4^{F_s}(t)$ can be related to static correlation functions, which, as discussed previously, do not show any long range order. Alternatively the peak is a clear signature of dynamic heterogeneity and shows that the dynamics is maximally correlated on time-scales of the order of the relaxation time. A rough estimation of the corresponding dynamical correlation length

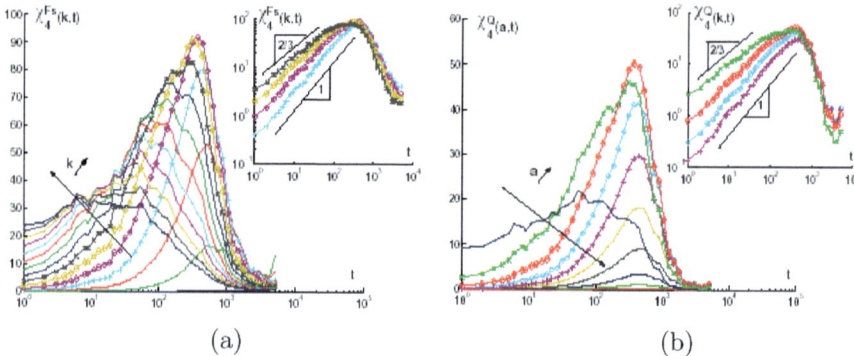

Fig. 4.18. Four-points correlations (**a**): $\chi_4^{F_s}(t)$ as a function of time for values of $k = 1, 3, \ldots, 29$. *Inset*: Log-Log plot for $k = 7, 9, 11, 13$. (**b**): $\chi_4^Q(t)$ as a function of time for values of $a = 0.05, 0.1, \ldots, 0.5$. *Inset*: Log-Log plot for $a = 0.1, 0.15, 0.2, 0.25$

is obtained by identifying the peak of $\chi_4^{F_s}(t)$, of the order of 100, to a correlated area $\pi \xi_{het}^2$, leading to a length $\xi_{het} \propto 6$ in agreement with the estimate of the previous section. Very similar results are found for $\chi_4^Q(t)$ as shown in Fig. 4.18 for $a = 0.05, 0.1, \ldots, 0.5$. It is interesting to note that, as found for glass-formers in [43], the main contribution to $\chi_4^Q(t)$ comes from the fluctuations of the self-part of Q. Indeed we checked that for small and intermediate a, $\chi_4^Q(t) \simeq \chi_4^{Q_s}(t)$ and only for $a \rangle 0.25$ one starts to see a difference. The growth of $\chi_4^{F_s}(t)$ (resp. $\chi_4^Q(t)$) before the peak seems to follow a power law with an exponent which depends on k (resp. a) and varies between 1 and 2/3. As discussed in [50] the form of $\chi_4^{F_s}$ and χ_4^Q provides interesting information on the mechanism behind dynamical heterogeneity. Such power-law behaviours with exponents between 1 and 2/3 suggest that the dynamic correlations cannot be induced by independent defects or free volume diffusion [50].

It would now be very interesting to have some insight on the spatial origin of the fluctuations evidenced by the computation of $\chi_4(t)$. One way to understand how these fluctuations relate to spatial heterogeneities of the dynamics is to decompose, say, $Q(a, t)$ in local contributions: $N\hat{Q}(a, t) = \rho \int dr \hat{q}_a(r, t)$ where $\hat{q}_a(r, t) = 1/\bar{\rho} \int dr' \delta\rho(r, t) w_a(r - r') \delta\rho(r', 0)$.

Using this last expression one finds that $\chi_4^Q(t) = \rho \int dr G_4^Q(r, t)$ where $G_4^Q(r, t) = \langle [\hat{q}_a(r, t) - \langle \hat{q}_a(r, t) \rangle][\hat{q}_a(0, t) - \langle \hat{q}_a(0, t) \rangle] \rangle$. This last expression states that $\chi_4^Q(t)$ is nothing but the mean value over the sample of the spatial correlations among the local temporal correlation. It clearly shows that a large value of $\chi_4^Q(t)$ has to be related to long range spatial correlations of $G_4^Q(r, t)$, which is the spatio-temporal representation of the local temporal correlations.

Figure 4.19 presents a grey-scale plot of the self-part $\hat{q}_{as}(r, t) = \sum_i \delta(r - r_i(0)) w_a(r_i(t) - r_i(0))$ for $t = 154, 435, 1113, 2526$ and $a = 0.15$. By definition $\hat{q}_{as}(r, t)$ measures in a coarse grained way the local mobility: if the particle that was close to r at $t = 0$ moved away more than a in the time interval t

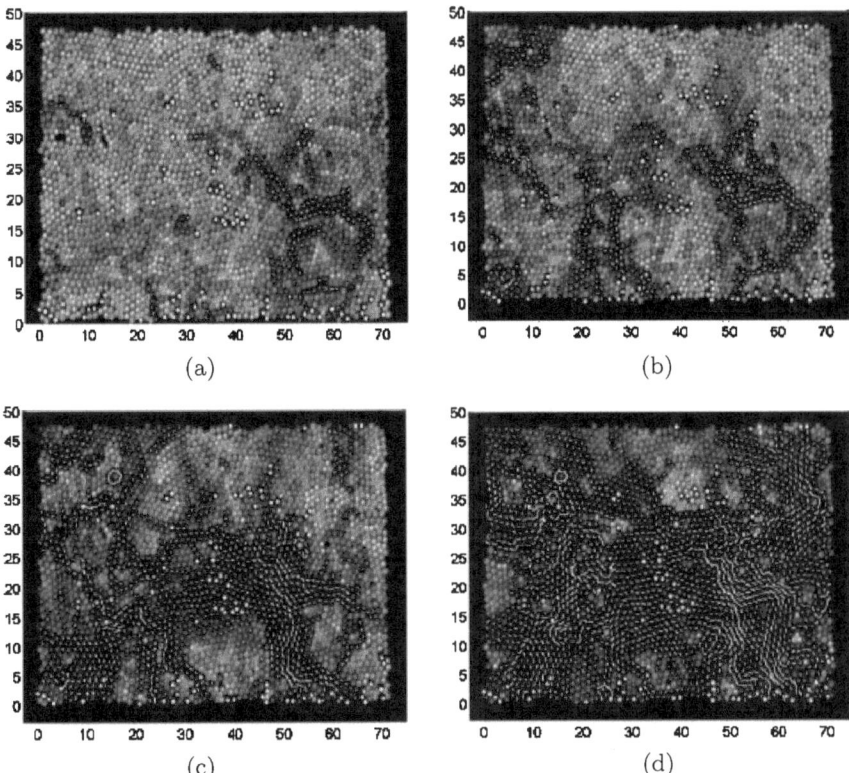

Fig. 4.19. *Grey-scale* plot of $\hat{q}_{a_s}(r,t)$, at $t = 154, 435, 1113, 2526$ from *top* to *bottom* in a grey-scale ($a = 0.15$). *Black* regions correspond to lower values of $\hat{q}_{a_s}$. The displacements of the particles during the interval of time t are plotted in *yellow*. The *yellow dots* are particles that have been lost during tracking

then $\hat{q}_{a_s}(r,t) \simeq 0$. The yellow lines in Fig. 4.19 are the particle displacements in the time interval t. The four chosen time intervals correspond from top to bottom to short-times, relaxation times, moderate long times, long-times. At short-times ($t = 154$) only few particles have moved and from Fig. 4.19 it appears that they do so in a string-like fashion. On larger times ($t = 435, 1113$) the relaxed regions are ramified and finally, at very long time ($t = 2526$) the overall majority of the particles has moved substantially but there remain few (rather large) regions not yet relaxed. These findings, similar to the ones found in simulation of super-cooled liquids [41,43] and experiments on colloidal suspensions [32] suggest that the mobility is organized in clusters, which are the direct visual evidence of the dynamical heterogeneities.

To further quantify the heterogeneities, we estimate how large is the mobility difference between fast and slow grains. Figure 4.20(a) displays the self part of the Van-Hove correlation function, i.e. the probability distribution of the grains displacements amplitudes for $t = 438$ (corresponding to the

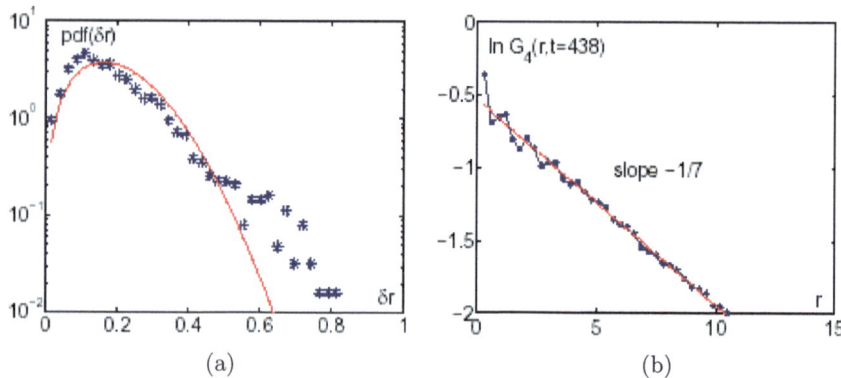

Fig. 4.20. (**a**): Self part of the Van Hove correlation function after angular integration at $t = 438$; the *continuous line* is the pdf obtained assuming a Gaussian distribution. (**b**): $\ln(G_4(r, 438))$ as a function of r; the *straight line* is a linear fit

maximum of $\chi_4^Q(t)$). It clearly demonstrates the excess of fast and slow grains compared to the distribution obtained when assuming a Gaussian process (in continuous line). The fast grains are roughly five time faster than the slow ones. Furthermore, we obtain $G_4(r, 438)$ by computing the radial autocorrelation of $\hat{q}_{as}(r, t)$, and averaging over ten realizations. Figure 4.20(b) shows that $G_4(r, 438)$ decays exponentially over a characteristic dynamical length $\xi = 7$, in agreement with the value obtained from the peak of $\chi_4^{F_s}$.

4.3.3 Partial Conclusion

In this second part of the lecture, we have seen that the analogy observed at the macroscopic level between dense granular media close to the jamming transition and super-cooled liquids close to the glass transition has indeed microscopic grounds. Despite the difference in the driving mechanisms – a mechanical instead of a thermal forcing –, the diffusion properties of a single particle and the collective relaxation of the system share very strong similarities including the existence of a dynamical length increasing at the transition.

At the root of these very strong similarities is the physical nature of the transition. In the case of thermal systems close to the glass transition, the dynamics is dominated by the complex shape of the multidimensional potential energy landscape. The thermal activation being weaker and weaker, the system spend more and more time in meta-stable states. Eventually the system does not equilibrate on experimental time scales and falls out of equilibrium. In the case of the dense granular systems under gentle forcing, the grains rearrange among mechanically stable configurations which are the equivalent of the meta-stable states. When the external forcing is decreased, or when the density is increased, the grains rearrangements become more and more difficult to produce. In both cases, the relaxation evolves towards a global structural relaxation involving collective behaviours and characterized by dynamical heterogeneities.

As a matter of fact, the analogy is so strong that the glass transition can be seen as a specific case of jamming transition as suggested by Liu and Nagel [68]. The interest of such a unifying view is double. First, as we shall see in the last part of the present lecture, theories developed in the field of glasses have inspired interesting development in the field of granular media. Second, as we saw in this part, granular media can be seen reciprocally as a very convenient experimental system for studying the microscopic features of the structural relaxation close to the glassy state.

4.4 Looking for a Statistical Description

As just stated, one of the key ingredients of the non-trivial phenomenology observed in both granular media and thermal glasses is the large number of microscopic meta-stable states, among which the system hops during its slow dynamical evolution. In the context of glassy systems, Stillinger et al. [51, 52], introduced the concept of inherent structures, namely the potential local minima. Following the ideas of Goldstein [53], the phase space trajectory of the system can be described as successive steps among the potential basins. The entropy is then claimed to be separable into one vibrational part accounting for the vibrational modes around the minima and one configurational part accounting for the numerous inherent structures. In the thermodynamical analogy proposed by Edwards [54], that we shall discuss in the following, the mechanically stable states of a granular packing are given a similar role to that of the inherent states and called "blocked states".

The natural question that immediately arises is that of the weight of these configurations in a given experiment and how they encode the specificity of the dynamics. Various forms of the fluctuation-dissipation relation have been generalized to out of equilibrium situations of thermal systems by Cugliandolo, Kurchan and collaborators [2,3]. Such generalizations lead to the definition of an effective temperature for the long-time behaviour of glassy systems. The existence of such an effective temperature suggests for these systems some kind of "ergodicity" in the dynamics among the meta-stable states. Extending these ideas to the case of granular media as suggested by Kurchan [55,56] may provide some validity to Edwards" assumption that all blocked configurations in the jammed state are equiprobable leading to the so called "Edwards' ensemble".

However the situation is far from being clear. Let us recall some of the remarks made by Bouchaud [57] in his lecture in Les Houches, to motivate the last part of the present lecture:

- Despite phenomenological analogies between temperature and gentle tapping, one should keep in mind that tapping is a long-wavelength excitation, whereas temperature is thought to give rise to very short wavelength excitation. Accordingly detailed balance and activated process ideas might need to be reconsidered.

• The choice of the microscopic variables is already not obvious. Also, dealing with continuous variables such as the contact forces for instance, one has to assume the uniformity of the a priori measure on the forces as done on the canonical variables (position and momentum) when building the microcanonical ensemble for particles. However, in the latter case, the procedure is justified by the Liouville theorem resulting from the Hamiltonian dynamics. In the case of granular media no physical prescription has been proposed yet.

In this part, we shall present Edwards' proposal, discuss how and whether they can be tested experimentally and finally produce some recent results on free volume statistics inside a bi-dimensional granular packing.

4.4.1 Edwards' Proposal

In the statistical physics of Hamiltonian systems [58,59], the microscopic configurations $\mathcal{C}$ are described by the canonical variables prescribed by the Liouville's theorem, the momenta and positions $\mathcal{C}(p_i, q_i)$ of all particles. In the case of an isolated system with total energy E, one obtains as a stationary state of the Liouville's equation, a uniform equilibrium probability density over the micro-states of energy E. Accordingly for a system defined by its Hamiltonian $H(p_i, q_i)$:

$$P(\mathcal{C}(p_i, q_i)) = \frac{1}{\Omega(E)}\delta(H(p_i, q_i){-}E) \text{ with } \Omega(E) = \int \prod_i dp_i dq_i \delta(H(p_i, q_i){-}E).$$
(4.27)

The entropy at equilibrium and the temperature are then given following the construction presented in Subsect. 4.2.2 by:

$$S(E) = -\sum_{\mathcal{C}} P(\mathcal{C}) \ln(P(\mathcal{C})) = \ln \Omega(E) \quad \text{and} \quad \beta = \frac{1}{T} = \frac{\partial \ln \Omega(E)}{\partial E}. \quad (4.28)$$

Behind this very elegant formalism stand a few but essential properties of Hamiltonian systems. We have already mentioned the prescription for the appropriate microscopic variables (p_i, q_i), by the Liouville's theorem, which derives itself from the canonical structure of the equations of Hamilton. One must also consider the symmetries such as the time-reversal and the time-translational invariances, the latter giving rise to the conservation of energy. Finally, assuming that the uniform distribution is the true distribution of the system is given by the ergodic hypothesis.

Consider now a granular media close to the jammed state. In Edwards' description [54,60,61], it is first assumed that the volume $\mathcal{V}$ is the key macroscopic quantity governing the behaviour of the system. Then, it is assumed that the statistics is completely dominated by the "blocked configurations", which are claimed to all have the same statistical weight. Hence the probability of a configuration $\mathcal{C}$ in a system of fixed volume V is:

$$P(\mathcal{C}) = \frac{1}{\Omega(V)}\Theta(\mathcal{C})\delta(V(\mathcal{C}) - V) \quad \text{with} \quad \Omega(V) = \int d\mathcal{C}\Theta(\mathcal{C})\delta(\mathcal{V} - V). \quad (4.29)$$

where $\Theta(\mathcal{C})$ is a constraint to restrict the configurations to the "blocked states". An analogous entropy and the corresponding analogue of the temperature, named the "compactivity" are then given by

$$S(V) = -\sum_{\mathcal{C}} P(\mathcal{C})\ln(P(\mathcal{C})) = \ln \Omega(V) \quad \text{and} \quad \frac{1}{X} = \frac{\partial \ln \Omega(V)}{\partial V}. \quad (4.30)$$

Given the very strong properties of the Hamiltonian systems which support the equilibrium statistical description, Edwards' proposal looks at first sight a rather crude analogy and at least calls for a few comments.

- Let us discuss first the choice of the volume. It is a natural extensive macroscopic quantity and it clearly plays a crucial role in the rearrangements of the grains among jammed configurations. However, it should be shown that it is conserved by the dynamics, a key ingredient for the above construction as illustrated in the first part of this lecture. The total volume accessible to the grains can be fixed, such as in the shear experiment conducted in Saclay. Assuming then some tiling of the space accessible to the grains in a given experiment, the grains' rearrangements can be described by a redistribution of the volume among the grains. In this sense, there is indeed a local conservation of the volume. Even if the total volume is not fixed, as in most tapping experiments, on can check that the system is large enough to ensure sufficiently small fluctuations of its volume. In such a case, provided that the grains rearrangements do not cascade to the free surface of the packing, the system can serve as a reservoir of volume for a sub-system, which then has to be described in the canonical formalism. Yet, one sees that one important hypothesis is to have enough local redistribution of the volume.
- The choice of the microscopic variables, as already mentioned, is extremely ambiguous. There is no general prescription neither for the minimal list of relevant physical quantities, nor for the choice of the appropriate variables to describe them. Ignoring physical quantities will falsify the computation of the density of states. Having no prescription for the correct choice of variables induces an irreducible ambiguity since the uniform measure for continuous variables is not conserved under a change of variables.
- Time-reversal symmetry and the ergodic hypothesis are crucial for assuming a uniform distribution among the accessible configurations. Given the existence of dissipation and the very slow compaction of granular media under gentle tapping there is little chance to observe time-reversal symmetry in the general case. Furthermore, even for a stationary dynamics,

checking the existence of micro-reversibility in a real system, is out of reach of experimental investigations.

Altogether Edwards' description is a challenging proposal, the implementation of which is far from being obvious and which calls for experimental and numerical validations. Despite some clear examples where Edwards' approach fails [62], various checks have been made so far in mean field models of the glass-transition [63], in schematic finite-dimensional models with kinetic constraints [64,65], in spin-glass models with a-thermal driving between the blocked states [66] and finally in a few more realistic models of particle deposition [67] or MD simulations of shear driven granular media [69]. Reviewing these studies is out of the scope of this lecture. It should be stated, however, that in most of these works, the accent is put on the validation of the uniform measure over the blocked states. A given model being chosen, its dynamics is computed at constant volume. Blocked states are identified and dynamical averages of macroscopic quantities are compared with averages over the blocked states assuming equal weights. The issue of the proper choice of variables to describe a granular media is not considered. Finally most of these models use Monte-Carlo algorithm to generate the dynamics so that implicitly, the dynamics is reminiscent of an Hamiltonian kind of dynamics. In the last section, we shall discuss what can be tested experimentally in Edwards' proposal and present recent results obtained in this direction.

4.4.2 Experimental Test of Edwards' Proposal?

Volume Fluctuations

Formula (4.29) gives the probability of a given configuration of volume V, according to Edwards' proposal. In order to investigate its validity, let us assume that the probability distribution over the blocked configurations is *not* uniform, but given by a density $f(\mathcal{C})$. Formula (4.29) then turns into:

$$P_\mu(\mathcal{C}) = \frac{1}{Z_\mu(V)} f(\mathcal{C})\Theta(\mathcal{C})\delta(V(\mathcal{C}) - V) \quad \text{with}$$

$$Z_\mu(V) = \int d\mathcal{C} f(\mathcal{C})\Theta(\mathcal{C})\delta(V(\mathcal{C}) - V) \tag{4.31}$$

As for usual thermodynamics, it is uneasy to study an isolated system since the experimentally measurable quantities of interest are then fixed from the outside. The usual way to proceed is to consider a subsystem, free to exchange volume with a reservoir, that is a system in the canonical situation. We have seen in the first part of this lecture, Subsect. 4.2.2, equation (4.12), that *provided that both $f(\mathcal{C})$ and $\Theta(\mathcal{C})$ factorize for any partition of the system,* the canonical probability distribution can be written:

$$P_c(\mathcal{C}) = \frac{1}{Z_c(X)} f(\mathcal{C})\Theta(\mathcal{C}) \exp -(V(\mathcal{C})/X) \quad \text{with} \quad \frac{1}{X} = \frac{\partial \ln Z_\mu(V)}{\partial V}\bigg|_{V^*},$$

$$\tag{4.32}$$

where V^* is the most probable value of the volume of the reservoir.

This probability distribution is still out of reach of experimental and even numerical investigations, since it requires to sample all microscopic configurations. However, as seen in Subsect. 4.2.2, equation (4.15), the usual thermodynamical equalities remain valid, so that the fluctuations of the volume can be related to the average volume by:

$$\langle V^2 \rangle - \langle V \rangle^2 = X^2 \frac{\partial \langle V \rangle}{\partial X} . \tag{4.33}$$

Inverting this relation, one can in principle extract from the simultaneous measure of $\langle V \rangle$ and $\langle V^2 \rangle$ the dependence of the compactivity on the volume $X(V)$. This is precisely what has been done by Schröter et al. [21] in the Austin experimental set-up presented on Fig. (4.3d) and that we shall now discuss in more details.

In their work Schröter et al. [21] use a periodic train of flow pulses in a fluidized bed. A column of glass beads in water is expanded by an upward stream of water until it reaches a homogeneously fluidized state, and then the flow is switched off. The fluidized bed forms a sediment of volume fraction Φ, which depends in a reproducible way on the flow rate of the pulse. This forcing results in a history independent steady state where the volume exhibits Gaussian fluctuations around its average value. The history independence is demonstrated by ramping up and down in flow rate; both the averaged volume fraction Φ_{avg} and the standard deviation σ_Φ depend only on the flow rate of the last flow pulse, not the earlier history of the bed.

As shown on Fig. 4.21(a) the variation of σ_Φ with Φ_{avg} is well fitted by a parabola with a minimum for some specific value of the averaged volume fraction. Relating this minimum of the fluctuations to a maximum in the number of statistically independent spatial regions at the moment of solidification, the authors suggest the following explanation. For smaller volume fraction, the sample is more fragile and local rearrangements induce large reorganizations. For larger volume fraction, the free volume becomes smaller, the system is more jammed and any local rearrangement requires a large reorganization of the packing.

Using the relation (4.33), the authors derive the following relation

$$\frac{\lambda \rho}{m} \int_{\Phi_{RLP}}^{\Phi} d\phi \left(\frac{\phi}{\sigma_\phi} \right)^2 = \frac{1}{X(\Phi)} \tag{4.34}$$

where it has been assumed in the spirit of Edwards' proposal that $X(\Phi_{RLP}) = \infty$ and that Φ_{RLP} is obtained in the limit of very large flow rate. Note that in the present lecture λ, the equivalent of the Boltzmann constant, has been fixed to one. This relation leads to the dependence of the compactivity X on the averaged volume fraction Φ_{avg} displayed on Fig. 4.21(b).

The above results are the very first experimental measurements of the so-called compactivity. Unfortunately, in the absence of a theoretical prediction for the dependence of the compactivity on the volume, they do not check

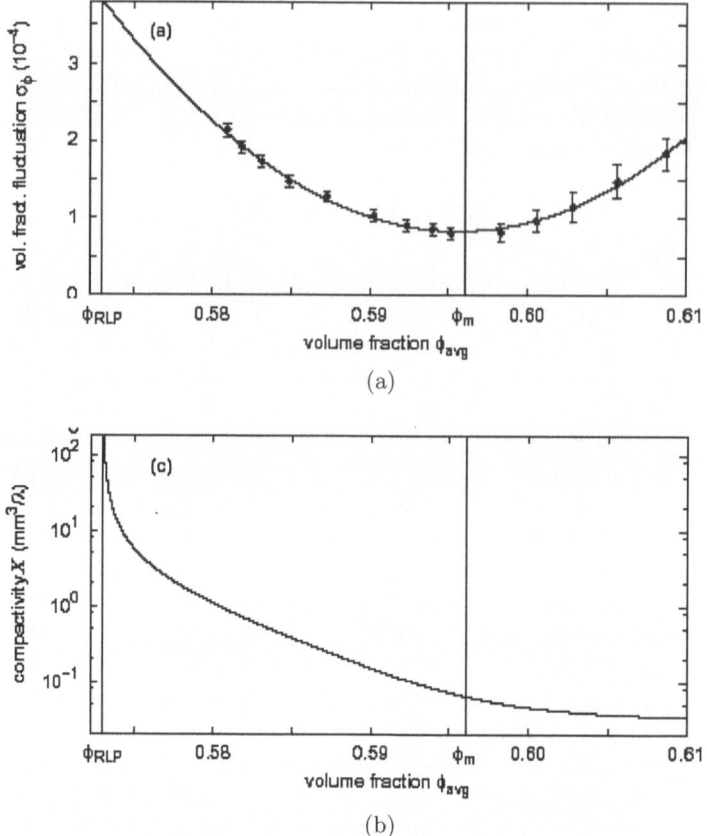

Fig. 4.21. Fluctuations and compactivity in Austin experiment (**a**): Volume fluctuations as a function of the mean volume. (**b**): Compactivity as a function of the mean volume

Edwards' proposal in anyway. As a matter of fact, for any system, in any situation, it is always possible to first measure the averaged value and the fluctuations of any given macroscopic observable V, then *define* X assuming a thermodynamical relation such as (4.33) and obtain $X(V)$.

Free Volume Distributions

To go one step further, one might think of investigating the full probability distribution of the volume, not only at the scale of the packing but for subsystems of increasing sizes. From the canonical probability distribution (Eq. 4.32), one readily computes the probability of observing a volume V in a subsystem of N grains:

$$P_c(V_N) = \int d\mathcal{C} P_c(\mathcal{C})\delta(V(\mathcal{C}) - V_N) = \frac{Z_\mu(V_N)}{Z_c(X)} \exp -V_N/X , \qquad (4.35)$$

Apart from the exponential weight, most of the information about the system lies in the pre-factor dependence on V_N. Hence, one crucial step to go further is to precise what are the variables which describe the microscopic configurations. Given the role played by the volume, it could be natural to consider the volumes w_i associated to each grain through some tiling of the space, as suggested by Edwards. However, it is clear that the choice of such a set of variables is not sufficient since it does not include the forces at the contacts. As a consequence one will have to include a density of state $\rho(w_1, w_2, \ldots, w_N)$ in the description and the microcanonical partition function will read:

$$Z_\mu(V_N) = \int \prod_i dw_i \rho(w_i) f(w_i) \Theta(w_i) \delta \left(\sum_i w_i - V_N \right) . \qquad (4.36)$$

One already sees that without a theoretical prescription for the density of states ρ, a formidable task to achieve, there is little chance to test the measure $f(w_i)$. Let us take an example to make this last point more precise. Consider a system for which $f(w_i) = \prod_i w_i^{\eta-1}$ [15, 16] and for which $\rho(\lambda w_1, \lambda w_2, \ldots, \lambda w_N) = \lambda^{\gamma N} \rho(w_1, w_2, \ldots, w_N)$. Then introducing the adimensionalized volumes ω_i defined by $w_i = \omega_i V_N/N = \omega_i \nu_N$, one obtains:

$$Z_\mu(V_N) = (V/N)^{N-1}(V/N)^{\gamma N}(V/N)^{N(\eta-1)}$$
$$\int \prod_i d\omega_i \rho(\omega_i) \Theta(\omega_i) \delta \left(\sum_i \omega_i - N \right) , \qquad (4.37)$$

and thereby

$$P_c(\nu_N) = \frac{A(N)}{Z_c(X)} \nu_N^{(\gamma+\eta)N-1} \exp -N\nu_N/X . \qquad (4.38)$$

This last expression shows clearly that the details of the microcanonical measure (here the value of η) cannot be distinguished from the specific properties of the density of state (here the value of γ). In particular the uniform measure ($\eta = 1$) does not emerge as a special case.

However, the microscopic physics of the system remains fully embedded in the microcanonical partition function and therefore, it is still of interest to investigate its shape and in particular to evaluate its dependence on the system size. This task has been conducted by Da-Cruz et al. [70] in the case of a bi-dimensional packing. The experimental set up (Fig. 4.22a) consists in a rectangular glass container which contains 5000 nickel plated brass cylindrical spacers of two different diameters $d_s = 4$ mm and $d_l = 5$ mm in equal number. In the following d_s has been chosen as the unit length. The cell is half filled with a single layer of such hard disks mixed together, resulting in an homogeneous and disordered bi-dimensional packing. The cell is mounted on a horizontal axis and rotated around this axis in such a way that the grains fall from one side to the other every half cycle. The experimental procedure is the following. The cell starts in a vertical position and is rotated one cycle,

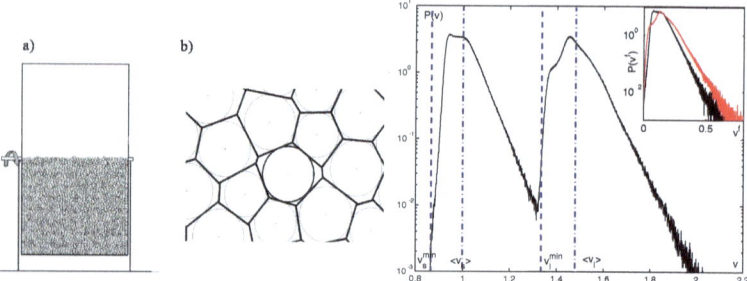

Fig. 4.22. (a) Experimental set-up and sketch of the modified Voronoi tessellation; **(b)** Distribution of the Voronoi cells area. Vertical *dashed lines*: minimal Voronoi cell area. Vertical *dash-dotted lines*: conditional average Voronoi cell area. Inset: distributions of the free volume conditioned by the grain size; *(dark)*: small grains; *(grey)*: large grains

at a constant speed of one cycle per minute. During this cycle, the grains fall from one side to the other and then back to the initial side. The engine is stopped, the system allowed to reach a mechanically stable state, and a picture of the bulk is taken. 15000 of such cycles are performed. The pictures hence taken display on average 300 grains. For each picture the centers of the spacers are located and their Voronoi diagram is computed (Fig 4.22b), taking into account the bidispersity of the assembly. One then collects the area of the cells along with the type, position and index of the associated grains. Out of these raw data, the statistical distribution of the free volumes occupied first by one grain, then by clusters of an increasing number of neighbouring grains are extracted and analyzed.

Figure 4.22(b) displays the distribution of the Voronoi cell areas. The distribution displays two peaks centered on $\langle v_s \rangle = 1.00$ (resp. $\langle v_l \rangle = 1.49$), the averaged area occupied by the small, (resp. the large) grains computed independently. Also indicated on the figure are the minimal values that a Voronoi cell can possibly take – the closest regular hexagon- for each type of grain, $v_s^{min} = \sqrt{3}/2 \simeq 0.866$ and $v_l^{min} = \sqrt{3}/2(d_l/d_s)^2 \simeq 1.35$. Both peaks present a well defined exponential tail, which is easily isolated when considering the distributions of the free volume ($v_{s,l}^f = v - v_{s,l}^{min}$), for each type of grain as shown on the inset of Fig. 4.22(b). Note that these exponential tails and the associated characteristic free volumes should not be interpreted as a signature of the Gibbs weight, and thereby as some kind of validation of Edwards' hypothesis as sometime suggested in the literature. At least one should consider larger subsystems.

Accordingly, let us turn to the free volume distributions for clusters of neighbouring grains. Figure 4.23(a) displays the distribution of the free volume per grain inside clusters of N neighbouring grains, $v_N^f = N^{-1}\Sigma_{i=1}^{N}v_i^f$. The authors choose to describe the distributions of the free volume per grain inside

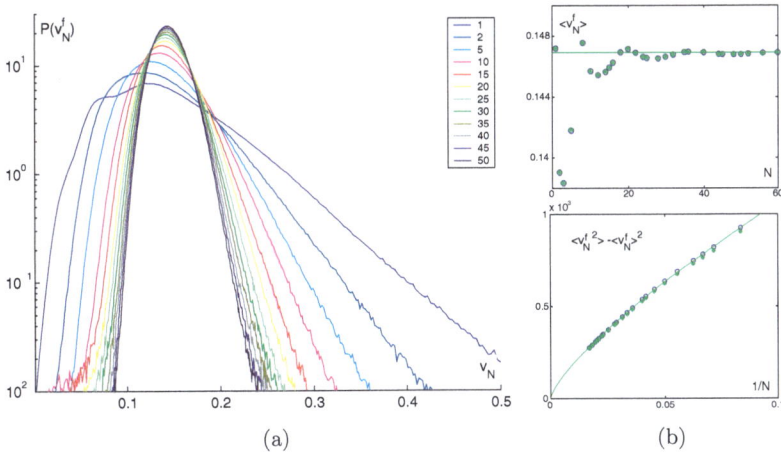

Fig. 4.23. Free volume statistics. (**a**): distributions of the free volume per grain inside clusters of N grains; the larger N, the narrower the distribution. (**b**): dependence on N of the first (*top*) and second (*bottom*) moments of the free volume distributions (○): computed from the data; (∗): extracted from the fit of the distributions by a Gamma law; (*plain line*): fit of their dependence on N

a cluster of N neighbouring grains by a Gamma law of parameters η_N and X_N as suggested by the the the shape of the distributions, the example discussed here above and the expected convergence towards the gaussian law:

$$P(v_N^f) = \frac{1}{X_N^{\eta_N}\, \Gamma(\eta_N)} (v_N^f)^{\eta_N-1} e^{-v_N^f/X_N} . \qquad (4.39)$$

where Γ is the Euler Gamma function. Once chosen the form of the distributions, one computes their first two moments and obtain η_N and X_N, through the relations $\langle v_N^f \rangle = \eta_N X_N$ and $\langle v_N^{f\,2} \rangle - \langle v_N^f \rangle^2 = \eta_N X_N^2$. As expected, $\langle v_N^f \rangle$ rapidly evolves towards a constant (Fig. 4.23(b)-top). On the contrary $\langle v_N^{f\,2} \rangle - \langle v_N^f \rangle^2$ varies like $N^{-\alpha}$ with $\alpha = 0.75 \pm 0.0025$, in contrast with the $1/N$ dependence expected for independent variables (Fig. 4.23(b)-bottom). Altogether, the distribution of the free volume per grain inside clusters of N grains is well described by a Gamma law, the parameters of which exhibit the following dependences on N : $\eta_N = \eta_{\text{eff}} N^\alpha$ and $X_N = X_{\text{eff}} N^{-\alpha}$, with $\eta = 3.5$ and $X = 0.041$.

Altogether, rewriting the above Gamma law in the limit of large N, one obtains that the logarithm of the distribution of the free volume per grain inside clusters of N grains scales as $N^\alpha g(v, \eta_{\text{eff}}, X_{\text{eff}})$ with $g(v) = \eta(\ln(v/(\eta X)) - v/(\eta X) + 1)$, $\alpha \simeq 3/4$, $\eta_{\text{eff}} \simeq 7/2$ and $X_{\text{eff}} = 0.041$. Finally, one can also write the distribution of the free volume per grain inside clusters of N grains as:

$$P(v) = \frac{1}{Z(N, \eta, X)} e^{-N^\alpha \left(\frac{v}{X} - s(v) \right)}, \quad \text{with} \quad s(v) = \eta \ln(v), \qquad (4.40)$$

and thereby $\frac{1}{X} = \frac{\partial s}{\partial v}\Big|_{\langle v \rangle}$, an exact result given the Gamma law distribution and more generally expected from a saddle point calculation in the large N limit.

This central result deserve a few comments: First, the observed non extensive factor N^α is presumably the evidence of long range correlations between the free volumes of individual grains. Indeed, in the presence of correlations decaying with the distance r as $1/r^\gamma$, one has in two dimensions, for $\gamma < 2$, that the second moment of the average of N centered random variable scales like $N^{-\gamma/2}$. In the present case, we would thus infer the existence of long range correlations decaying like $1/r^{3/2}$. If the existence of such long range correlations is confirmed, then the thermodynamical description will have to take them into account in order to define properly the extensive and the associated intensive parameters. For instance, the use of the relation (4.33) as done by Schröter et al. leads to the definition of a compactivity, which depends on the system size! Also, the existence of such long range correlations may invalidate the hypothesis of a local conservation of the volume. Second the above analysis has allowed to define two effective parameters which characterize the probability distribution of the free volume for one grain. How do they relate to Edwards' compactivity? Do they equilibrate between subsystems put into contact? Is it still possible to define a thermometer in the most general sense? Many questions remain open.

To conclude this part, testing Edwards hypothesis appears to be extremely challenging. Whereas much of the focus is usually put on the uniformity assumption for the probability distribution of the blocked states, we have seen here that in practice it is hard to distinguish it from another factorizable distribution, until one has a full microscopic description of the systems and its dynamics. Conversely, a lot can be learned from the investigation of the probability distributions of various macroscopic variables. It is of major interest to understand how many intensive parameters are necessary to describe these distributions and whether they equilibrate between subsystems in contact. Identifying these parameters would be a major step in the thermodynamical description of granular systems.

4.5 Conclusion and Perspectives

In this lecture we have tried to discuss both theoretical ideas and experimental results on the thermodynamics of granular media and its statistical grounds. After having made precise the concepts of thermal vs. a-thermal systems, we have reviewed the experimental evidences of the strong similarities between the granular media close to the jamming transition and super-cooled liquids close to the glass transition, at both the macroscopic and the microscopic scale. Finally, we have discussed in details Edwards' proposal for a statistical description of jammed granular media and illustrated the kind of experimental study, which are conducted in this spirit.

As recalled in the introduction, the understanding of granular media and a-thermal systems in general is far from being completed. Many of the ideas exposed here will change; many experimental results will find new interpretations. Let us still stress one more time, what we believe are the main messages of this lecture in the present state of knowledge.

In the first part, it was shown that the definition of a temperature or an equivalent is actually not related to the scale of the particles, but to the existence of an extensive conserved quantity. In the second part, it has been observed that the idea of a unified description for the glass and the jamming transition has indeed strong evidences at the scale of the individual particles. Finally, we saw that from an experimental point of view, testing the uniformity of the measure over the blocked configurations is a chimera, until a full microscopic description of the system is provided. However, in the meantime looking for relevant extensive and intensive thermodynamical parameters is a key step for achieving a thermodynamical description of non-hamiltonian systems. In this matter we have stressed that one must be careful with potential long range correlations and associated non-extensivity.

Finally, let us suggest some further developments in the field. In the first part, we have seen how to define a thermodynamical equivalent of the temperature for stationary non-hamiltonian dynamics with a conserved quantity. Kurchan has proposed to extend the definition of the effective temperature obtained in the glassy regime for thermal glasses, to the case of a-thermal systems [55]. It would be of great interest to relate both approaches. One way for instance would be to study glassy regimes in a modified version of the model introduced by Bertin et al [15]. Given the strength of the similarities between granular media close to the jamming transition and the super-cooled liquids close to the glass transition, and given the rather easy access to the details of the particles dynamics in the case of the granular media, it would be of great benefit to further investigate the mechanisms underlying the development of the dynamical heterogeneities. Finally, given the possibility of extracting intensive parameters from the free volume distributions inside a granular packing, it is now a priority to test whether some of these parameters equilibrate between subsystems.

Acknowledgments

The content of these lecture notes owes a lot to my collaborators, without whom none of the present ideas and results would have been developed: G. Marty, F. Da Cruz, F. Lechenault, S. Deboeuf, E. Bertin, G. Biroli. Several papers written with them have inspired the present notes. I want to thank all the members of the "Glassy Work group" in Saclay, in particular, J.P. Bouchaud,

E. Vincent, G. Biroli, as well as J. Kurchan, J.M. Lück, C. Godrèche and A. Barrat for many inspiring discussions. I am grateful to the GDR-Midi, which provides an enriching and intellectually challenging context. I want to thank A. Coniglio and his team for having invited me to exchange our views on the topics. Finally let me express my gratitude to M. Henkel and M. Pleimling for giving me the opportunity to give this lecture. Also a special thank to E. Bertin and F. Lechenault for their careful rereading.

References

1. J. Casas-Vasquez and D. Jou. *Rep. Prog. Phys.*, **66**, 1937 (2003).
2. L. F. Cugliandolo, J. Kurchan, and L. Peliti. *Phys. Rev. E*, **55**, 3898 (1997).
3. L. F. Cugliandolo and J. Kurchan. *Phys. Rev. Lett.*, **71**, 173 (1993).
4. J. L. Barrat and W. Kob. *Europhys. Lett.*, **46**, 637 (1999).
5. L. Berthier and J. L. Barrat. *Phys. Rev. Lett.*, **89**(095702) (2002).
6. A. Crisanti and F. Ritort. *J. Phys A*, Math. Gen., **36**(R181) (2003).
7. G. Parisi. *Phys. Rev. Lett.*, **79**, 3660 (1997).
8. F. Sciortino and P. Tartaglia. *Phys. Rev. Lett.*, **86**, 107 (2001).
9. B. Abou and F. Gallet. Probing an nonequilibrium einstein relation in an aging colloidal glass. *Phys. Rev. Lett.*, **93**(160603) (2004).
10. L. Bellon, S. Ciliberto, and C. Laroche. *Europhys. Lett.*, **53**, 511 (2001).
11. G. D'Anna, P. Mayor, A. Barrat, V. Loreto, and F. Nori. Observing brownian motion in vibration-fluidized granular matter. *Nature*, **424**, 909–911 (2003).
12. T. S. Grigera and N. E. Israeloff. *Phys. Rev. Lett.*, **83**, 5038 (1999).
13. D. Herisson and M. Ocio. *Phys. Rev. Lett.*, **88**(257202) (2002).
14. C. Song, P. Wang, and H. A. Makse. Experimental measurement of an effective temperature for jammed granular materials. *Proc. Nat. Acad. Sci.*, **102**, 2299–2304 (2005).
15. E. Bertin, O. Dauchot, and M. Droz. Temperature in nonequilibrium systems with conserved energy. *Phys. Rev. Lett.*, **93**, 230601 (2004).
16. E. Bertin, O. Dauchot, and M. Droz. Non-equilibrium temperatures in steady-state systems with conserved energy. *Phys. Rev. E*, **71**, 046140 (2005).
17. J. B. Knight, C. G. Fandrich, C. N. Lau, H. M. Jaeger, and S. R. Nagel. Density relaxation in a vibrated granular media. *Phys. Rev. E*, **51**, 3957–3963 (1995).
18. P. Philippe and D. Bideau. Compaction dynamics of a granular medium under vertical tapping. *Europhys. Lett.*, **60**, 677–683 (2002).
19. P. Philippe and D. Bideau. *Granular medium undervertical tapping*, Change of compaction and convection dynamics around the liftoff threshold. *Phys. Rev. Lett.*, **91**, 104302 (2003).
20. M. Nicolas, P. Duru, and O. Pouliquen. Compaction of a granular material under cyclic shear. *Eur. Phys. J. E*, **3**, 309–314 (2000).
21. M. Schröter, D.I. Goldman, and H.L. Swinney. Stationnary state volume fluctuations in a granular medium. *Phys. Rev. E*, **71**(030301(R)) (2005).
22. G. D'Anna and G. Gremaud. The jamming route to the glass state in a weakly perturbed granular media. *Nature*, **413**, 407–409 (2001).
23. A. Kabla and G. Debregeas. Contact dynamics in a gentlyvibrated granular pile. *Phys. Rev. Lett.*, **92**(035501) (2004).

24. R. Kohlrausch. *Pogg. Ann. Phys. Chem.*, **91**, 179 (1854).

25. G. Williams and D.C. Watts. *Trans. Faraday Soc.*, **66**, 80 (1970).

26. T. Boutreux and P. G. de Gennes. *Physica A*, **244**, 59 (1997).

27. E.R. Nowak, J.B. Knight, E. Ben-Naim, H.M. Jaeger, and S.R. Nagel. Density fluctuations in vibrated granular materials. *Phys. Rev. E*, **57(2)**, 1971 (1998).

28. C. Josserand, A. V. Tkachenko, D. M. Mueth, and H. M. Jaeger. Memory effect in granular materials. *Phys. Rev. Lett.*, **85**, 3632–3635 (2000).

29. O. Pouliquen, M. Belzons, and M. Nicolas. Fluctuating particle motion during shear induced granular compaction. *Phys. Rev. Lett.*, **91**, 014301 (2003).

30. G. Marty and O. Dauchot. Subdiffusion and cage effect in a sheared granular material. *Phys. Rev. Lett.*, **94**, 015701 (2005).

31. O. Dauchot, G. Marty, and G. Biroli. Dynamical heterogeneity close to the jamming transition in a sheared granular material. `cond-mat/0507152` submitted to Phys. Rev. Lett. (2005).

32. E. R. Weeks and D.A. Weitz. Properties of cage rearrangements observed near the colloidal glass transition. *Phys. Rev. Lett.*, **89**, 095704 (2002).

33. B. Doliwa and A. Heuer. *Phys. Rev. Lett.*, **80**, 4915 (1998).

34. B. Doliwa and A. Heuer. *J. Phys. Cond. Mat.*, **11**, A227 (1999).

35. M.M. Hurley and P. Harrowell. *Phys. Rev. E*, **52**, 1694 (1995).

36. E. R. Weeks, J.C. Crocker, A.C. Levitt, A. Schofield, and D.A. Weitz. *Science*, **287**, 627 (2000).

37. C. Donati, J. F. Douglas, W. Kob, S. J. Plimpton, P. H. Poole, and S. C. Glotzer. *Phys. Rev. Lett.*, **80**, 2338 (1998).

38. A. Heuer. *Phys. Rev. E*, **56**, 730 (1997).

39. W. Kob and H. C. Andersen. *Phys. Rev. E*, **51**, 4626–4641 (1995).

40. W. Kob and H. C. Andersen. *Phys. Rev. E*, **52**, 4134–4153 (1995).

41. H.C. Andersen. *PNAS*, **102**, 6686 (2005).

42. L. Berthier. *Phys. Rev. E*, **69**(020201(R)) (2004).

43. N. Lacevic, F.W. Starr, T.B. Schroeder, and S.C. Glotzer. *J. Chem. Phys*, **119**, 7372 (2003).

44. S. Whitelam, L. Berthier, and J. P. Garrahan. *Phys. Rev. Lett.*, **92**, 185705 (2004).

45. M.A. Ediger. *Annu. Rev. Phys. Chem.*, **51**, 99 (2000).

46. R. Stinchcombe, A. Lefèvre, L. Berthier. cond-mat/0410741 (2003).

47. C. Donati, S. Franz, S.C. Glotzer, and G. Parisi. *J. Non-Cryst. Sol.*, **307**, 215–224 (2002).

48. S. Franz and G. Parisi. *J. Phys. Cond. Matt.*, **12**, 6353 (2000).

49. L.F. Cugliandolo and J.L. Iguain. *Phys. Rev. Lett.*, **85**, 3448 (2000).

50. C. Toninelli, M. Wyart, G. Biroli, L. Berthier, and J.-P. Bouchaud. *Phys. Rev. E*, **71**(041505) (2005).

51. F.H. Stillinger and T.A. Weber. *Phys. Rev. A*, **25**, 978 (1982).

52. F.H. Stillinger and T.A. Weber. *Science*, **225**, 983 (1984).

53. M. Goldstein. *J. Chem. Phys.*, **51**, 3728 (1969).

54. S.F. Edwards and R.B.S. Oakeshott. Theory of powders. *Physica A*, **157**, 1080–1090 (1989).

55. J. Kurchan. Emergence of macroscopic temperatures in systems that are not thermodynamical microscopically, towards a thermodynamical description of slow granular rheology. *J. Phys. Condens. Matter*, **29**, 6611–6617 (2000).

56. J. Kurchan. In A.J. Liu and S.R. Nagel, editors, *Jamming and Rheology*, Constrained dynamics on Microscopic and Macroscopics scales, pp. 72–79, London, 2001. Taylor and Francis).

57. J.P. Bouchaud. *Granular media*, some ideas from statistical physics. lecture given in Les Houches (2002).

58. B. Diu, C. Guthmann, D. Lederer, and B. Roulet. Hermann, Paris (1989).

59. L.D. Landau and E.M. Lifshitz. Pergamon, New-York (1970).

60. S. F. Edwards. The role of entropy in the specification of a powder. In A. Mehta, editor, *Granular Matter, An Interdisciplinary approach*, pp. 121–140, New-York, 1994. Springer-Verlag).

61. S.F. Edwards and D.V. Grinev. *Phys. Rev. Lett.*, **82**(5397) (1999).

62. C. Godrèche and J.M. Lück. Metastability in zero-temperature dynamics, Statistics of attractors. *J. Phys. Cond. Matt.*, 2005. to appear, `cond-mat/0412077`).

63. R. Monasson. The structural glass transition and the entropy of the metastable states. *Phys. Rev. Lett.*, **75**(15), 2847 (1995).

64. A. Barrat, J. Kurchan, V. Loreto, and M. Sellitto. Edwards' measures for powders and glasses. *Phys. Rev. Lett.*, **85**, 5034–5037 (2000).

65. A. Barrat, J. Kurchan, V. Loreto, and M. Sellitto. Edwards' measures, a thermodynamic construction for dense granular media and glasses. *Phys. Rev. E*, **63**, 51301 (2001).

66. D.S. Dean and A. Lefèvre. Steady state behavior of mechanically perturbed spin glasses and ferromagnets. *Phys. Rev. E*, **64**, 046110 (2001).

67. J. J. Brey, A. Prados, and B. Sanchez-Rey. Thermodynamic description in a simple model for granular compaction. *Physica A*, **275**, 310–324 (2000).

68. A. J. Liu and S. R. Nagel. Jamming is not just cool anymore. *Nature*, **396**, 21–22 (1998).

69. H. A. Makse and J. Kurchan. Testing the thermodynamic approach to granular matter with a numerical model of a decisive experiment. *Nature*, **415**, 614–616 (2002).

70. F. Da Cruz, F. Lechenault, O. Dauchot, and E. Bertin. Free volume distributions inside a bidimensional granular medium. In *Powders and Grains 2005, Stuttgart* (2005).

5

Introduction to Simulation Techniques

W. Janke

Institut für Theoretische Physik and Centre for Theoretical Sciences (NTZ),
Universität Leipzig, Augustusplatz 10/11,
D-04109 Leipzig, Germany
wolfhard.janke@itp.uni-leipzig.de

These lectures give an introduction to Monte Carlo simulations of classical statistical physics systems and their statistical analysis. After briefly recalling a few elementary properties of phase transitions, the concept of importance sampling Monte Carlo methods is discussed and illustrated by a few standard local update algorithms (Metropolis, heat-bath, Glauber). Then emphasis is placed on thorough analyses of the generated data paying special attention to the choice of estimators, autocorrelation times and statistical error analysis. This leads to the phenomenon of critical slowing down at continuous phase transitions. For illustration purposes, only the two-dimensional Ising model will be needed. To overcome the slowing-down problem, non-local cluster algorithms have been developed which will be discussed next. Then the general tool of reweighting techniques will be explained. This paves the way to introduce simulated and parallel tempering methods which are very useful for simulations of complex, possibly disordered systems. Finally, also the important alternative approach using multicanonical ensembles is briefly outlined.

5.1 Introduction

The statistical mechanics of complex physical systems poses many hard problems which are very difficult if not impossible to solve by purely analytical methods. Numerical simulation techniques will therefore be indispensable tools on our way to a better understanding of systems such as (spin) glasses and disordered magnets, or of the huge field of biologically motivated problems such as protein folding, to mention only a few important classical problems. Quantum statistical problems in condensed matter or the broad field of elementary particle physics and quantum gravity are other major applications.

The numerical tools commonly employed can be roughly divided into molecular dynamics (MD) and Monte Carlo (MC) simulations. With the still ongoing advances in computer technology – according to Moore's law, since about

W. Janke: *Introduction to Simulation Techniques*, Lect. Notes Phys. **716**, 207–260 (2007)
DOI 10.1007/3-540-69684-9_5   © Springer-Verlag Berlin Heidelberg 2007

1950, every 5 years a factor of 10 is gained in computing speed – both approaches can be expected to gain even more importance in the future than they have already today. In the past few years the predictive power of especially the MC approach was in addition considerably enhanced by the discovery of greatly improved simulation algorithms. Not all of them are already well enough understood to be applicable to really complex physical systems. But, as a first step, it is gratifying to note that at least for relatively simple spin systems, orders of magnitude of computing time can be saved by these refinements. The purpose of these lecture notes is to give a concise introduction to what is feasible today. For further reading, there are quite a few recent textbooks [1–4] available which treat some of the material discussed here in more depth. In particular, in these books one can also find recent applications to physically relevant systems which are purposely omitted in this short introduction.

For illustration purposes, we shall rather confine ourselves to the simplest spin models, the Ising and Potts models. From a theoretical point of view, also spin systems are still of current interest since they provide the possibility to compare completely different approaches such as field theory, series expansions, and simulations. They are also the ideal testing ground for general concepts such as universality, scaling or finite-size scaling, where even today some new features can still be discovered. And last but not least, they have found a revival in slightly disguised form in quantum gravity and conformal field theory, where they serve as idealized "matter" fields on Feynman graphs or fluctuating manifolds.

The rest of these lecture notes is organized as follows. In Sect. 5.2, the definitions of Ising and Potts models are recalled and some standard observables (specific heat, magnetization, susceptibility, correlation functions,...) are briefly discussed. Next the most characteristic properties of phase transitions, scaling properties and the definition of critical exponents are summarized. In Sect. 5.3, the basic method underlying all importance sampling Monte Carlo simulations is described. The following Sect. 5.4 is first devoted to a short discussion of the initial non-equilibrium period and ageing phenomena, and then in Sect. 5.5 a fairly detailed account of statistical error analysis in equilibrium is given which also includes temporal correlation effects. The latter highlight the problems of critical slowing down at a continuous phase transition and phase coexistence with exponentially large flipping times at a first-order transition. One very successful solution of the former problem are non-local cluster algorithms which are described in Sect. 5.6. In Sect. 5.7 we discuss reweighting techniques which quite naturally lead to the tempering update algorithms explained in Sect. 5.8. These algorithms may be viewed as dynamical reweighting methods that can circumvent exponentially large flipping times and proved to be very successful for the simulation of complex, disordered systems. The alternative method of multicanonical ensembles is only very briefly discussed in Sect. 5.9, with emphasis on similarities and

differences to tempering methods. Finally, Sect. 5.10 contains a few conclud-
ing remarks.

5.2 Models and Phase Transitions

5.2.1 Models and Observables

Most of the simulation techniques introduced below can be illustrated for the
simple Ising spin model whose partition function is defined as [5]

$$Z = \sum_{\{\sigma_i\}} \exp(-H/k_B T) , \tag{5.1}$$

with

$$H = -J \sum_{\langle ij \rangle} \sigma_i \sigma_j - h \sum_i \sigma_i , \qquad \sigma_i = \pm 1 . \tag{5.2}$$

Here T is the temperature and h is an external magnetic field, k_B is Boltzmann's constant, the spins σ_i are assumed (for simplicity) to live on the sites
i of a D-dimensional cubic lattice of volume $V = L^D$, and the symbol $\langle ij \rangle$
indicates that the lattice sum runs over all 2D nearest-neighbor pairs. In all
examples discussed below, periodic boundary conditions will be assumed.

Standard observables are the internal energy per site, $e = E/V$, with
$E = -d \ln Z/d\beta \equiv \langle H \rangle$, and the specific heat,

$$C/k_B = \frac{de}{d(k_B T)} = \beta^2 \left(\langle H^2 \rangle - \langle H \rangle^2 \right) /V , \tag{5.3}$$

where $\beta \equiv 1/k_B T$. In the following we always use units in which $k_B \equiv 1$
and $J \equiv 1$. On finite lattices the magnetization and susceptibility are usually
defined as

$$m = M/V = \langle |\mu| \rangle , \qquad \mu = \sum_i \sigma_i / V , \tag{5.4}$$

$$\chi = \beta V \left(\langle \mu^2 \rangle - \langle |\mu| \rangle^2 \right) . \tag{5.5}$$

In the high-temperature phase one often employs the fact that the magnetization vanishes in the infinite volume limit and considers

$$\chi' = \beta V \langle \mu^2 \rangle . \tag{5.6}$$

Similarly, the spin-spin correlation function,

$$G(\boldsymbol{x}_i - \boldsymbol{x}_j) = \langle \sigma_i \, \sigma_j \rangle - \langle \sigma_i \rangle \langle \sigma_j \rangle , \tag{5.7}$$

then simplifies to $G(\boldsymbol{x}_i - \boldsymbol{x}_j) = \langle \sigma_i \, \sigma_j \rangle$, where $\boldsymbol{x}_i$ measures the position of the
lattice sites i which are numbered, say, in a lexicographical order. At large

distances, $G(\boldsymbol{x}) \propto \exp(-|\boldsymbol{x}|/\xi)$ decays exponentially. Its decay rate defines the correlation length

$$\xi = - \lim_{|\boldsymbol{x}| \to \infty} |\boldsymbol{x}| / \ln G(\boldsymbol{x}) , \tag{5.8}$$

which strictly speaking depends on the (discrete) orientation of $\boldsymbol{x}$. For definiteness we will hence always consider correlations along one of the main lattice directions.

In vanishing external field the Ising model exhibits a continuous phase transition in temperature for all dimensions $D \geq 2$. The two-dimensional (2D) model is self-dual, relating its behaviour for high temperatures $T = 1/\beta$ to that at $T^* = 1/\beta^*$ in the low-temperature phase, where

$$\sinh(2\beta) \sinh(2\beta^*) = 1 . \tag{5.9}$$

Under the mild assumption of a *single* phase-transition point, this fixes already the critical temperature to be

$$\sinh(2\beta_c) = 1 \quad \text{or} \quad \beta_c = \ln(1 + \sqrt{2})/2 = 0.440\,686\ldots , \quad T_c = 2.269\,185\ldots . \tag{5.10}$$

The exact solution by Onsager [6–8] in 1944 yields the free energy, internal energy, specific heat etc. in zero external field for the general case of anisotropic couplings J_x, J_y. Also the exact result for the magnetization below T_c in zero field (again for general J_x, J_y) was first announced by Onsager at a conference in Florence 1949 [9]. The first published derivation was given three years later by Yang [10] in 1952, for the special case $J_x = J_y = J$, and subsequently generalized to arbitrary J_x, J_y by Chang [11] in the same year. Even the correlation length in arbitrary directions is known analytically [7,8]. Along the coordinate axes of a square lattice, the formula takes a surprisingly simple form,

$$\xi_d(\beta) = \frac{1}{2(\beta^* - \beta)} \quad (\beta < \beta_c) , \tag{5.11}$$

$$\xi_o(\beta^*) = \xi_d(\beta)/2 \quad (\beta^* > \beta_c) , \tag{5.12}$$

where β and β^* are the dual couplings defined in (5.9). The susceptibility, however, is still not exactly known, even though highly accurate approximations could be derived [7,8,12]. In 3D even for the simple Ising model, no exact solutions are available. Numerical work, high-temperature series expansions and field-theoretical considerations provide, however, very precise estimates. From 4D on, the so-called upper critical dimension, mean-field behaviour starts to be become qualitatively correct, albeit in 4D only up to multiplicative logarithmic corrections. The critical temperature is for all finite dimensions $D \geq 3$ only approximately known, approaching $T_c = 2D$ in the mean-field limit $D \to \infty$.

A simple generalization of the Ising model is the q-state Potts model [13] whose Hamiltonian in zero external field is given by

$$H_{\text{Potts}} = -J \sum_{\langle ij \rangle} \delta_{\sigma_i \sigma_j} \,, \qquad \sigma_i \in 1, \dots, q \,, \qquad (5.13)$$

which is equivalent to the Ising model for $q = 2$. The 2D Potts model is exactly known from self-duality, $(\exp(\beta) - 1)(\exp(\beta^*) - 1) = q$, to exhibit at $\beta_c = \ln(1 + \sqrt{q})$ a second-order transition for $q \leq 4$ and a first-order transition for all $q \geq 5$ [14,15]. At the transition point (but only there), a couple of exact results are available for both types of transitions, including the free energy, internal energy and specific heat [14,15] as well as the correlation length ξ_d in the disordered phase and the related interface tension σ_{od} between the ordered and disordered phase [16]. In 3D, numerical work suggests for all $q \geq 3$ a first-order transition which rapidly becomes stronger with increasing q.

5.2.2 Phase Transitions

The most interesting aspect of a system's phase diagram is the region where cooperation effects may cause a phase transition, e.g., from a disordered phase at high temperatures to an ordered phase at low temperatures as in the paradigmatic Ising model. To predict the properties of this most challenging region of a phase diagram as accurately as possible is one of the major objectives of all statistical mechanics approaches, including numerical computer simulation studies. The theory of phase transitions is a very broad subject described comprehensively in many textbooks (see, e.g., Refs. [17–20]). Here we only roughly classify them into *first-order* and *second-order* (or, more generally, continuous) phase transitions, and give a very brief summary of those properties that are most relevant for numerical simulations.

Some characteristic features of the thermodynamic behaviour at first- and second-order phase transitions are sketched in Fig. 5.1. Most phase transitions in Nature are of first order [21–24]. The best known example is the field-driven transition in magnets at temperatures below the Curie point, while the paradigm of a temperature-driven first-order transition experienced every day is ordinary melting [25,26]. Simple models sharing such a behaviour are the Ising and Potts models defined in (5.2) and (5.13). In general, first-order phase transitions are characterized by *discontinuities* in the order parameter (the jump Δm of the magnetization m in Fig. 5.1), or the energy (the latent heat Δe), or both. This reflects the fact that, at the transition temperature T_0, two (or more) phases can coexist. In the example of a magnet at low temperatures the coexisting phases are the phases with positive and negative magnetization, while at the melting transition they are the solid (ordered) and liquid (disordered) phases. The correlation length in the coexisting pure phases is usually finite. Consequently also the specific heat and the susceptibility do not diverge in the pure phases. Mathematically there are, however, superimposed delta-function like singularities associated with the jumps of e and m.

In these lecture notes we will mainly consider second-order phase transitions, which are characterized by a *divergent* correlation length at the transition point. The growth of correlations as one reaches the critical region from

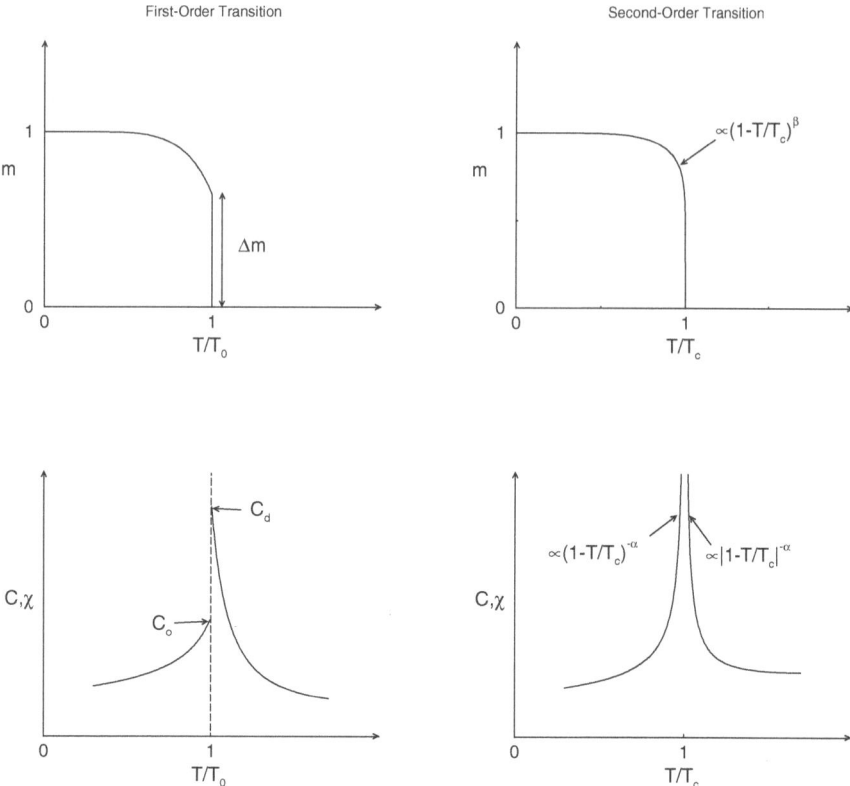

Fig. 5.1. The characteristic behaviour of the magnetization, m, specific heat, C, and susceptibility, χ, at first- and second-order phase transitions

high temperatures is illustrated in Fig. 5.2, where six typical spin configurations of the 2D Ising model on a 100×100 lattice are shown. One clearly observes the emerging larger and larger domains or clusters which eventually start percolating the system when the critical point is approached. While this apparently gives an intuitive picture of what happens near criticality, some care is necessary with the interpretation of such plots since the domains or clusters visible in Fig. 5.2 are so-called *geometrical* clusters, whose fractal and percolation properties do *not* encode the proper thermodynamic *critical* behaviour. Rather, they carry information on a closely related *tricritical* point [27]. The proper Fortuin-Kasteleyn clusters encoding the *critical* properties of the model can be constructed by a stochastic rule implied by the Fortuin-Kasteleyn representation of Potts models. These clusters, which are always smaller than the geometrical ones, form also the basis for cluster-update algorithms discussed later in Sect. 5.6.

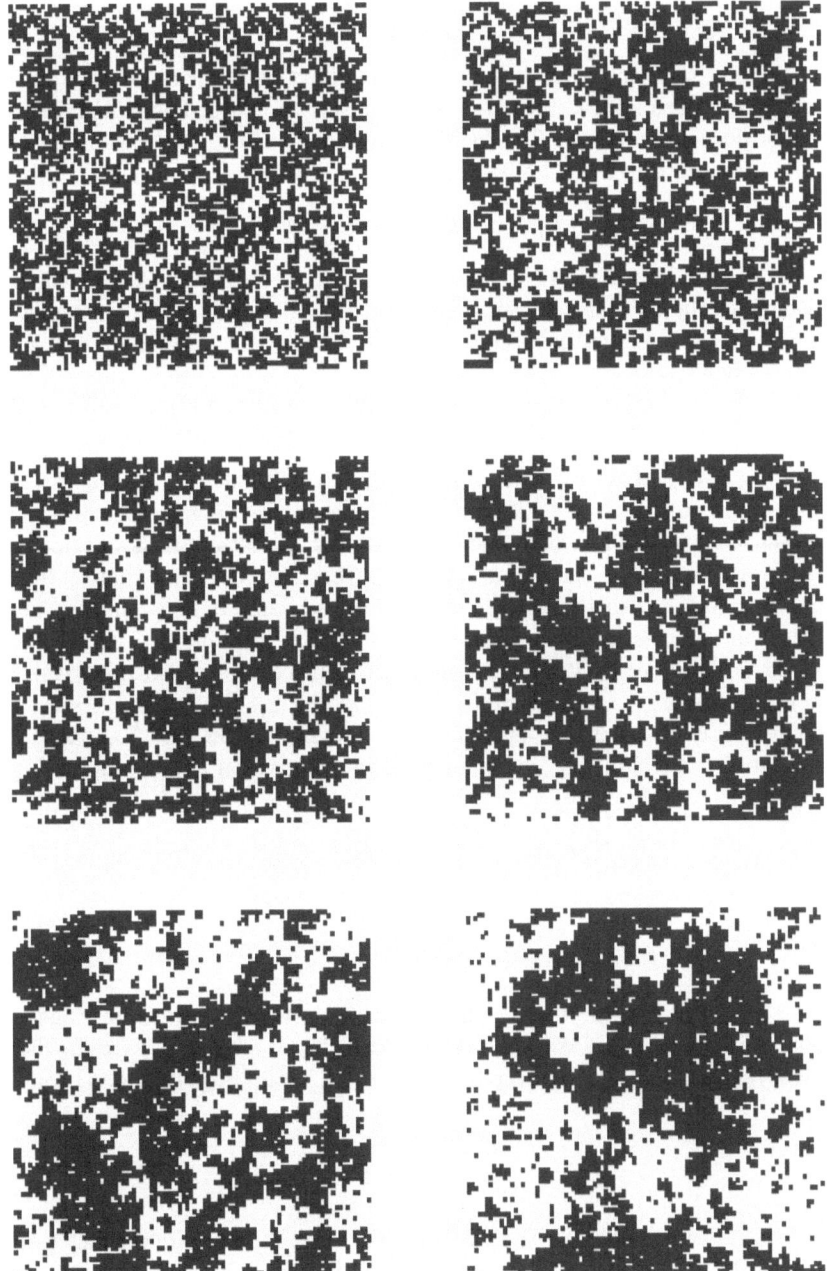

Fig. 5.2. From high temperatures (*upper left*) to the critical region (*lower right*), characterized by large spatial correlations. Shown are actual 2D Ising configurations for a 100×100 lattice at $\beta/\beta_c = 0.50$, 0.70, 0.85, 0.90, 0.95, and 0.98

Temperature Scaling

For an infinite correlation length, thermal fluctuations are equally important on all length scales, and one therefore expects power-law singularities in thermodynamic functions. The leading singularity of the correlation length is usually parameterized in the high-temperature phase as

$$\xi = \xi_{0+} |1 - T/T_c|^{-\nu} + \dots \quad (T \geq T_c) , \tag{5.14}$$

where the ... indicate sub-leading corrections (analytical as well as confluent). This defines the critical exponent ν and the critical amplitude ξ_{0+} on the high-temperature side of the transition. In the low-temperature phase one expects a similar behaviour,

$$\xi = \xi_{0-} (1 - T/T_c)^{-\nu} + \dots \quad (T \leq T_c) , \tag{5.15}$$

with the same critical exponent ν but a different critical amplitude $\xi_{0-} \neq \xi_{0+}$.

An important feature of second-order phase transitions is that due to the divergence of ξ the short-distance details of the Hamiltonian should not matter. This is the basis of the *universality* hypothesis which states that all (short-ranged) systems with the same symmetries and same dimensionality should exhibit similar singularities governed by one and the same set of critical exponents. For the amplitudes this is not true, but certain amplitude ratios are also universal.

The singularities of the specific heat, magnetization (for $T < T_c$), and susceptibility are similarly parameterized by the critical exponents α, β, and γ, respectively,

$$C = C_{\text{reg}} + C_0 |1 - T/T_c|^{-\alpha} + \dots , \tag{5.16}$$

$$m = m_0 (1 - T/T_c)^{\beta} + \dots , \tag{5.17}$$

$$\chi = \chi_0 |1 - T/T_c|^{-\gamma} + \dots , \tag{5.18}$$

where C_{reg} is a regular background term, and the amplitudes are again in general different on the two sides of the transition, cf. Fig. 5.1. Right at the critical temperature T_c, two further exponents δ and η are defined through

$$m \propto h^{1/\delta} , \tag{5.19}$$

$$G(\boldsymbol{r}) \propto r^{-d+2-\eta} . \tag{5.20}$$

The critical exponents for the 2D and 3D Ising model and the 2D q-state Potts model with $q = 3$ and 4 are collected in Table 5.1.

Finite-Size Scaling

For systems of finite size, as in any numerical simulation, the correlation length cannot diverge, and also the divergences in all other quantities are

Table 5.1. Critical exponents of the 2D q-state Potts model with $q = 2, 3$ and 4, and the 3D Ising model. All 2D exponents are exactly known [14, 15], while for the 3D Ising model the "world-average" for ν and γ calculated in Ref. [28] is quoted. The other exponents follow from the hyperscaling relation $\alpha = 2 - D\nu$, and the scaling relations $\beta = (2 - \alpha - \gamma)/2$, $\delta = \gamma/\beta + 1$, and $\eta = 2 - \gamma/\nu$

Model	ν	α	β	γ	δ	η
2D Ising	1	0 (log)	1/8	7/4	15	1/4
3D Ising	0.630 05(18)	0.109 85	0.326 48	1.237 17(28)	4.7894	0.036 39
2D $q = 3$ Potts	5/6	1/3	1/9	13/9	14	4/15
2D $q = 4$ Potts	2/3	2/3	1/12	7/6	15	1/2

then rounded and shifted [29–32]. This is illustrated in Fig. 5.3, where the specific heat of the 2D Ising model on various $L \times L$ lattices is shown. The curves are computed from the exact solution of Kaufman [33] for any $L_x \times L_y$ lattice with periodic boundary conditions (see also Ferdinand and Fisher [34]).

Near T_c the role of ξ in the scaling formulas is then taken over by the linear size of the system, L. By rewriting

$$|1 - T/T_c| \propto \xi^{-1/\nu} \longrightarrow L^{-1/\nu} , \tag{5.21}$$

we see that at T_c the scaling laws (5.16)–(5.18) are replaced by the *finite-size scaling* (FSS) Ansätze,

$$C = C_{\text{reg}} + aL^{\alpha/\nu} + \dots , \tag{5.22}$$

$$m \propto L^{-\beta/\nu} + \dots , \tag{5.23}$$

$$\chi \propto L^{\gamma/\nu} + \dots . \tag{5.24}$$

In general these scaling laws are valid in the vicinity of T_c as long as the scaling variable $x = (1 - T/T_c)L^{1/\nu}$ is kept fixed [29–32]. In particular this is true for the locations T_{max} of the (finite) maxima of thermodynamic quantities such as the specific heat or susceptibility, which are expected to scale with the system size as

$$T_{\text{max}} = T_c(1 - x_{\text{max}}L^{-1/\nu} + \dots) . \tag{5.25}$$

In this more general formulation the scaling law for, e.g., the susceptibility reads

$$\chi(T, L) = L^{\gamma/\nu} f(x) . \tag{5.26}$$

By plotting $\chi(T, L)/L^{\gamma/\nu}$ vs the scaling variable x, one thus expects that the data for different T and L fall onto a kind of master curve. This is a nice way to demonstrate the scaling properties visually.

Similar considerations for first-order phase transitions show that also the delta function like singularities, originating from phase coexistence, are smeared out for finite systems [35–39]. They are replaced by narrow peaks

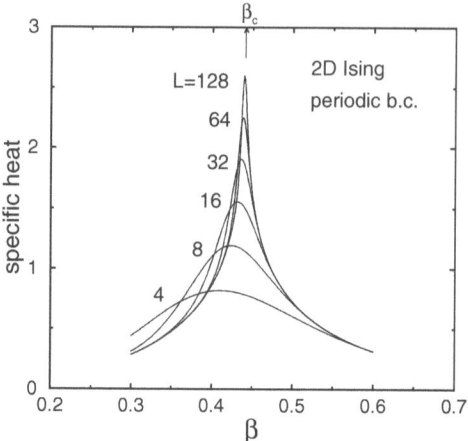

Fig. 5.3. Finite-size scaling behaviour of the specific heat of the 2D Ising model on $L \times L$ lattices. The critical point is indicated by the arrow on the top axis

whose height (width) grows proportional to the volume (1/volume) with a displacement of the peak location from the infinite-volume limit proportional to 1/volume [16, 40–44].

5.3 The Monte Carlo Method

Let us now discuss how the expectation values in (5.3)–(5.7) can be computed numerically. A direct summation of the partition function is impossible, since already for the Ising model with only two possible states per site the number of terms would be enormous: $2^{2500} \approx 10^{750}$ for a modestly large 50×50 lattice![1] Also a naive random sampling of the spin configurations does not work. Here the problem is that the relevant region in the high-dimensional phase space is relatively narrow and hence too rarely hit by random sampling. The solution to this problem is known since long: One has to use the *importance sampling* technique [45].

5.3.1 Importance Sampling

The basic idea of importance sampling is that one does not pick configurations at random, but draws them directly according to their Boltzmann weight,

$$P^{\text{eq}}(\{\sigma_i\}) \propto \exp\left(-\beta H(\{\sigma_i\})\right) . \tag{5.27}$$

---

[1] For comparison, the estimated number of protons in the Universe is 10^{80}.

In more mathematical terms one sets up a Markov chain,

$$\cdots \xrightarrow{W} \{\sigma_i\} \xrightarrow{W} \{\sigma_i'\} \xrightarrow{W} \{\sigma_i''\} \xrightarrow{W} \cdots \; ,$$

with a transition probability W satisfying the conditions

(a) $W(\{\sigma_i\} \longrightarrow \{\sigma_i'\}) \geq 0$ for all $\{\sigma_i\}, \{\sigma_i'\}$,   (5.28)

(b) $\sum_{\{\sigma_i'\}} W(\{\sigma_i\} \longrightarrow \{\sigma_i'\}) = 1$ for all $\{\sigma_i\}$,   (5.29)

(c) $\sum_{\{\sigma_i\}} W(\{\sigma_i\} \longrightarrow \{\sigma_i'\}) P^{\mathrm{eq}}(\{\sigma_i\}) = P^{\mathrm{eq}}(\{\sigma_i'\})$ for all $\{\sigma_i'\}$. (5.30)

From (5.30) we see that the desired Boltzmann distribution P^{eq} is a fixed point of W. A somewhat simpler sufficient condition is *detailed balance*,

$$P^{\mathrm{eq}}(\{\sigma_i\})W(\{\sigma_i\} \longrightarrow \{\sigma_i'\}) = P^{\mathrm{eq}}(\{\sigma_i'\})W(\{\sigma_i'\} \longrightarrow \{\sigma_i\}) \; . \qquad (5.31)$$

By summing over $\{\sigma_i\}$ and using (5.29), the more general condition (5.30) follows. After an initial equilibration period (cf. Sect. 5.4), expectation values can then be estimated as an arithmetic mean over the Markov chain, e.g.,

$$E = \langle H \rangle = \sum_{\{\sigma_i\}} H(\{\sigma_i\}) P^{\mathrm{eq}}(\{\sigma_i\}) \approx \frac{1}{N} \sum_{j=1}^{N} H(\{\sigma_i\}_j) \; , \qquad (5.32)$$

where $\{\sigma_i\}_j$ denotes the spin configuration at "time" j. A more detailed exposition of the mathematical concepts underlying any Markov chain Monte Carlo algorithm can be found in many textbooks and reviews [1–4, 29, 46, 47].

5.3.2 Local Update Algorithms

The required Markov chain properties can be satisfied with many different concrete update rules. These can be roughly divided into *local* and *non-local* algorithms. While non-local algorithms such as multigrid schemes or the cluster-update methods to be discussed later in Sect. 5.6 may considerably improve the performance of the simulations, they are more specialized and hence usually not automatically applicable to a given arbitrary physical system. This is why the conceptually much simpler local algorithms continue to be very important.

Metropolis Algorithm

The most flexible update prescription is the standard Metropolis algorithm [48] where the Markov chain is realized by *locally* updating the degrees of freedom step by step. This works for discrete and continuously varying degrees of freedom, and for lattice and off-lattice formulations. Examples for lattice formulations range from our simple, paradigmatic Ising model, over freely rotating Heisenberg spins to field theories such as the Ginzburg-Landau model. Also all kinds of lattice gauge theories and even non-perturbative formulations of quantum gravity can be simulated with this method. Off-lattice formulations cover a huge range of physical phenomena. Prominent examples are simulations of fluids, polymers and proteins, to name only a few important applications. Depending on the problem at hand, the degrees of freedom may be spins, field values or gauge potentials, or particle positions in space. There is also no principle restriction on the form of the interactions which may be short- or long-ranged or even of mean-field type.

If E and E' denote the energy before and after the proposed local update, respectively, then the probability to accept this proposal is given by [48]

$$W(\{\sigma_i\} \longrightarrow \{\sigma_i'\}) = \begin{cases} 1 & E' < E \\ \exp\left[-\beta(E' - E)\right] & E' \geq E, \end{cases} \tag{5.33}$$

where the proposed new spin configuration $\{\sigma_i'\}$ differs from $\{\sigma_i\}$ only by a single flipped spin. More compactly, this may also be written as

$$W(\{\sigma_i\} \longrightarrow \{\sigma_i'\}) = \min\{1, \exp\left[-\beta(E' - E)\right]\}. \tag{5.34}$$

If the energy is lowered by the proposed update, it is thus always accepted. On the other hand, when the energy would be increased for the new configuration, the update has still to be accepted with a certain probability in order to ensure the proper treatment of entropic contributions – in thermal equilibrium, the *free* energy is minimized and not the energy. Only in the limit of zero temperature, $\beta \longrightarrow \infty$, this probability tends to zero and the MC algorithm degenerates to a minimization algorithm for the energy functional. With some additional refinements, this is the basis of the simulated annealing technique [49], which is often applied to hard optimization and minimization problems.

To show that the detailed balance condition (5.31) is indeed satisfied, we first consider the case that the proposed spin update lowers the energy, $E' < E$. In this case, the l.h.s. of (5.31) becomes $\exp(-\beta E) \times 1 = \exp(-\beta E)$. On the r.h.s. we have to take into account that the reverse move would increase the energy, $E > E'$, with E now playing the role of the "new" energy. Hence now the second line of (5.33) with E and E' interchanged is relevant, such that the r.h.s. of (5.31) becomes $\exp(-\beta E') \times \exp(-\beta(E - E')) = \exp(-\beta E)$, proving the equality of the two sides of the detailed balance condition. In the case that the proposed spin update increases the energy, $E' < E$, a similar reasoning leads to $\exp(-\beta E) \times \exp(-\beta(E' - E)) = \exp(-\beta E') = \exp(-\beta E') \times 1$.

Even though this "proof" looks rather like a tautology, it is indeed nontrivial, as one can easily convince oneself by replacing the r.h.s. of the Metropolis rule (5.33) by some general function $f(E' - E)$. The detailed balance condition then reads $\exp(-\beta E)f(E' - E) = \exp(-\beta E')f(E - E')$. With $\Delta E \equiv E' - E$ this can be recast into the form

$$g(\Delta E) \equiv \exp(\beta\Delta E/2)f(\Delta E) = \exp(-\beta\Delta E/2)f(-\Delta E) = g(-\Delta E) , \quad (5.35)$$

showing that $g(\Delta E) \equiv \exp(\beta\Delta E/2)f(\Delta E)$ can be quite a general function which, however, must be even in ΔE, $g(\Delta E) = g(-\Delta E)$. The simplest choice $g(\Delta E) = $ const., leads to $f(\Delta E) = $ const. $\exp(-\beta\Delta E/2)$. While this would satisfy detailed balance, it is still not a permissible choice because the r.h.s. of (5.33) should admit the interpretation as a probability. For a given model, this can often be repaired by considering the allowed range of ΔE and introducing a suitable normalization factor [1]. Requiring thus that $0 \leq f(\Delta E) \leq 1$, we see that $0 \leq g(\Delta E) \leq \exp(\beta\Delta E/2)$. Choosing just the upper bound, $g(\Delta E) = \exp(\beta\Delta E/2)$ for $\Delta E < 0$ and applying a (non-differentiable) symmetrization to define $g(\Delta E)$ for $\Delta E \geq 0$, we end up with

$$g(\Delta E) = \exp(\beta\Delta E/2)\min\{1,\exp(-\beta\Delta E)\} = \begin{cases} \exp(\beta\Delta E/2) & \Delta E < 0 \\ \exp(-\beta\Delta E/2) & \Delta E \geq 0 , \end{cases}$$
$$(5.36)$$

implying $f(\Delta E) = \min\{1,\exp(-\beta\Delta E)\}$ – which is nothing but the Metropolis rule (5.34).

How is the Metropolis update rule (5.33) implemented in practice? Since the possible values of the transition probability W are restricted to values between 0 and 1, one first draws a uniformly distributed random number $r \in [0,1)$. Then, if $W \leq r$, the proposed update is accepted, and otherwise it is rejected and one continues with the next spin. In words this is easy to state. In practice, however, "drawing a random number" in a computer program is a pretty involved mathematical problem [50]. Since in most applications a huge number of random numbers is required (for, say, 1 million sweeps through a 2D Ising lattice of size $1000 \times 1000 = 10^6$ already 10^{12}) and each random number usually occupies 8 Bytes, it is neither practical nor feasible to store physically generated, "truly" random (whatever that really means ...) events on a hard disk. Also, reading them from the hard disk into the computer memory would be far too slow. Therefore, one uses in MC computer simulations so-called "pseudo-random number generators", or short RNGs, which use deterministic rules to produce (more or less) uniformly distributed numbers, whose values are "very hard" to predict. In other words, given a finite sequence of subsequent pseudo-random numbers, it should be almost impossible to predict the next one or to even guess the deterministic rule underlying their generation. The "goodness" of a RNG is thus measured by the difficulty to derive its underlying deterministic rule. Related requirements are the absence of trends (correlations) and a very long period. Furthermore, a RNG should be portable among different computer platforms, and it should yield reproducible results for testing purposes.

There are many different ways how the degrees of freedom to be updated can be chosen. They may be picked at random or according to a random permutation, which can be updated every now and then. But also a simple fixed lexicographical (sequential) order is permissible. In lattice models one may also update first all odd and then all even sites, which is the usual choice in vectorized codes. A so-called sweep is completed when on the average[2] for all degrees of freedom an update was proposed. The qualitative behaviour of the update algorithm is not sensitive to these details, but its quantitative performance does depend on the choice of update scheme.

The big merit of this simple algorithm is its flexibility which allows the application to a great variety of physical systems. The main drawback of this and most other *local* update algorithms (one exception is the overrelaxation method [51–54]) is that it is plagued by large autocorrelation times which severely limit the statistical accuracy achievable with a given computer budget as will be explained in detail in Sect. 5.5.

Heat-Bath Algorithm

This algorithm is only applicable to lattice models and at least in its most straightforward form only to discrete degrees of freedom with a few allowed states. The new value of the selected variable at site i_0 is determined by testing all its possible states in the "heat-bath" of its (fixed) neighbors (i.e., 4 on a square lattice and 6 on a simple-cubic lattice with nearest-neighbor interactions):

$$W(\{\sigma_i\} \longrightarrow \{\sigma_i'\}) = \frac{e^{-\beta H(\{\sigma_i'\}}}{\sum_{\sigma_{i_0}'} e^{-\beta H(\{\sigma_i'\}}} , \qquad (5.37)$$

which obviously satisfies the detailed balance condition (5.31) since

$$e^{-\beta H(\{\sigma_i\}} \frac{e^{-\beta H(\{\sigma_i'\}}}{\sum_{\sigma_{i_0}'} e^{-\beta H(\{\sigma_i'\}}} = e^{-\beta H(\{\sigma_i'\}} \frac{e^{-\beta H(\{\sigma_i\}}}{\sum_{\sigma_{i_0}} e^{-\beta H(\{\sigma_i\}}} . \qquad (5.38)$$

Due to the summation over all local states, special tricks are necessary when each degree of freedom can take many different states, and only in special cases the heat-bath method can be efficiently generalized to continuous degrees of freedom. In the special case of the Ising model with only two states per spin, (5.37) may be written more explicitly as

$$W(\{\sigma_i\} \longrightarrow \{\sigma_i'\}) = \frac{e^{-\beta \sigma_{i_0}' E_{i_0}}}{e^{\beta E_{i_0}} + e^{-\beta E_{i_0}}} , \qquad (5.39)$$

where $\sigma_{i_0} E_{i_0}$ is the energy of the spin at site i_0 in the state σ_{i_0}, that is $E_{i_0} = -J \sum_j \sigma_j - h$, where j runs over all sites interacting with site i_0 and h is the external magnetic field. The energy difference $\Delta E = E_{\text{new}} - E_{\text{old}}$ can be expressed as $\Delta E = (\sigma_{i_0}' - \sigma_{i_0}) E_{i_0}$, since by definition no other interactions are affected by the spin value at site i_0.

Let us now assume that before the update $\sigma_{i_0} = +1$. The probability that the spin is flipped to $\sigma_{i_0}' = -1$ is then

---

[2] This is only relevant when the random update order is chosen.

$$W(\sigma_{i_0} \longrightarrow -\sigma_{i_0}) = \frac{e^{\beta E_{i_0}}}{e^{\beta E_{i_0}} + e^{-\beta E_{i_0}}} . \tag{5.40}$$

Since in this case $\Delta E = -2E_{i_0}$, the flip probability can be equivalently written as

$$W(\sigma_{i_0} \longrightarrow -\sigma_{i_0}) = \frac{e^{-\beta \Delta E/2}}{e^{\beta \Delta E/2} + e^{-\beta \Delta E/2}} . \tag{5.41}$$

This is also true in the other case where initially $\sigma_{i_0} = -1$. The heat-bath probability of a flip to $\sigma'_{i_0} = +1$ is then $e^{-\beta E_{i_0}} / (e^{\beta E_{i_0}} + e^{-\beta E_{i_0}})$, but since the energy difference now reads $\Delta E = +2E_{i_0}$, we again arrive at the flip probability (5.41).

The order of updating the individual variables can be done as for the Metropolis algorithm (random, sequential, . . .).

Glauber Algorithm

This update procedure [55], named after the 2005 Nobel Laureate Roy J. Glauber of Harvard University[3], is conceptually similar to the Metropolis algorithm in that one also here proposes locally a change for a single degree of freedom and then accepts this update proposal with a certain probability. For the Ising model with spins $\sigma_i = \pm 1$ this update rule is often written as

$$W(\sigma_i \longrightarrow -\sigma_i) = \frac{1}{2} \left[1 + \sigma_i \tanh(\beta E_i) \right] , \tag{5.42}$$

where as before $\sigma_i E_i$ is the energy of the ith spin in the current "old" state, that is $E_i = -J \sum_j \sigma_j - h$.

Since $\sigma_i = \pm 1$ and using the point symmetry of the tanh-function, one may rewrite $\sigma_i \tanh(\beta E_i)$ as $\tanh(\sigma_i \beta E_i)$. In a local spin flip $\sigma_i \longrightarrow -\sigma_i$, only the energy contributions collected in E_i are affected, and we obtain again $\Delta E = E_{\text{new}} - E_{\text{old}} = -2\sigma_i E_i$ for the total energy change due to the proposed flip. Hence we can rewrite (5.42) as

$$W(\sigma_i \longrightarrow -\sigma_i) = \frac{1}{2} \left[1 - \tanh(\beta \Delta E/2) \right] . \tag{5.43}$$

In this representation, the acceptance probability is explicitly seen to depend only on the total energy change – similar to the Metropolis case. In this form it is thus possible to generalize the Glauber update rule from the Ising model

---

[3] Half of the Nobel Prize in Physics 2005 was awarded to Roy J. Glauber for his outstanding theoretical contributions to what is called today "Quantum Optics", with his seminal papers dating back to the year 1963 [Phys. Rev. Lett. **10**, 84 (1963); Phys. Rev. **130**, 2529 (1963); *ibid.* **131**, 2766 (1963)], when also his paper [55] on dynamical properties of the Ising model appeared. The other half of the 2005 Prize is shared by John L. Hall of the University of Colorado and Theodor W. Hänsch of Ludwig-Maximilians-Universität Munich.

with only two states per spin to any general model that can be simulated with the Metropolis procedure. Also detailed balance is straightforward to prove.

By using trivial identities for hyperbolic functions, (5.43) can be further recast to read

$$W(\sigma_i \longrightarrow -\sigma_i) = \frac{1}{2} \left[\frac{\cosh(\beta \Delta E/2) - \sinh(\beta \Delta E/2)}{\cosh(\beta \Delta E/2)} \right] = \frac{e^{-\beta \Delta E/2}}{e^{\beta \Delta E/2} + e^{-\beta \Delta E/2}} .$$
(5.44)

Notice that this agrees with the flip probability (5.41) of the heat-bath algorithm for the Ising model, i.e., heat-bath updates for the special case of a 2-state model and the Glauber update algorithm are identical.

The Glauber (or equivalently heat-bath) update algorithm for the Ising model is also of theoretical interest since in this case the MC (pseudo-) dynamics can be calculated analytically – albeit only in one dimension [55]. For two and higher dimensions no exact solutions are known.

5.4 Initial Non-Equilibrium Period and Ageing

The initial equilibration or thermalization period, in general, is a non-trivial non-equilibrium process which is of interest it its own right. Long suspected to be a consequence of the slow dynamics of glassy systems only, the phenomenon of ageing for example has also been found in the phase-ordering kinetics of simple ferromagnets such as the Ising model. To study this effect numerically, we only need the methods introduced so far since most theoretical concepts assume a *local* spin-flip dynamics as realized by one the three update algorithms discussed above. Similarly to the concept of universality classes in equilibrium, all three algorithms should yield qualitatively similar results, being representatives of what is commonly referred to as dynamical Glauber universality class.

Let us assume that we pick as the initial configuration of the Markov chain a completely disordered state. If the simulation is run at a temperature $T > T_c$, equilibration will, in fact, be fast and nothing spectacular happens. If we choose instead to do the simulation right at T_c or at a temperature $T < T_c$, the situation is, however, quite different. In the latter two cases one speaks of a "quench", since the starting configuration is in a statistical sense far away from a typical equilibrium configuration at temperature T. This is easiest to understand for temperatures $T < T_c$, where the typical equilibrium state consists of homogeneously ordered configurations. After the quench, local regions of parallel spins start forming domains or clusters, and the non-equilibrium dynamics of the system is governed by the movement of the domain walls. In order to minimize their surface energy, the domains grow and straighten their surface. This mechanism is illustrated in Fig. 5.4 for the 2D Ising and 3-state Potts model, showing the time evolution after a quench to $T < T_c$ from an initially completely disordered state. This leads

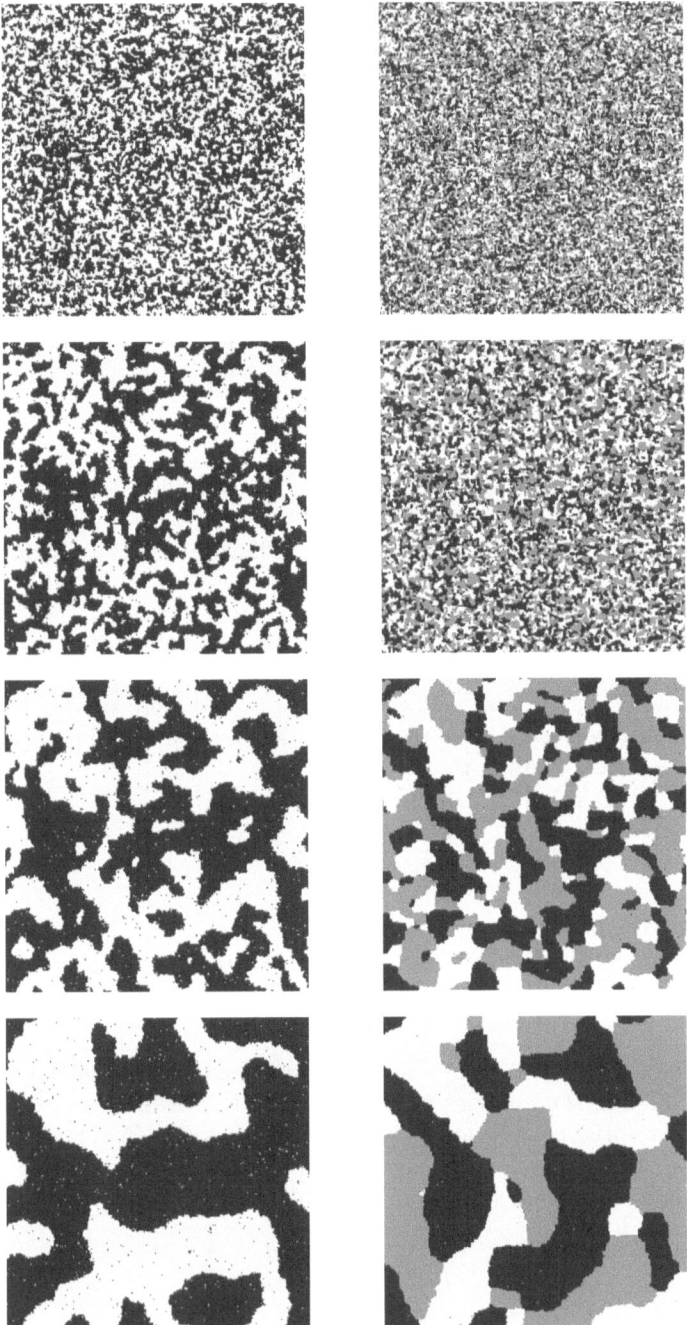

Fig. 5.4. Phase-ordering with progressing MC "time" (*from top to bottom*) of initially disordered spin configurations for the 2D Ising model at $T = 1.5 \approx 0.66\,T_c$ (*left*) and the 2D 3-state Potts model at $T = 0.4975 \approx T_c/2$ (*right*) (from Ref. [75])

to a growth law for the typical correlation length scale of the form $\xi \sim t^{1/z}$, where t is the time (measured in units of sweeps) elapsed since the quench. In the case of a simple ferromagnet like the Ising- or q-state Potts model with a non-conserved scalar order parameter, the dynamical exponent can be found exactly as $z = 2$ [56], according to diffusion or random-walk arguments. Right at the transition temperature, critical dynamics (for a recent review, see Ref. [57]) plays the central role and the dynamical exponent takes the somewhat larger non-trivial value $z \approx 2.17$ [58]. To equilibrate the whole system, ξ must approach the system size L, so that the typical relaxation time for equilibration scales as

$$\tau_{\text{relax}} \sim L^z \; . \tag{5.45}$$

Note that this implies in the infinite-volume limit $L \to \infty$ that true equilibrium can never be reached.

Since $1/z < 1$, the relaxation process after the quench happens on a growing time scale. This can be revealed most clearly by measurements of two-time quantities $f(t, s)$ with $t > s$, which no longer transform time-translation invariantly as they would do for small perturbations in equilibrium, where f would be a function of the time difference $t - s$ only. Instead, in phase-ordering kinetics, two-time quantities depend non-trivially on the *ratio* t/s of the two times. The dependence of the relaxation on the so-called "waiting time" s is the notional origin of ageing: older samples respond more slowly.

Commonly considered two-time quantities are the two-time autocorrelation function (in q-state Potts model notation)

$$C(t, s) = \frac{1}{q - 1} \left(\frac{q}{V} \sum_{i=1}^{V} \left[\delta_{\sigma_i(t), \sigma_i(s)} \right]_{\text{av}} - 1 \right) \tag{5.46}$$

and the two-time response function

$$R(t, s) = \left. \frac{\delta [\sigma_i(t)]_{\text{av}}}{\delta h(s)} \right|_{h=0} , \tag{5.47}$$

where $h(s)$ is the amplitude of a small spatially random external field which is switched off after the waiting time s and $[\dots]_{\text{av}}$ denotes an average over different random initial configurations (and random fields in (5.47)). In computer simulation studies it is more convenient to consider the integrated response or thermoremanent magnetization (TRM) [59],

$$\rho(t, s) = T \int_0^s du R(t, u) = \frac{T}{h} M_{\text{TRM}}(t, s) \; . \tag{5.48}$$

Dynamical scaling arguments predict the scaling forms (for reviews see, e.g., Refs. [60, 61])

$$C(t, s) = s^{-b} f_C(t/s) , \qquad R(t, s) = s^{-1-a} f_R(t/s) , \tag{5.49}$$

with scaling functions f_C and f_R which approach for large values of the scaling variable $x \equiv t/s$ the power-law behaviour

$$f_C(x) \to x^{-\lambda_C/z} \,, \qquad f_R(x) \to x^{-\lambda_R/z} \quad (x \gg 1) \,. \qquad (5.50)$$

In phase-ordering kinetics after a quench to $T < T_c$, $b = 0$ and $z = 2$ [56]. As the other exponents depend on the dimensionality of the considered system, we shall focus here on two dimensions only, where for the Ising model, it is commonly accepted that $\lambda_C = \lambda_R = 5/4$. The value of the remaining exponent a, however, is more controversial [62]. In the literature there are strong claims for $a = 1/z = 1/2$ [60, 63], but also $a = 1/4$ [64] has been conjectured.

Extending the symmetry considerations to *local* scale invariance in analogy to conformal invariance [65], even the explicit form of the scaling function $f_R(x)$ has been predicted [66, 67],

$$f_R(x) = r_0 x^{1+a-\lambda_R/z}(x-1)^{-1-a} \,, \qquad (5.51)$$

where r_0 is a normalization constant. The integration over the response function in (5.48) leads for the thermoremanent magnetization to the scaling form [64, 68–70]

$$\rho(t, s) = r_0 s^{-a} f_M(t/s) + r_1 s^{-\lambda_R/z} g_M(t/s) \,, \qquad (5.52)$$

where some care is necessary in dealing with cross-over effects leading to the second term which can be argued to take the explicit form $g_M(x) \approx x^{-\lambda_R/z}$. The first term follows directly from the integration over the explicit expression for $f_R(x)$ in (5.51) which results in a hypergeometric function [67, 69],

$$f_M(x) = x^{-\lambda_R/z} \, {}_2F_1(1+a, \lambda_R/z - a; \lambda_R/z - a + 1; 1/x) \,. \qquad (5.53)$$

Due to the linear combination of scaling functions in (5.52) with s-dependent prefactors, the scaling properties cannot be tested easily. One therefore usually subtracts first the correction term $\propto g_M(x)$ and then considers $f_M(x)$. While the two-time autocorrelation function $C(t, s)$ is conceptually and in particular computationally the much simpler quantity, local scale invariance predictions are much harder to derive for $C(t, s)$ than for $R(t, s)$. The expression for $f_C(x)$ contains combinations of hypergeometric and incomplete Gamma functions, depending on three additional undetermined constants apart from a normalization factor [71].

In computer simulations one proceeds as follows. One prepares many independent disordered start configurations of the order of a few hundred to a few thousand and monitors for each of them the time evolution after the quench to $T < T_c$ or, in critical relaxation [57], to $T = T_c$. Here it is important to make sure that the time evolutions are statistically independent of each other. In practice this means that different random number sequences have to be used for each sample.[4] The final result (for each time s and t) is then an average over these samples.

For the 2D and 3D Ising model, extensive numerical tests of the scaling predictions have been performed by Henkel, Pleimling and collaborators [69–71], showing a very good agreement with the almost parameter-free analytical expressions. To check the generality of the scaling arguments, we extended this work in a recent MC study [75,76] to more general q-state Potts models in two dimensions. Figures 5.5 and 5.6 compare our numerical results for the 2D Ising ($q = 2$) and 3-state Potts model after a quench to $T = 1.5 \approx 0.66\,T_c$ (in Ising model normalization) and $T = 0.4975 \approx T_c/2$, respectively (assuming in both cases $\lambda_C = \lambda_R$ with $\lambda_C \approx 1.25$ [75, 76] and $a = 1/z = 1/2$). We see that the two models behave very similarly during ageing, i.e., also for the 3-state Potts model the scaling predictions (5.49) are well satisfied. Moreover, the explicit analytical predictions for $f_M(t/s)$ in (5.53) and (the more complicated one) for $C(t, s)$ as given in Ref. [71] relying on local scale invariance are both in excellent agreement with the MC data. For details of the numerical set-up, see Refs. [75, 76], where also additional simulations of the 2D 8-state Potts model are described that give similarly good results.

5.5 Statistical Analysis of Monte Carlo Data

About a decade ago most of the statistical analysis methods discussed in this section were still quite cumbersome since due to disk-space limitations they usually had to be applied "on the flight" during the simulation. In particular dynamical aspects of a given model are usually not easy to predict beforehand such that the guess of reasonable analysis parameters was quite difficult. The situation has changed dramatically when it became affordable to store hundreds of megabytes on hard-disks. Since then a simulation study can clearly be separated into "raw data generation" and "data analysis" parts. The interface between these two parts should consist of time series of measurements of the relevant physical observables taken during the actual simulations. In principle there are no limitations on the choice of observables $\mathcal{O}$ which could be, for example, the energy H or the magnetization μ. Once the system is in equilibrium (which, in general, is non-trivial to assure), we simply save $\mathcal{O}_j \equiv \mathcal{O}[\{\sigma_i\}_j]$ where j labels the measurements. Given these data files one can perform detailed error analyses; in particular adapting parameters to a specific situation is now straightforward and very fast.

---

[4] If the same sequence of random numbers, i.e., the same dynamics, would be used for all samples with different start configurations, then one would study the phenomenon of "damage spreading" [72–74], where one basically asks how likely it is that two initially different configurations merge after some time into the same state when evolving under exactly the *same* dynamics.

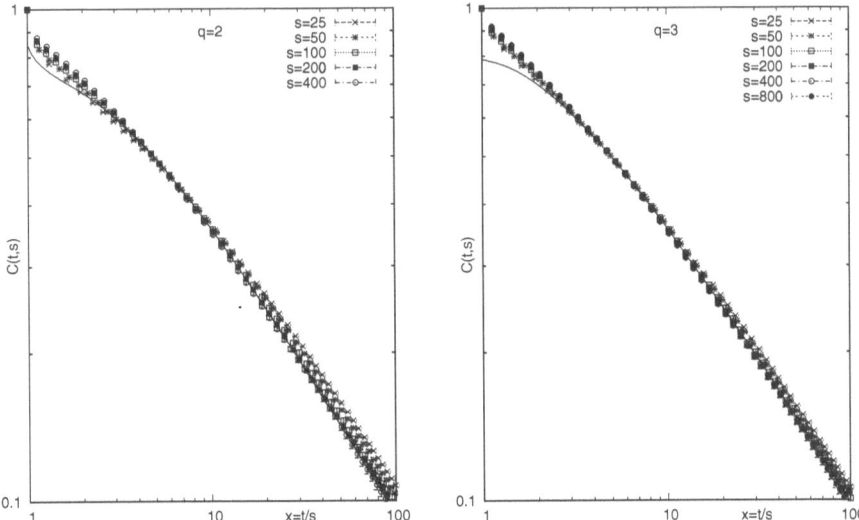

Fig. 5.5. Two-time autocorrelation function for the 2D Ising ($q = 2$) (*left*) and 3-state Potts (*right*) models [75,76]. The different data symbols correspond to different waiting times s. The *solid lines* show the fits to the scaling prediction [71] based on local scale invariance

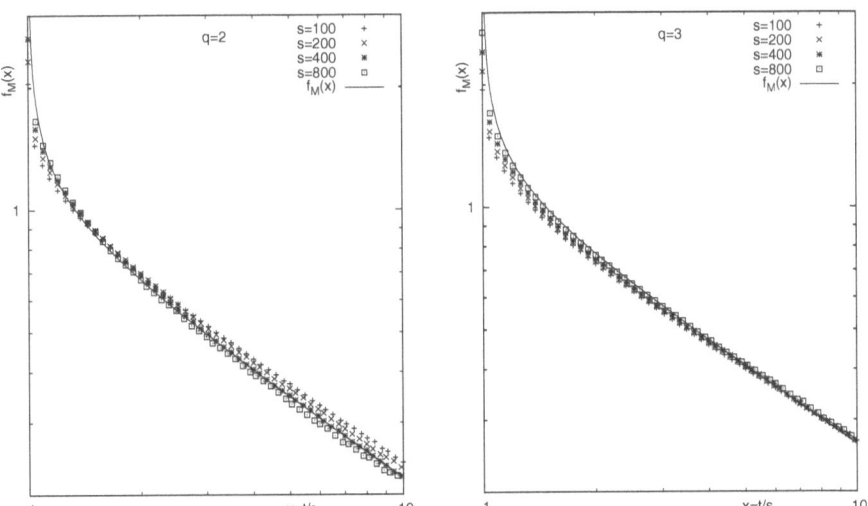

Fig. 5.6. Leading contribution $f_M(x)$ to the thermoremanent magnetization (5.48) and (5.52) for the 2D Ising ($q = 2$) (*left*) and 3-state Potts (*right*) models [75,76]. The different data symbols correspond to different waiting times s. The *solid lines* show the local scale invariance prediction (5.53) which depends only a single normalization parameter

5.5.1 Estimators

If the time series data result from an importance sampling MC simulation, the expectation value $\langle \mathcal{O} \rangle$ can be estimated as a simple arithmetic mean over the Markov chain,

$$\overline{\mathcal{O}} = \frac{1}{N} \sum_{j=1}^{N} \mathcal{O}_j \, , \tag{5.54}$$

where we assume that the time series collects, after an appropriate equilibration period, a total of N measurements. Conceptually it is important to distinguish between the expectation value $\langle \mathcal{O} \rangle$ and the mean value $\overline{\mathcal{O}}$, which is an estimator for the former. While $\langle \mathcal{O} \rangle$ is an ordinary number and represents the exact thermal average (which is only for very few models known), the estimator $\overline{\mathcal{O}}$ is still a *random* number, fluctuating around the theoretically expected value. Of course, in practice this is a "virtual" concept as one does not probe the fluctuations of the mean value directly since this would require repeating the whole MC simulation many times. However, one can estimate its variance,

$$\sigma_{\overline{\mathcal{O}}}^2 = \langle [\overline{\mathcal{O}} - \langle \overline{\mathcal{O}} \rangle]^2 \rangle = \langle \overline{\mathcal{O}}^2 \rangle - \langle \overline{\mathcal{O}} \rangle^2 \, , \tag{5.55}$$

from the statistical properties of individual measurements $\mathcal{O}_j$ in a single MC run.

5.5.2 Uncorrelated Measurements and Central-Limit Theorem

For simplicity, let us first make the unrealistic assumption that the N subsequent measurements $\mathcal{O}_j$ are all completely uncorrelated (as would be true only in simple sampling). Then the relation between the two variances would simply be

$$\sigma_{\overline{\mathcal{O}}}^2 = \sigma_{\mathcal{O}_j}^2 / N \, , \tag{5.56}$$

where $\sigma_{\mathcal{O}_j}^2 = \langle \mathcal{O}_j^2 \rangle - \langle \mathcal{O}_j \rangle^2$ is the variance of the individual measurements. A further, milder assumption is, of course, that the simulation is already in equilibrium so that time-translation invariance over the Markov chain is satisfied. Equation (5.56) is true for any distribution $\mathcal{P}(\mathcal{O}_j)$ of the $\mathcal{O}_j$. For the energy or magnetization the latter distributions are often plotted as physically directly relevant histograms (see, e.g., Fig. 5.14(b) below) whose squared width $(= \sigma_{\mathcal{O}_j}^2)$ is proportional to the specific heat or susceptibility, respectively.

Whatever form the distribution $\mathcal{P}(\mathcal{O}_j)$ assumes (which, in fact, is already often close to Gaussian because the $\mathcal{O}_j$ are usually lattice averages over many degrees of freedom), by the central limit theorem the distribution of the mean value is Gaussian, at least for uncorrelated data in the asymptotic limit of large N. The variance of the mean, $\sigma_{\overline{\mathcal{O}}}^2$, is the squared width of this (N dependent) distribution which is usually taken as the "one-sigma" squared error, $\epsilon_{\overline{\mathcal{O}}}^2 \equiv \sigma_{\overline{\mathcal{O}}}^2$, and quoted together with the mean value $\overline{\mathcal{O}}$. Under the assumption of a Gaussian distribution for the mean, the interpretation is that about 68% of all simulations under the same conditions would yield a mean value in the range $[\overline{\mathcal{O}} - \sigma_{\overline{\mathcal{O}}}, \overline{\mathcal{O}} + \sigma_{\overline{\mathcal{O}}}]$. For a "two-sigma" interval which also is sometimes used, this percentage goes up to about 95.4%, and for a "three-sigma" interval which is rarely quoted, the confidence level is higher than 99.7%.

5.5.3 Correlated Measurements and Autocorrelation Times

In "real life" things become more involved since when using importance sampling update algorithms subsequent measurements are necessarily correlated in time [77–79]. Inserting (5.54) into (5.55), one obtains

$$\sigma_{\overline{\mathcal{O}}}^2 = \langle \overline{\mathcal{O}}^2 \rangle - \langle \overline{\mathcal{O}} \rangle^2 = \frac{1}{N^2} \sum_{i,j=1}^{N} \langle \mathcal{O}_i \mathcal{O}_j \rangle - \frac{1}{N^2} \sum_{i,j=1}^{N} \langle \mathcal{O}_i \rangle \langle \mathcal{O}_j \rangle , \qquad (5.57)$$

and by collecting diagonal and off-diagonal terms one arrives at

$$\sigma_{\overline{\mathcal{O}}}^2 = \frac{1}{N^2} \sum_{i=1}^{N} \left(\langle \mathcal{O}_i^2 \rangle - \langle \mathcal{O}_i \rangle^2 \right) + \frac{1}{N^2} \sum_{i \neq j}^{N} \left(\langle \mathcal{O}_i \mathcal{O}_j \rangle - \langle \mathcal{O}_i \rangle \langle \mathcal{O}_j \rangle \right) . \qquad (5.58)$$

The first term is identified as the variance of the individual measurements multiplied with $1/N$. In the second sum we first use the symmetry $i \leftrightarrow j$ to reduce the summation to $\sum_{i \neq j}^{N} = 2 \sum_{i=1}^{N} \sum_{j=i+1}^{N}$. Reordering the summation and using time-translation invariance (assuming that equilibrium has already been reached, cf. the previous Sect. 5.4) we finally get

$$\sigma_{\overline{\mathcal{O}}}^2 = \frac{1}{N} \left[\sigma_{\mathcal{O}_i}^2 + 2 \sum_{k=1}^{N} \left(\langle \mathcal{O}_1 \mathcal{O}_{1+k} \rangle - \langle \mathcal{O}_1 \rangle \langle \mathcal{O}_{1+k} \rangle \right) \left(1 - \frac{k}{N} \right) \right] , \qquad (5.59)$$

where, due to the last factor, the $k = N$ term may trivially be kept in the summation. Factoring out $\sigma_{\mathcal{O}_i}^2$, this can be written as

$$\epsilon_{\overline{\mathcal{O}}}^2 \equiv \sigma_{\overline{\mathcal{O}}}^2 = \frac{\sigma_{\mathcal{O}_i}^2}{N} 2\tau'_{\mathcal{O},\mathrm{int}} , \qquad (5.60)$$

where we have introduced the (proper) *integrated* autocorrelation time

$$\tau'_{\mathcal{O},\mathrm{int}} = \frac{1}{2} + \sum_{k=1}^{N} A(k) \left(1 - \frac{k}{N} \right) , \qquad (5.61)$$

with

$$A(k) = \frac{\langle \mathcal{O}_i \mathcal{O}_{i+k} \rangle - \langle \mathcal{O}_i \rangle \langle \mathcal{O}_i \rangle}{\langle \mathcal{O}_i^2 \rangle - \langle \mathcal{O}_i \rangle \langle \mathcal{O}_i \rangle} \qquad (5.62)$$

denoting the normalized autocorrelation function ($A(0) = 1$).

For large time separations k the autocorrelation function decays exponentially ($a = $ const.),

$$A(k) \overset{k \to \infty}{\longrightarrow} a e^{-k/\tau_{\mathcal{O},\mathrm{exp}}} , \tag{5.63}$$

which defines the *exponential* autocorrelation time $\tau_{\mathcal{O},\mathrm{exp}}$. Since in any meaningful simulation study $N \gg \tau_{\mathcal{O},\mathrm{exp}}$, $A(k)$ in (5.61) is already exponentially small before the correction term in parentheses becomes important. For simplicity this correction is hence usually omitted (as is the "prime" of $\tau'_{\mathcal{O},\mathrm{int}}$ in (5.61)) and one employs the following definition for the *integrated* autocorrelation time:

$$\tau_{\mathcal{O},\mathrm{int}} = \frac{1}{2} + \sum_{k=1}^{N} A(k) . \tag{5.64}$$

The notion "integrated" derives from the fact that this may be interpreted as a trapezoidal discretization of the (approximate) integral $\tau_{\mathcal{O},\mathrm{int}} \approx \int_0^N \mathrm{d}k A(k)$. Notice that, in general, $\tau_{\mathcal{O},\mathrm{int}}$ (and also $\tau'_{\mathcal{O},\mathrm{int}}$) is different from $\tau_{\mathcal{O},\mathrm{exp}}$. In fact, one can show [80] that $\tau_{\mathcal{O},\mathrm{int}} \leq \tau_{\mathcal{O},\mathrm{exp}}$ in realistic models. Only if $A(k)$ is a pure exponential, the two autocorrelation times, $\tau_{\mathcal{O},\mathrm{int}}$ and $\tau_{\mathcal{O},\mathrm{exp}}$, coincide (up to minor corrections for small $\tau_{\mathcal{O},\mathrm{int}}$, see Eq. (5.86) below) [79].

Close to a critical point, the autocorrelation time scales for an infinite system typically as

$$\tau_{\mathcal{O},\mathrm{int}} \propto \tau_{\mathcal{O},\mathrm{exp}} \propto \xi^z , \tag{5.65}$$

where z is the dynamical critical exponent. For local algorithms, $z \approx 2$, which can be understood by a random-walk argument. Since $\xi \propto |T - T_c|^{-\nu} \to \infty$ when $T \to T_c$, also τ diverges when the critical point is approached. This leads to the phenomenon of *critical slowing down* at a continuous phase transition. In a finite system with extent L, ξ is basically replaced by L and

$$\tau_{\mathcal{O},\mathrm{int}} \propto \tau_{\mathcal{O},\mathrm{exp}} \propto L^z . \tag{5.66}$$

Non-local update algorithms such as multigrid schemes or in particular the cluster methods discussed later in Sect. 5.6 can reduce the value of the dynamical critical exponent z significantly, albeit in a model-dependent fashion.

At a first-order phase transition the "slowing-down" problem is even more severe, but the mechanism is completely different. Here, a finite system close to the (pseudo-) transition point can flip between the coexisting pure phases by crossing a two-phase region. Relative to the weight of the pure phases, this region of state space is strongly suppressed by an additional Boltzmann factor $\exp(-2\sigma L^{d-1})$, where σ denotes the interface tension between the coexisting phases, L^{d-1} is the (projected) "area" of the interface and the factor 2 accounts for periodic boundary conditions, which enforce for simple topological reasons always an even number of interfaces [16]. Whatever update algorithm is used, the time spent for crossing this highly suppressed rare-event region scales proportional to the inverse of this interfacial Boltzmann factor, i.e., the autocorrelation time behaves as

$$\tau \propto e^{2\sigma L^{d-1}} .\tag{5.67}$$

This exponential increase of autocorrelations with system size at a first-order phase transition is often described in the literature as *supercritical slowing down* (even though, strictly speaking, nothing is "critical" here). This type of slowing-down problem can be overcome in part by means of tempering and multicanonical methods also discussed later in Sects. 5.8 and 5.9.

As far as the accuracy of MC data is concerned, the important point of Eq. (5.60) is that due to temporal correlations of the measurements the statistical error $\epsilon_{\overline{\mathcal{O}}} \equiv \sqrt{\sigma_{\overline{\mathcal{O}}}^2}$ on the MC estimator $\overline{\mathcal{O}}$ is enhanced by a factor of $\sqrt{2\tau_{\mathcal{O},\text{int}}}$. This can be rephrased by writing the statistical error similar to the uncorrelated case as $\epsilon_{\overline{\mathcal{O}}} = \sqrt{\sigma_{\mathcal{O}_j}^2/N_{\text{eff}}}$, but now with a parameter

$$N_{\text{eff}} = N/2\tau_{\mathcal{O},\text{int}} \leq N ,\tag{5.68}$$

describing the *effective* statistics. This shows more clearly that only every $2\tau_{\mathcal{O},\text{int}}$ iterations the measurements are approximately uncorrelated and gives a better idea of the relevant effective size of the statistical sample. In view of the scaling behaviour of the autocorrelation time in (5.65) or (5.66) respectively (5.67), it is obvious that without extra care this effective sample size may become very small close to a continuous or first-order phase transition. Since some quantities (e.g., the specific heat or susceptibility) can severely be underestimated if the effective statistics is too small [81], any serious simulation should therefore provide at least a rough order-of-magnitude estimate of autocorrelation times.

5.5.4 Bias

For a better understanding of the latter point, let us consider as a specific example the specific heat, $C = \beta^2 V (\langle e^2 \rangle - \langle e \rangle^2) = \beta^2 V \sigma_{e_i}^2$. The standard estimator for the variance is

$$\hat{\sigma}_{e_i}^2 = \overline{e^2} - \overline{e}^2 = \overline{(e - \overline{e})^2} = \frac{1}{N} \sum_{i=1}^{N} (e_i - \overline{e})^2 .\tag{5.69}$$

What is the *expected* value of $\hat{\sigma}_{e_i}^2$? To answer this question, we subtract and add $\langle \overline{e} \rangle^2$,

$$\langle \hat{\sigma}_{e_i}^2 \rangle = \langle \overline{e^2} - \overline{e}^2 \rangle = \langle \overline{e^2} \rangle - \langle \overline{e} \rangle^2 - \left(\langle \overline{e}^2 \rangle - \langle \overline{e} \rangle^2 \right) ,\tag{5.70}$$

and then use the previously derived result: The first two terms on the r.h.s. of (5.70) just give $\sigma_{e_i}^2$, and the second two terms in parentheses yield $\sigma_{\overline{e}}^2 = \sigma_{e_i}^2 2\tau_{e,\text{int}}/N$, as calculated in (5.60). Combining these two results we arrive at

$$\langle \hat{\sigma}_{e_i}^2 \rangle = \sigma_{e_i}^2 \left(1 - \frac{2\tau_{e,\text{int}}}{N} \right) = \sigma_{e_i}^2 \left(1 - \frac{1}{N_{\text{eff}}} \right) \neq \sigma_{e_i}^2 .\tag{5.71}$$

The estimator $\hat{\sigma}^2_{e_i}$ as defined in (5.69) thus systematically underestimates the true value by a term of the order of $\tau_{e,\text{int}}/N$. Such an estimator is called *weakly biased* ("weakly" because the statistical error $\propto 1/\sqrt{N}$ is asymptotically larger than the systematic bias; for medium or small N, however, also prefactors need to be carefully considered).

We thus see that for large autocorrelation times or equivalently small effective statistics N_{eff}, the bias may be quite large. Since $\tau_{e,\text{int}}$ scales quite strongly with the system size for local update algorithms, some care is necessary in choosing the run time N. Otherwise the FSS of the specific heat and thus the determination of the *static* critical exponent α/ν could be completely spoiled by the temporal correlations!

As a side remark we note that even in the completely uncorrelated case the estimator (5.69) is biased, $\langle \hat{\sigma}^2_{e_i} \rangle = \sigma^2_{e_i} (1 - 1/N)$, since with our conventions in this case $\tau_{e,\text{int}} = 1/2$ (some authors use a different convention in which τ more intuitively vanishes in the uncorrelated case; but this has certain disadvantages in other formulas). In this case one can (and usually does) define a bias-corrected estimator,

$$\hat{\sigma}^2_{e_i,\text{corr}} = \frac{N}{N-1}\hat{\sigma}^2_{e_i} = \frac{1}{N-1}\sum_{i=1}^{N}(e_i - \bar{e})^2 , \qquad (5.72)$$

which obviously satisfies $\langle \hat{\sigma}^2_{e_i,\text{corr}} \rangle = \sigma^2_{e_i}$. For the squared error on the mean value, this leads to the error formula $\epsilon^2_{\bar{e}} = \hat{\sigma}^2_{\bar{e},\text{corr}} = \hat{\sigma}^2_{e_i,\text{corr}}/N = \frac{1}{N(N-1)}\sum_{i=1}^{N}(e_i - \bar{e})^2$, i.e., to the celebrated replacement of one of the $1/N$-factors by $1/(N-1)$ "due to one missing degree of freedom". Note that in the case of correlated data, a similar construction is at best approximately possible since the bias in (5.71) depends on the a priori unknown autocorrelation time $\tau_{e,\text{int}}$.

5.5.5 Numerical Estimation of Autocorrelation Times

The above considerations show that not only for the error estimation but also for the computation of static quantities themselves it is important to have control over autocorrelations. Unfortunately, it is very difficult to give reliable a priori estimates, and an accurate numerical analysis is often too time consuming. As a rough estimate it is about ten times harder to get precise information on dynamic quantities than on static quantities like critical exponents. A (weakly biased) estimator $\hat{A}(k)$ for the autocorrelation function is obtained by replacing in (5.62) the expectation values (ordinary numbers) by mean values (random variables), e.g., $\langle \mathcal{O}_i \mathcal{O}_{i+k} \rangle$ by $\overline{\mathcal{O}_i \mathcal{O}_{i+k}}$. With increasing separation k the relative variance of $\hat{A}(k)$ diverges rapidly. To get at least an idea of the order of magnitude of $\tau_{\mathcal{O},\text{int}}$ and thus the correct error estimate (5.60), it is useful to record the "running" autocorrelation time estimator

$$\hat{\tau}_{\mathcal{O},\text{int}}(k_{\max}) = \frac{1}{2} + \sum_{k=1}^{k_{\max}} \hat{A}(k) \, , \tag{5.73}$$

which approaches $\tau_{\mathcal{O},\text{int}}$ in the limit of large $k_{\max}$ where, however, its statistical error increases rapidly. As a compromise between systematic and statistical errors, an often employed procedure is to determine the upper limit $k_{\max}$ self-consistently by cutting off the summation once $k_{\max} \geq 6\hat{\tau}_{\mathcal{O},\text{int}}(k_{\max})$, where $A(k) \approx e^{-6} \approx 10^{-3}$. In this case an a priori error estimate is available [79, 82, 83],

$$\epsilon_{\tau_{\mathcal{O},\text{int}}} = \tau_{\mathcal{O},\text{int}} \sqrt{\frac{2(2k_{\max} + 1)}{N}} \approx \tau_{\mathcal{O},\text{int}} \sqrt{\frac{12}{N_{\text{eff}}}} \, . \tag{5.74}$$

For a 5% relative accuracy one thus needs at least $N_{\text{eff}} \approx 5\,000$ or $N \approx 10\,000\,\tau_{\mathcal{O},\text{int}}$ measurements. As an order of magnitude estimate consider the 2D Ising model with $L = 100$ simulated with a local update algorithm. The integrated autocorrelation time for this example is of the order of $L^2 \approx 100^2$ (ignoring an priori unknown prefactor of "order unity" which depends on the considered quantity), thus implying $N \approx 10^8$. Since in each sweep L^2 spins have to be updated and assuming that each spin update takes about 0.1 μsec, we end up with a total time estimate of about 10^5 seconds $\approx$ 1 CPU-day to achieve this accuracy.

Another possibility is to approximate the tail end of $A(k)$ by a single exponential as in (5.63). Summing up the small k part exactly, one finds [84]

$$\tau_{\mathcal{O},\text{int}}(k_{\max}) = \tau_{\mathcal{O},\text{int}} - ce^{-k_{\max}/\tau_{\mathcal{O},\exp}} \, , \tag{5.75}$$

where c is a constant. The latter expression may be used for a numerical estimate of both the exponential and integrated autocorrelation times [84].

5.5.6 Binning Analysis

It should be clear by now that ignoring autocorrelation effects can lead to severe underestimates of statistical errors. Applying the full machinery of autocorrelation analysis discussed above, however, is often too cumbersome. On a day by day basis the following binning analysis is much more convenient (though somewhat less accurate). By grouping the N original time-series data into N_B non-overlapping bins or blocks of length k (such that[5] $N = N_B k$), one forms a new, shorter time series of block averages,

$$\mathcal{O}_j^{(B)} \equiv \frac{1}{k} \sum_{i=1}^{k} \mathcal{O}_{(j-1)k+i} \, , \qquad j = 1, \ldots, N_B \, , \tag{5.76}$$

---

[5] Here we assume that N was chosen cleverly. Otherwise one has to discard some of the data and redefine N.

which by choosing the block length $k \gg \tau$ are almost uncorrelated and can thus be analyzed by standard means. The mean value over all block averages obviously satisfies $\overline{\mathcal{O}^{(B)}} = \overline{\mathcal{O}}$ and their variance can be computed according to the unbiased estimator (5.72), leading to the squared statistical error of the mean value,

$$\epsilon_{\overline{\mathcal{O}}}^2 \equiv \sigma_{\overline{\mathcal{O}}}^2 = \sigma_B^2/N_B = \frac{1}{N_B(N_B-1)} \sum_{j=1}^{N_B} (\mathcal{O}_j^{(B)} - \overline{\mathcal{O}^{(B)}})^2 \ . \tag{5.77}$$

By comparing with (5.60) we see that $\sigma_B^2/N_B = 2\tau_{\mathcal{O},\text{int}}\sigma_{\mathcal{O}_i}^2/N$. Recalling the definition of the block length $k = N/N_B$, this shows that one may also use

$$2\tau_{\mathcal{O},\text{int}} = k\sigma_B^2/\sigma_{\mathcal{O}_i}^2 \tag{5.78}$$

for the estimation of $\tau_{\mathcal{O},\text{int}}$. Estimates of $\tau_{\mathcal{O},\text{int}}$ obtained in this way are often referred to as "blocking τ" or "binning τ".

5.5.7 Jackknife Analysis

But even if the data are completely uncorrelated in time, one still has to handle the problem of error estimation for quantities that are not directly measured in the simulation but are computed as a non-linear combination of "basic" observables. This problem can either be solved by error propagation or by using the Jackknife method [85, 86] where instead of considering rather small blocks of length k and their fluctuations as in the binning method, one forms N_B large Jackknife blocks $\mathcal{O}_j^{(J)}$ containing all data but the j'th block of the previous binning method,

$$\mathcal{O}_j^{(J)} = \frac{N\overline{\mathcal{O}} - k\mathcal{O}_j^{(B)}}{N-k} \ , \qquad j = 1, \ldots, N_B \ . \tag{5.79}$$

Each of the Jackknife blocks thus consists of $N-k$ data, i.e., it contains almost as many data as the original time series. When non-linear combinations of basic variables are estimated, the bias is hence comparable to that of the total data set (typically $1/(N-k)$ compared to $1/N$). The N_B Jackknife blocks are, of course, trivially correlated because one and the same original data enter in $N_B - 1$ different Jackknife blocks. This trivial correlation caused by re-using the original data over and over again has nothing to do with temporal correlations. As a consequence the Jackknife block variance σ_J^2 will be much smaller than the variance estimated in the binning method. Because of the trivial nature of the correlations, however, this reduction can be corrected by multiplying σ_J^2 with a factor $(N_B - 1)^2$, leading to

$$\epsilon_{\overline{\mathcal{O}}}^2 \equiv \sigma_{\overline{\mathcal{O}}}^2 = \frac{N_B - 1}{N_B} \sum_{j=1}^{N_B} (\mathcal{O}_j^{(J)} - \overline{\mathcal{O}^{(J)}})^2 \ . \tag{5.80}$$

To summarize this section, any realization of a Markov chain, i.e., MC update algorithm, is characterized by autocorrelation times which enter directly in the statistical errors of MC estimates. Since temporal correlations always increase the statistical errors, it is a very important issue to develop MC update algorithms that keep autocorrelation times as small as possible. This is the reason why cluster and other non-local algorithms are so important.

5.5.8 A Simplified Model: The Bivariate Gaussian Time Series

A useful "gauge model" for all the statistical analysis tools discussed so far is the bivariate Gaussian time series which allows for fairly simple exact solutions. Once the numerical routines reproduce the exact answers for this artificial time series, it is almost certain that they also work properly for "true" time series generated by a MC simulation. The bivariate Gaussian time series is generated by the recursion

$$e_0 = e_0' ,$$
$$e_i = \rho e_{i-1} + \sqrt{1 - \rho^2} e_i' , \qquad i \geq 1 , \tag{5.81}$$

where $0 \leq \rho < 1$ and the e_i' are *independent*, identically distributed (often abbreviated as "i.i.d.") Gaussian random variables satisfying $\langle e_i' \rangle = 0$ and $\langle e_i' e_j' \rangle = \delta_{ij}$. By iterating the recursion (5.81) it is then easy to see that $\langle e_i \rangle = 0$, $\langle e_i^2 \rangle = 1$ and

$$e_k = \rho e_{k-1} + \sqrt{1 - \rho^2} e_k' = \rho^k e_0 + \sqrt{1 - \rho^2} \sum_{l=1}^{k} \rho^{k-l} e_l' , \tag{5.82}$$

so that

$$A(k) = \langle e_0 e_k \rangle = \rho^k \equiv e^{-k/\tau_{\exp}} . \tag{5.83}$$

In this simplified model the autocorrelation function is thus a pure exponential with an exponential autocorrelation time given by

$$\tau_{\exp} = -1/\ln \rho . \tag{5.84}$$

It should be stressed that in realistic situations a purely exponential decay can only be expected asymptotically for large k where the slowest mode dominates. For smaller time separations usually also many other modes contribute whose autocorrelation time is smaller.

The visual appearance of uncorrelated and correlated data with $\tau_{\exp} = 10$ and 50 is depicted in Figs. 5.7(a)–(c) where in each case one percent of the total "MC time" evolution consisting of 100 000 consecutive "measurements" according to the rule (5.81) is shown. Despite the quite distinct temporal evolutions, histogramming the time series leads to the same Gaussian distribution within error bars, as it should, cf. Fig. 5.7(d). The corresponding autocorrelation functions $A(k)$ are shown in Fig. 5.8(a).

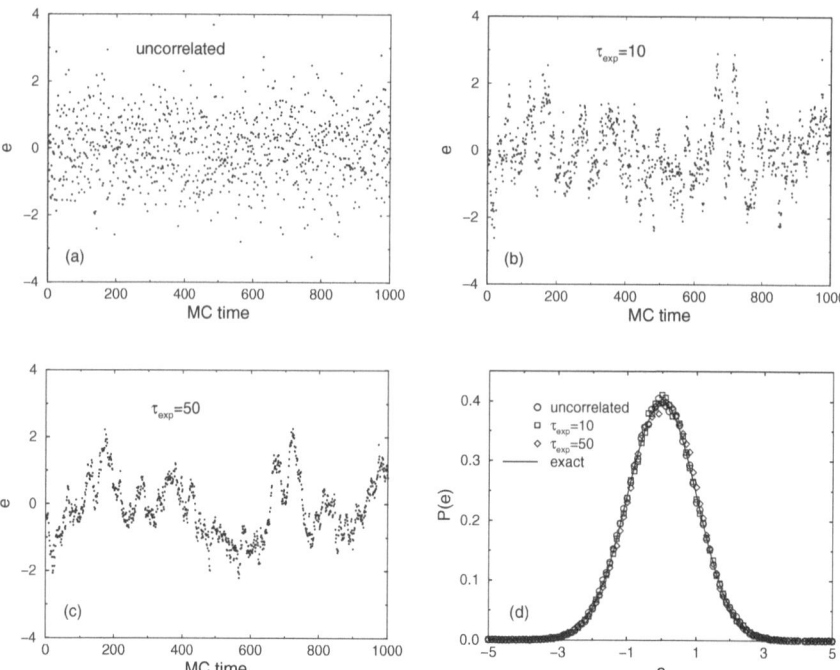

Fig. 5.7. "MC time" evolution according to the bivariate Gaussian process (5.81) (only the first percent shown) in (**a**) the uncorrelated case, (**b**) with $\tau_{\exp} = 10$, and (**c**) with $\tau_{\exp} = 50$. All three time evolutions with a total of $100\,000$ consecutive "measurements" lead to the same Gaussian histogram shown in (**d**)

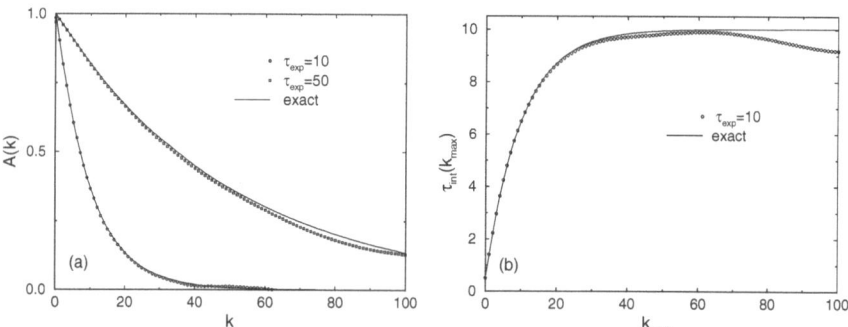

Fig. 5.8. (**a**) Autocorrelation functions and (**b**) integrated autocorrelation time for $\tau_{\exp} = 10$ on the basis of $100\,000$ "measurements" in comparison with exact results for the bivariate Gaussian model shown as the *solid lines*

The integrated autocorrelation time can also be calculated exactly,

$$\tau_{int} = \frac{1}{2} + \sum_{k=1}^{\infty} A(k) = \frac{1}{2}\frac{1+\rho}{1-\rho} = \frac{1}{2}\text{cth}(1/2\tau_{exp}) \tag{5.85}$$

$$= \tau_{exp}\left[1 + \frac{1}{12\tau_{exp}^2} + \mathcal{O}(1/\tau_{exp}^4)\right] . \tag{5.86}$$

This shows that for a purely exponential autocorrelation function to a very good approximation, $\tau_{int} \approx \tau_{exp}$, which would immediately follow from $\tau_{int} \approx \int_0^{\infty} dk A(k) = \tau_{exp}$.

As explained in the last section, one usually truncates the summation in (5.85) self-consistently at about $k_{max} = 6\tau_{int}$ ($\approx 6\tau_{exp}$) since $A(k)$ becomes very noisy for large time separations. Observing that (5.85) is nothing but a geometric series, also the resulting correction can be calculated exactly,

$$\tau_{int}(k_{max}) \equiv \frac{1}{2} + \sum_{k=1}^{k_{max}} A(k) = \frac{1}{2}\text{cth}(1/2\tau_{exp})\left[1 - \frac{2e^{-(k_{max}+1)/\tau_{exp}}}{1 + e^{-1/\tau_{exp}}}\right] \tag{5.87}$$

$$= \tau_{int}\left\{1 - [1 - \tanh(1/2\tau_{exp})]e^{-k_{max}/\tau_{exp}}\right\} \tag{5.88}$$

$$\approx \tau_{int}\left[1 - \left(1 - \frac{1}{2\tau_{exp}}\right)e^{-k_{max}/\tau_{exp}}\right] \qquad (\tau_{exp} \gg 1) , \tag{5.89}$$

showing that with increasing k_{max} the asymptotic value of $\tau_{int} \equiv \tau_{int}(\infty)$ is approached exponentially fast. This is illustrated in Fig. 5.8(b) for the bivariate Gaussian time series with $\tau_{exp} = 10$. Here we also see that for too large k_{max} the estimate for $\tau_{int}(k_{max})$ can deviate quite substantially from the exact value due to its divergent variance. The usually employed self-consistent cutoff would be around $6\tau_{exp} = 60$ where $\tau_{int}(k_{max}) \approx 9.89$.

Let us now turn to the binning analysis by decomposing as in (5.76) the total number of measurements N into N_B non-overlapping blocks of length k ($N = N_B k$). In our simple example, the expected value of the block averages is, of course, zero, $\langle e_{B,n} \rangle = \frac{1}{k}\sum_{i=1}^{k}\langle e_{(n-1)k+i}\rangle = 0$. The variance of the block variables is hence just the expectation value of $e_{B,n}^2$,

$$\sigma_B^2 = \langle e_{B,n}^2 \rangle = \frac{1}{k^2}\sum_{i,j=1}^{k}\rho^{|i-j|} = \frac{1}{k^2}\left[k + 2\sum_{i=1}^{k}\sum_{j=1}^{i-1}\rho^{i-j}\right]$$

$$= \frac{1}{k}\left[1 + \frac{2\rho}{1-\rho} - \frac{2\rho}{k}\frac{1-\rho^k}{(1-\rho)^2}\right] . \tag{5.90}$$

Recalling (5.85) this can be rewritten as

$$k\sigma_B^2 = 2\tau_{int}\left[1 - \frac{\tau_{int}}{k}\left(1 - e^{-k/\tau_{exp}}\right)/\cosh^2(1/2\tau_{exp})\right] \tag{5.91}$$

$$\approx 2\tau_{int}\left[1 - \frac{\tau_{exp}}{k}\left(1 - e^{-k/\tau_{exp}}\right)\right] \qquad (\tau_{exp} \gg 1) , \tag{5.92}$$

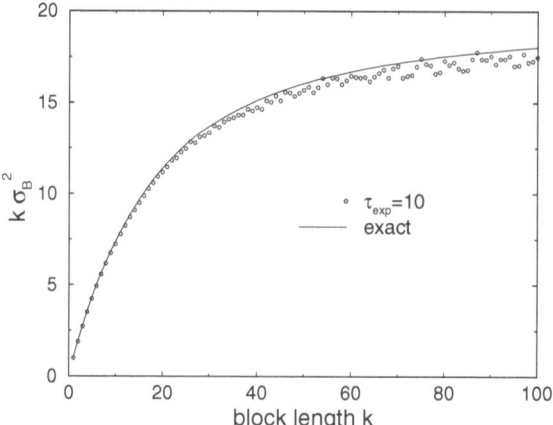

Fig. 5.9. Binning analysis of 100 000 "measurements" in the bivariate Gaussian model with $\tau_{\exp} = 10$. The *solid line* shows the exact result

showing that with increasing block length k the asymptotic value $2\tau_{\mathrm{int}}$ is approached according to a power law. For an illustration see Fig. 5.9.

5.5.9 Applications to the 2D Ising Model

In this section the autocorrelation and error analysis is illustrated for the 2D Ising model which albeit still very simple exhibits already some effects also seen in more complicated systems. The simulations are done with the Metropolis update algorithm for a 16×16 square lattice with periodic boundary conditions at the infinite-volume critical point $\beta_c = \ln(1 + \sqrt{2})/2 \approx 0.440\,686\,793\,4\ldots$. The spins were updated in sequential order by proposing always a spin flip[6] and accepting or rejecting this proposal according to (5.33). The raw data of the simulation are collected in a time-series file, storing 1000000 measurements of the energy and magnetization taken after each sweep over the lattice, after discarding (quite generously) the first 200000 sweeps to equilibrate the system.

The last 500 sweeps of the time evolution of the energy are shown in Fig. 5.10(a), which should be compared with the Gaussian model time series in Figs. 5.7(b) and (c). Using the complete time series the autocorrelation functions were computed according to (5.62). The only difference to the analysis of the simplified model is that instead of using the Gaussian data one now reads in the Ising model time series – the analysis program is exactly the

---

[6] If the spins are updated in sequential order, but a spin flip is proposed with only 50% probability, the temporal correlations are much larger ($\tau_{e,\mathrm{int}} \approx 27$) [87]. This quite unusual update procedure was (inadvertently) chosen in Ref. [87], because always proposing a spin flip with sequential update order does *not* work properly for the 1D model with its only two nearest neighbours.

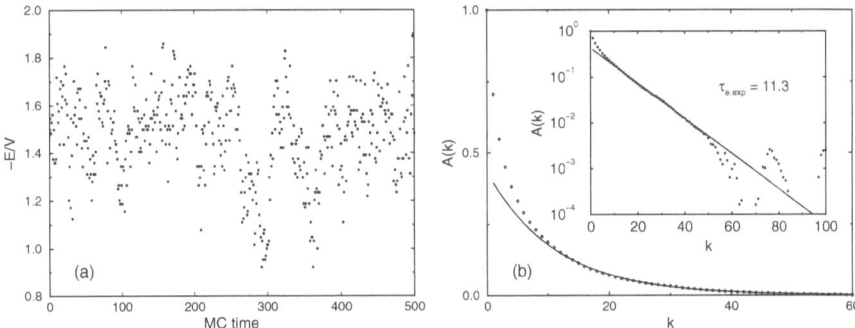

Fig. 5.10. (**a**) Part of the time evolution of the energy $e = E/V$ for the 2D Ising model on a 16×16 lattice at β_c and (**b**) the resulting autocorrelation function. The inset shows the same data on a logarithmic scale, revealing the fast initial drop for very small k and the noisy behaviour for large k. The *solid lines* show a fit to the ansatz $A(k) = a \exp(-k/\tau_{e,\exp})$ in the range $10 \leq k \leq 40$ with $\tau_{e,\exp} = 11.3$ and $a = 0.432$

same. The result for the energy autocorrelations is shown in Fig. 5.10(b). On the linear-log scale of the inset we clearly see the asymptotic linear behaviour of $\ln A(k)$. A linear fit of the form (5.63), $A(k) = a \exp(-k/\tau_{e,\exp})$, in the range $10 \leq k \leq 40$ yields an estimate for the *exponential* autocorrelation time of $\tau_{e,\exp} \approx 11.3$. Apart from the noise for large k, which is also present in the simplified model for finite statistics, the main difference to the artificial data of the simplified model lies in the small k behaviour. For the Ising model we clearly notice an initial fast drop, corresponding to faster relaxing modes, before the asymptotic behaviour sets in. This is, in fact, the generic behaviour of autocorrelation functions in realistic models.

Once the autocorrelation function is known, it is straightforward to sum up the integrated autocorrelation time. The result for the energy is depicted in Fig. 5.11(a), yielding an estimate of $\tau_{e,\mathrm{int}} \approx 5.93$. The binning analysis shown in Fig. 5.11(b) gives a consistent result as it should. Note that due to the initial fast drop of $A(k)$ the exponential autocorrelation time $\tau_{e,\exp} \approx 11.3$ is much larger than the integrated autocorrelation time $\tau_{e,\mathrm{int}} \approx 5.93$, which is in accord with the general inequality [80] quoted above.

5.6 Cluster Algorithms

The main drawback of local update algorithms is their pronounced critical slowing down at a continuous phase transition where temporal correlations diverge (thermodynamic limit) or become very large (finite-size scaling region): $\tau \propto \xi^z$ or $\propto L^z$ with $z \approx 2$. Since excitations on all length scales become important at T_c, it is intuitively clear that some sort of non-local updates should alleviate this problem. While it was clear since long that clusters or

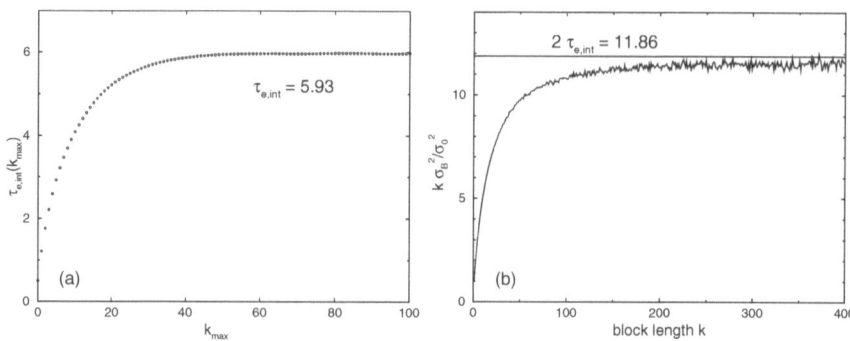

Fig. 5.11. (a) Integrated autocorrelation time approaching $\tau_{e,\mathrm{int}} \approx 5.93$ for large upper cutoff $k_{\max}$ and (b) binning analysis for the energy of the 2D Ising model on a 16×16 lattice at β_c. The *horizontal line* in (b) shows $2\tau_{e,\mathrm{int}}$ with $\tau_{e,\mathrm{int}}$ read off from (a)

droplets should play a central role in such an update, it took until 1987 before Swendsen and Wang [88] proposed a legitimate cluster update procedure for Potts models. Soon after Wolff [89] discovered the so-called single-cluster variant and developed a generalization to $O(n)$-symmetric spin models. By now cluster updates have been derived for many other models as well [90], but they are still less general applicable than local update algorithms of the Metropolis type. We therefore start again with the Ising model where (as for more general Potts models) the prescription for a cluster-update algorithm can be easily read off from the equivalent Fortuin-Kasteleyn representation [91–94],

$$Z = \sum_{\{\sigma_i\}} \exp\left(\beta \sum_{\langle ij\rangle} \sigma_i \sigma_j\right) \tag{5.93}$$

$$= \sum_{\{\sigma_i\}} \prod_{\langle ij\rangle} e^\beta \left[(1-p) + p\delta_{\sigma_i\sigma_j}\right] \tag{5.94}$$

$$= \sum_{\{\sigma_i\}} \sum_{\{n_{ij}\}} \prod_{\langle ij\rangle} e^\beta \left[(1-p)\delta_{n_{ij},0} + p\delta_{\sigma_i\sigma_j}\delta_{n_{ij},1}\right] , \tag{5.95}$$

with

$$p = 1 - e^{-2\beta} . \tag{5.96}$$

Here the n_{ij} are bond variables which can take the values $n_{ij} = 0$ or 1, interpreted as "deleted" or "active" bonds. In the first line of this derivation we used the trivial fact that the product $\sigma_i\sigma_j$ of two Ising spins can only take the two values ± 1, so that $\exp(\beta\sigma_i\sigma_j) = x + y\delta_{\sigma_i\sigma_j}$ can easily be solved for x and y. And in the second line we made use of the "deep" identity $a + b = \sum_{n=0}^{1} (a\delta_{n,0} + b\delta_{n,1})$.

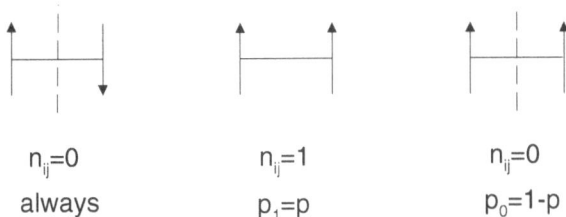

$n_{ij}=0$ $n_{ij}=1$ $n_{ij}=0$

always $p_1=p$ $p_0=1$-p

Fig. 5.12. Illustration of the bond variable update. The bond between unlike spins is always "deleted" as indicated by the *dashed line*. A bond between like spins is only "active" with probability $p = 1 - \exp(-2\beta)$. Only at zero temperature ($\beta \longrightarrow \infty$) stochastic and geometrical clusters coincide

Swendsen-Wang Cluster

According to (5.95) a cluster update sweep then consists of alternating updates of the bond variables n_{ij} for given spins with updates of the spins σ_i for a given bond configuration. In practice one proceeds as follows:

1. Set $n_{ij} = 0$ if $\sigma_i \neq \sigma_j$, or assign values $n_{ij} = 1$ and 0 with probability p and $1 - p$, respectively, if $\sigma_i = \sigma_j$, cp. Fig. 5.12.
2. Identify clusters of spins that are connected by "active" bonds ($n_{ij} = 1$).
3. Draw a random value ± 1 independently for each cluster (including one-site clusters), which is then assigned to all spins in a cluster.

Technically the cluster identification part is the most complicated step, but there are by now quite a few efficient algorithms available which can even be used on parallel computers. Vectorization, on the other hand, is only partially possible.

Notice the difference between the just defined *stochastic* clusters and *geometrical* clusters whose boundaries are defined by drawing lines through bonds between unlike spins. In fact, since in the stochastic cluster definition also bonds between like spins are "deleted" with probability $p_0 = 1 - p = \exp(-2\beta)$, stochastic clusters are on the average smaller than geometrical clusters. Only at zero temperature ($\beta \longrightarrow \infty$) p_0 approaches zero and the two cluster definitions coincide. As described above, the cluster algorithm is referred to as Swendsen-Wang (SW) or multiple-cluster update [88]. The distinguishing point is that the *whole* lattice is decomposed into stochastic clusters whose spins are assigned a random value $+1$ or -1. In one sweep one thus attempts to update all spins of the lattice.

Wolff Cluster

Shortly after the original discovery of cluster algorithms, Wolff [89] proposed a somewhat simpler variant in which only a single cluster is flipped at a time. This variant is therefore sometimes also called single-cluster algorithm. Here

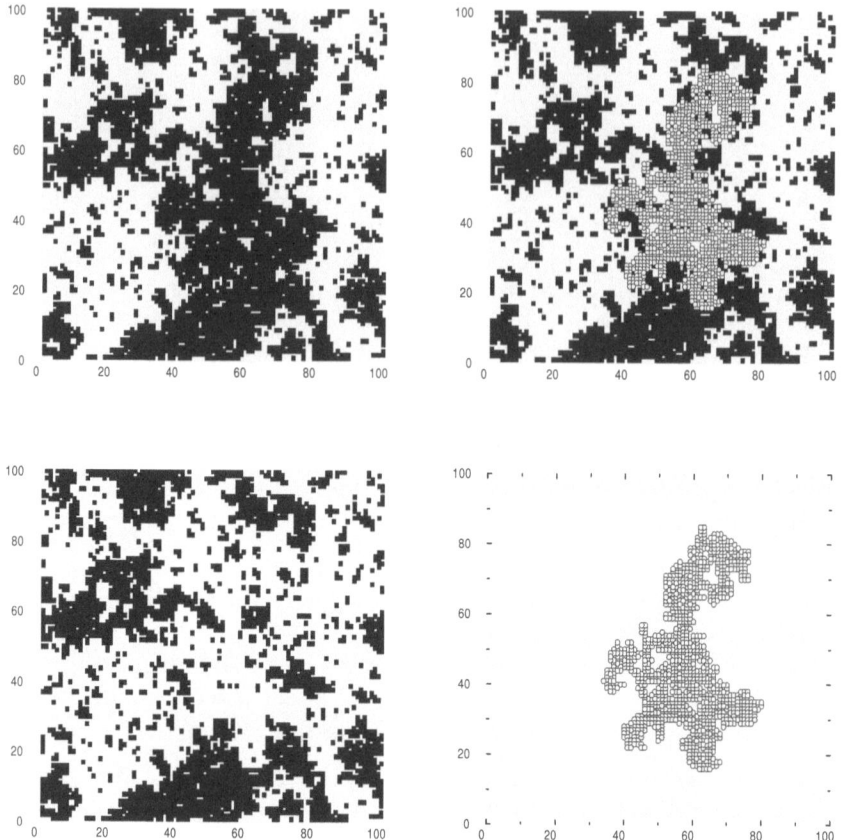

Fig. 5.13. Illustration of the Wolff cluster update, using actual simulation results for the 2D Ising model at $0.97 \times \beta_c$ on a 100×100 lattice. *Upper left*: Initial configuration. *Upper right*: The stochastic cluster is marked. *Lower left*: Final configuration after flipping the spins in the cluster. *Lower right*: The flipped cluster

one chooses a lattice site at random, constructs only the cluster connected with this site, and then flips all spins of this cluster. A typical example is shown in Fig. 5.13. In principle, one could also here choose for the new spin value $+1$ or -1 at random, but then nothing at all would be changed if one hits the current value of the spins.

Here a sweep consists of $V/\langle |C| \rangle$ single cluster steps, where $\langle |C| \rangle$ denotes the average cluster size. With this definition autocorrelation times are directly comparable with results from the Metropolis or Swendsen-Wang algorithm. Apart from being somewhat easier to program, Wolff's single-cluster variant is usually more efficient than the Swendsen-Wang multiple-cluster algorithm, especially in 3D. The reason is that with the single-cluster method, on the average, larger clusters are flipped.

Figure 5.13 also nicely illustrates the difference between *geometrical* and *stochastic FK* clusters as already pointed out in Sect. 5.2 in connection with Fig. 5.2. In the upper right configuration plot one clearly sees that the stochastic cluster is much smaller than the underlying black geometrical one. It is worth to emphasize again that only the stochastic FK clusters encode in their fractal and percolation properties the *critical* behaviour of the thermodynamic system. In the 3D Ising model, geometrical clusters do not even percolate at the proper critical temperature (but already at a about 2% smaller temperature). This is one of the reasons why early attempts to construct "cluster updates" (working with geometrical clusters) were not successful. In 2D, also geometrical clusters *do* percolate at T_c, but they are still not useful for algorithmic purposes because their fractal properties are not directly related to the *critical* behaviour of the thermodynamic system at hand. Rather they encode the properties of a *tricritical* point in a related model (the diluted $q = 1$ Potts model) [27]. For instance, while at the end of this section it will be shown that the average FK cluster size $\langle |C| \rangle_{\mathrm{FK}}$ is a so-called improved estimator for the Ising susceptibility, $\langle |C| \rangle_{\mathrm{FK}} = \chi'/\beta \propto L^{\gamma/\nu}$, and hence scales in 2D with the proper Ising model exponents $\gamma = 7/4 = 1.75$ and $\nu = 1$, one finds for the average geometrical cluster size $\langle |C| \rangle_{\mathrm{geo}} \propto L^{\gamma_{\mathrm{geo}}/\nu_{\mathrm{geo}}}$ with the exact exponent ratio $\gamma_{\mathrm{geo}}/\nu_{\mathrm{geo}} = 91/48 = 1.8958\ldots$. Note that both cluster quantities can be measured in the same MC simulation run [27].

Performance for the Ising Model

The advantage of cluster algorithms is most pronounced close to criticality where excitations on all length scales occur. A convenient performance measure is thus the dynamical critical exponent z (even though one should always check that the proportionality constant in $\tau \propto L^z$ is not exceedingly large, but this is definitely not the case here [95]). Some results on z are collected in Table 5.2, which allow us to conclude:

(1) Compared to local algorithms with $z \approx 2$, z is dramatically reduced for both cluster variants in 2D and 3D.
(2) In 2D, Swendsen-Wang and Wolff cluster updates are equally efficient, while in 3D, the Wolff update is clearly favourable.
(3) In 2D, the scaling with system size can hardly be distinguished from a very weak logarithmic scaling. Note that this is consistent with the Li-Sokal bound [96] for the Swendsen-Wang cluster algorithm of $\tau_{\mathrm{SW}} \geq C$ ($= C_0 + A \ln L$ for the 2D Ising model), implying $z_{\mathrm{SW}} \geq \alpha/\nu$ ($= 0$ for the 2D Ising model).
(4) Different observables (e.g., energy E and magnetization M) may yield quite different values for z when defined via the scaling behaviour of the integrated autocorrelation time.

Table 5.2. Dynamical critical exponents z for the 2D and 3D Ising model ($\tau \propto L^z$). The subscripts indicate the observables and method used ("exp" resp. "int": exponential resp. integrated autocorrelation time, "rel": relaxation, "dam": damage spreading)

Algorithm	2D	3D	Observable	Authors
Metropolis	2.1667(5)	–	$z_{M,\mathrm{exp}}$	Nightingale and Blöte [58]
	–	2.032(4)	z_{dam}	Grassberger [97]
	–	2.055(10)	$z_{M,\mathrm{exp}}$	Ito $et\ al.$ [98]
Swendsen-Wang cluster	0.35(1)	0.75(1)	$z_{E,\mathrm{exp}}$	Swendsen and Wang [88]
	0.27(2)	0.50(3)	$z_{E,\mathrm{int}}$	Wolff [95]
	0.20(2)	0.50(3)	$z_{\chi,\mathrm{int}}$	Wolff [95]
	$0(\log L)$	–	$z_{M,\mathrm{exp}}$	Heermann and Burkitt [99]
	0.25(5)	–	$z_{M,\mathrm{rel}}$	Tamayo [100]
Wolff cluster	0.26(2)	0.28(2)	$z_{E,\mathrm{int}}$	Wolff [95]
	0.13(2)	0.14(2)	$z_{\chi,\mathrm{int}}$	Wolff [95]
	0.25(5)	0.3(1)	$z_{E,\mathrm{rel}}$	Ito and Kohring [101]

Embedded Clusters

While it is quite easy to generalize the derivation (5.93)–(5.96) to q-state Potts models (because as in the Ising model each contribution to the energy, $\delta_{\sigma_i \sigma_j}$, can take only two different values), for O(n) spin models with Hamiltonian

$$H = -J \sum_{\langle ij \rangle} \boldsymbol{\sigma}_i \cdot \boldsymbol{\sigma}_j \; ; \quad \boldsymbol{\sigma}_i = (\sigma_{i,1}, \sigma_{i,2}, \ldots, \sigma_{i,n}) \; ; |\boldsymbol{\sigma}_i| = 1 \qquad (5.97)$$

one needs a new strategy for $n \geq 2$ [89, 102–104] (the case $n = 1$ degenerates again to the Ising model). Here the basic idea is to isolate Ising degrees of freedom by projecting the spins $\boldsymbol{\sigma}_i$ onto a randomly chosen unit vector $\boldsymbol{r}$,

$$\boldsymbol{\sigma}_i = \boldsymbol{\sigma}_i^{\parallel} + \boldsymbol{\sigma}_i^{\perp} \; ; \quad \boldsymbol{\sigma}_i^{\parallel} = \epsilon |\boldsymbol{\sigma}_i \cdot \boldsymbol{r}| \, \boldsymbol{r}; \quad \epsilon = \mathrm{sign}(\boldsymbol{\sigma}_i \cdot \boldsymbol{r}) \, . \qquad (5.98)$$

If this is inserted in the original Hamiltonian one ends up with an effective Hamiltonian

$$H = -\sum_{\langle ij \rangle} J_{ij} \epsilon_i \epsilon_j + \mathrm{const} \, , \qquad (5.99)$$

with positive random couplings $J_{ij} = J |\boldsymbol{\sigma}_i \cdot \boldsymbol{r}| |\boldsymbol{\sigma}_j \cdot \boldsymbol{r}| \geq 0$, whose Ising degrees of freedom ϵ_i can be updated with a cluster algorithm as described above.

Improved Estimators

A further advantage of cluster algorithms is that they lead quite naturally to so-called improved estimators which are designed to further reduce the statistical errors. Suppose we want to measure the expectation value $\langle \mathcal{O} \rangle$ of an observable $\mathcal{O}$. Then any estimator $\hat{\mathcal{O}}$ satisfying $\langle \hat{\mathcal{O}} \rangle = \langle \mathcal{O} \rangle$ is permissible. This does not determine $\hat{\mathcal{O}}$ uniquely since there are infinitely many other possible choices, $\hat{\mathcal{O}}' = \hat{\mathcal{O}} + \hat{\mathcal{X}}$, where the added estimator $\hat{\mathcal{X}}$ is assumed to have zero expectation, $\langle \hat{\mathcal{X}} \rangle = 0$. The variances of the estimators $\hat{\mathcal{O}}'$, however, can be quite different and are not necessarily related to any physical quantity (contrary to the standard mean-value estimator of the energy whose variance is proportional to the specific heat). It is exactly this freedom in the choice of $\hat{\mathcal{O}}$ which allows the construction of improved estimators.

For the single-cluster algorithm an improved "cluster estimator" for the spin-spin correlation function in the high-temperature phase, $G(\boldsymbol{x}_i - \boldsymbol{x}_j) \equiv \langle \boldsymbol{\sigma}_i \cdot \boldsymbol{\sigma}_j \rangle$, is given by [104]

$$\hat{G}(\boldsymbol{x}_i - \boldsymbol{x}_j) = n \frac{V}{|C|} \boldsymbol{r} \cdot \boldsymbol{\sigma}_i \, \boldsymbol{r} \cdot \boldsymbol{\sigma}_j \, \Theta_C(\boldsymbol{x}_i) \Theta_C(\boldsymbol{x}_j) \,, \tag{5.100}$$

where $\boldsymbol{r}$ is the normal of the mirror plane used in the construction of the cluster of size $|C|$ and $\Theta_C(\boldsymbol{x})$ is its characteristic function ($=1$ if $\boldsymbol{x} \in C$ and 0 otherwise). For the Fourier transform, $\tilde{G}(\boldsymbol{k}) = \sum_{\boldsymbol{x}} G(\boldsymbol{x}) \exp(-i\boldsymbol{k} \cdot \boldsymbol{x})$, this implies the improved estimator

$$\hat{\tilde{G}}(\boldsymbol{k}) = \frac{n}{|C|} \left[\left(\sum_{i \in C} \boldsymbol{r} \cdot \boldsymbol{\sigma}_i \cos \boldsymbol{k}\boldsymbol{x}_i \right)^2 + \left(\sum_{i \in C} \boldsymbol{r} \cdot \boldsymbol{\sigma}_i \sin \boldsymbol{k}\boldsymbol{x}_i \right)^2 \right] \,, \tag{5.101}$$

which, for $\boldsymbol{k} = \boldsymbol{0}$, reduces to an improved estimator for the susceptibility χ' in the high-temperature phase,

$$\hat{\tilde{G}}(\boldsymbol{0}) = \hat{\chi}'/\beta = \frac{n}{|C|} \left(\sum_{i \in C} \boldsymbol{r} \cdot \boldsymbol{\sigma}_i \right)^2 \,. \tag{5.102}$$

For the Ising model ($n = 1$) this reduces to $\chi'/\beta = \langle |C| \rangle$, i.e., the improved estimator of the susceptibility is just the average cluster size of the single-cluster update algorithm. For the XY and Heisenberg model one finds empirically that in two as well as in three dimensions $\langle |C| \rangle \approx 0.81 \chi'/\beta$ for $n = 2$ ([102, 108]) and $\langle |C| \rangle \approx 0.75 \chi'/\beta$ for $n = 3$ ([104, 109]), respectively.

It should be noted that by means of the estimators (5.100)–(5.102) a significant reduction of variance should only be expected outside the FSS region where the average cluster size is small compared to the volume of the system.

5.7 Reweighting Techniques

Even though the physics underlying reweighting techniques [110, 111] is extremely simple and the basic idea has been known since long (see the list of references in Ref. [111]), their power in practice has been realized only relatively late in 1988. The important observation by Ferrenberg and Swendsen [110,111] was that the best performance is achieved *near* criticality where histograms are usually broad. In this sense reweighting techniques are complementary to improved estimators.

5.7.1 Single-Histogram Technique

The single-histogram reweighting technique [110] is based on the following very simple observation. If we denote the number of states (spin configurations) that have the same energy E by $\Omega(E)$, the partition function at the simulation point $\beta_0 = 1/k_B T_0$ can always be written as[7]

$$Z(\beta_0) = \sum_{\{s\}} e^{-\beta_0 H(\{s\})} = \sum_E \Omega(E) e^{-\beta_0 E} \propto \sum_E P_{\beta_0}(E) \ , \qquad (5.103)$$

where we have introduced the unnormalized energy histogram (density)

$$P_{\beta_0}(E) \propto \Omega(E) e^{-\beta_0 E} \ . \qquad (5.104)$$

If we would normalize $P_{\beta_0}(E)$ to unit area, the r.h.s. would have to be divided by $\sum_E P_{\beta_0}(E) = Z(\beta_0)$, but the normalization will be unimportant in what follows. Let us assume we have performed a Monte Carlo simulation at inverse temperature β_0 and thus know $P_{\beta_0}(E)$. It is then easy to see that

$$P_\beta(E) \propto \Omega(E) e^{-\beta E} = \Omega(E) e^{-\beta_0 E} e^{-(\beta-\beta_0)E} \propto P_{\beta_0}(E) e^{-(\beta-\beta_0)E} \ , \quad (5.105)$$

i.e., the histogram at any point β can be derived, in principle, by *reweighting* the simulated histogram at β_0 with the exponential factor $\exp[-(\beta - \beta_0)E]$. Notice that in reweighted expectation values,

$$\langle f(E) \rangle(\beta) = \sum_E f(E) P_\beta(E) / \sum_E P_\beta(E) \ , \qquad (5.106)$$

the normalization of $P_\beta(E)$ indeed cancels. This gives for instance the energy $\langle e \rangle(\beta) = \langle E \rangle(\beta)/V$ and the specific heat $C(\beta) = \beta^2 V[\langle e^2 \rangle(\beta) - \langle e \rangle(\beta)^2]$, in principle, as a continuous function of β from a single MC simulation at β_0, where $V = L^d$ is the system size.

As an example of this reweighting procedure, using actual Swendsen-Wang cluster simulation data (with 5000 sweeps for equilibration and 50 000 sweeps for measurements) of the 2D Ising model at $\beta_0 = \beta_c = \ln(1 + \sqrt{2})/2 = 0.440\,686\ldots$ on a 16×16 lattice with periodic boundary conditions, the specific heat $C(\beta)$ is shown in Fig. 5.14(a) and compared with the curve obtained from the exact Kaufman solution [33,34] for finite $L_x \times L_y$ lattices. This clearly

---

[7] For simplicity we consider here only models with *discrete* energies. If the energy varies continuously, sums have to be replaced by integrals, etc. Also lattice size dependences are suppressed to keep the notation short.

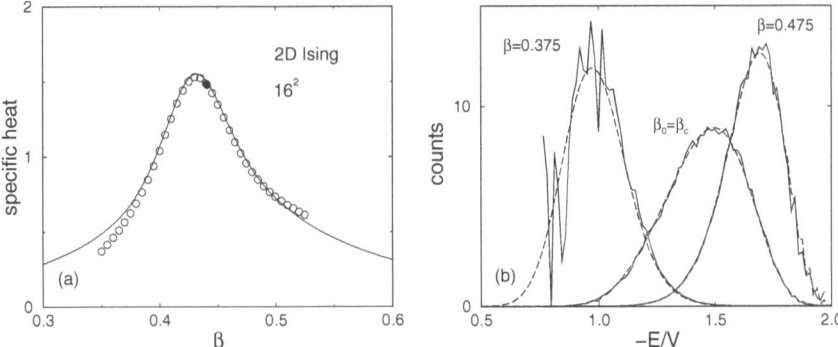

Fig. 5.14. (a) The specific heat of the 2D Ising model on a 16×16 square lattice computed by reweighting from a single MC simulation at $\beta_0 = \beta_c$, marked by the filled data symbol. The continuous line shows for comparison the exact solution of Kaufman [33,34]. (b) The corresponding energy histogram at β_0, and reweighted to $\beta = 0.375$ and $\beta = 0.475$. The *dashed lines* show for comparison the exact histograms obtained from Beale's [112] expression

demonstrates that, in practice, the β-range over which reweighting can be trusted is limited. The reason for this limitation are unavoidable statistical errors in the numerical determination of P_{β_0} using a MC simulation. In the tails of the histograms the relative statistical errors are largest, and the tails are exactly the regions that contribute most when multiplying $P_{\beta_0}(E)$ with the exponential reweighting factor to obtain $P_\beta(E)$ for β values far off the simulation point β_0. This is illustrated in Fig. 5.14(b) where the simulated histogram at $\beta_0 = \beta_c$ is shown together with the reweighted histograms at $\beta = 0.375 \approx \beta_0 - 0.065$ and $\beta = 0.475 \approx \beta_0 + 0.035$, respectively. Here the quality of the histograms can be judged by comparing with the curves obtained from Beale's [112] exact expression for $\Omega(E)$.

As a rule of thumb, the range over which reweighting should produce accurate results can be estimated by requiring that the peak location of the reweighted histogram should not exceed the energy value at which the input histogram had decreased to about one half or one third of its maximum value. In most applications this range is wide enough to locate from a single simulation, e.g., the specific-heat maximum by employing standard maximization routines to the continuous function $C(\beta)$. This is by far more convenient, accurate and faster than the traditional way of performing many simulations close to the peak of $C(\beta)$ and trying to determine the maximum by spline or least-squares fits.

For an analytical estimate of the reweighting range we now require that the peak of the reweighted histogram is within the width $\langle e \rangle(T_0) \pm \Delta e(T_0)$ of the input histogram (where a Gaussian histogram would have decreased to $\exp(-1/2) \approx 0.61$ of its the maximum value),

$$|\langle e \rangle(T) - \langle e \rangle(T_0)| \le \Delta e(T_0) \,, \tag{5.107}$$

where we have made use of the fact that for a not too asymmetric histogram $P_{\beta_0}(E)$ the maximum location approximately coincides with $\langle e \rangle(T_0)$. Recalling that the half width Δe of a histogram is related to the specific heat via $(\Delta e)^2 \equiv \langle (e - \langle e \rangle)^2 \rangle = \langle e^2 \rangle - \langle e \rangle^2 = C(\beta_0)/\beta_0^2 V$ and using the Taylor expansion $\langle e \rangle(T) = \langle e \rangle(T_0) + C(T_0)(T - T_0) + \ldots$, this can be written as $C(T_0)|T - T_0| \le T_0 \sqrt{C(T_0)/V}$ or

$$\frac{|T - T_0|}{T_0} \le \frac{1}{\sqrt{V}} \frac{1}{C(T_0)} \,. \tag{5.108}$$

Since $C(T_0)$ is known from the input histogram this is quite a general estimate of the reweighting range. For the example in Fig. 5.14 with $V = 16 \times 16$, $\beta_0 = \beta_c \approx 0.44$ and $C(T_0) \approx 1.5$, this estimate yields $|\beta - \beta_0|/\beta_0 \approx |T - T_0|/T_0 \le 0.04$, i.e., $|\beta - \beta_0| \le 0.02$ or $0.42 \le \beta \le 0.46$. By comparison with the exact solution we see that this is indeed a fairly conservative estimate of the reliable reweighting range.

If we only want to know the scaling behaviour with system size $V = L^D$, we can go one step further by considering three generic cases:

i)   *Off-critical*, where $C(T_0) \approx$ const., such that

$$\frac{|T - T_0|}{T_0} \propto V^{-1/2} = L^{-D/2} \,. \tag{5.109}$$

ii) *Critical*, where $C(T_0) \simeq a_1 + a_2 L^{\alpha/\nu}$, with a_1 and a_2 being constants, and α and ν denoting the standard critical exponents of the specific heat and correlation length, respectively. For $\alpha > 0$, the leading scaling behaviour becomes $|T - T_0|/T_0 \propto L^{-D/2} L^{-\alpha/2\nu}$. Assuming hyperscaling ($\alpha = 2 - D\nu$) to be valid, this simplifies to

$$\frac{|T - T_0|}{T_0} \propto L^{-1/\nu} \,, \tag{5.110}$$

i.e., the typical scaling behaviour of pseudo-transition temperatures in the finite-size scaling regime of a second-order phase transition [113]. For $\alpha < 0$, $C(T_0)$ approaches asymptotically a constant and the leading scaling behaviour of the reweighting range is as in the off-critical case.

iii) *First-order transitions*, where $C(T_0) \propto V$. This yields

$$\frac{|T - T_0|}{T_0} \propto V^{-1} = L^{-D} \,, \tag{5.111}$$

which is again the typical finite-size scaling behaviour of pseudo-transition temperatures close to a first-order phase transition [16].

If we also want to reweight other quantities such as the magnetization $\langle m \rangle$ we have to go one step further. The conceptually simplest way would be to store two-dimensional histograms $P_{\beta_0}(E, M)$ where $M = Vm$ is the total magnetization. We could then proceed in close analogy to the preceding case, and even reweighting to non-zero magnetic field h would be possible, which enters via the Boltzmann factor $\exp(\beta h \sum_i s_i) = \exp(\beta h M)$. However, the storage requirements may be quite high (of the order of V^2), and it is often preferable to proceed in the following way. For any function $g(M)$, e.g., $g(M) = M^k$, we can write

$$\langle g(M) \rangle = \sum_{\{s\}} g(M(\{s\})) e^{-\beta_0 H}/Z(\beta_0) = \sum_{E,M} \Omega(E, M) g(M) e^{-\beta_0 E}/Z(\beta_0)$$

$$= \sum_E \frac{\sum_M \Omega(E, M) g(M)}{\sum_M \Omega(E, M)} \sum_M \Omega(E, M) e^{-\beta_0 E}/Z(\beta_0) \ . \qquad (5.112)$$

Recalling that $\sum_M \Omega(E, M) e^{-\beta_0 E}/Z(\beta_0) = \Omega(E) e^{-\beta_0 E}/Z(\beta_0) = P_{\beta_0}(E)$ and defining the *microcanonical* expectation value of $g(M)$ at fixed energy E (sometimes denoted as a "list"),

$$\langle\langle g(M) \rangle\rangle(E) \equiv \frac{\sum_M \Omega(E, M) g(M)}{\sum_M \Omega(E, M)} \ , \qquad (5.113)$$

we arrive at

$$\langle g(M) \rangle = \sum_E \langle\langle g(M) \rangle\rangle(E) P_{\beta_0}(E) \ . \qquad (5.114)$$

Identifying $\langle\langle g(M) \rangle\rangle(E)$ with $f(E)$ in Eq. (5.106), the actual reweighting procedure is precisely as before. Mixed quantities, e.g. $\langle E^k M^l \rangle$, can be treated similarly. One caveat of this method is that one has to decide beforehand which "lists" $\langle\langle g(M) \rangle\rangle(E)$ one wants to store during the simulation, e.g., which powers k in $\langle\langle M^k \rangle\rangle(E)$ are relevant. An example for computing $\langle\langle |M| \rangle\rangle(E)$ and $\langle\langle M^2 \rangle\rangle(E)$ using the data of Fig. 5.14 is shown in Fig. 5.15.

An alternative and more flexible method is based on time series. Suppose we have performed a MC simulation at β_0 and stored the time series of N measurements $E_1, E_2, \ldots, E_N$ and $M_1, M_2, \ldots, M_N$. Then the most general expectation values at another inverse temperature β can simply be obtained from

$$\langle f(E, M) \rangle = \sum_{i=1}^{N} f(E_i, M_i) e^{-(\beta-\beta_0)E_i} / \sum_{i=1}^{N} e^{-(\beta-\beta_0)E_i} \ , \qquad (5.115)$$

i.e., in particular all moments $\langle E^k M^l \rangle$ can be computed. Notice that this can also be written as

$$\langle f(E, M) \rangle = \langle f(E, M) e^{-(\beta-\beta_0)E} \rangle_0 / \langle e^{-(\beta-\beta_0)E} \rangle_0 \ , \qquad (5.116)$$

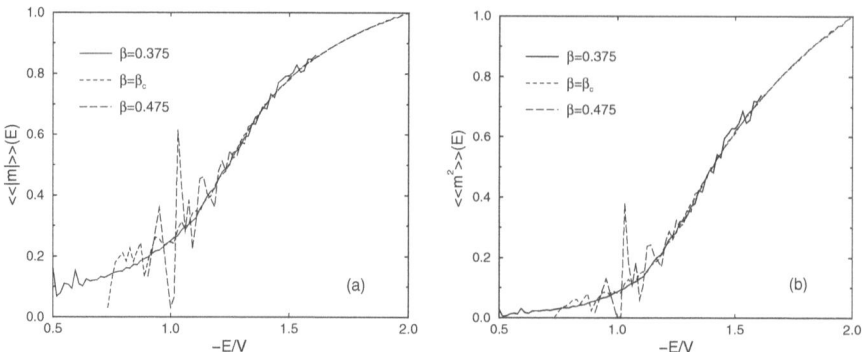

Fig. 5.15. Microcanonical expectation values for (**a**) the absolute magnetization and (**b**) the magnetization squared obtained from the 2D Ising model simulations shown in Fig. 5.14

where the subscript 0 refers to expectation values taken at β_0. Another very important advantage of the last formulation is that it works without any systematic discretization error also for continuously distributed energies and magnetizations.

As nowadays hard-disk space is no real limitation anymore, it is advisable to store time series in any case. This guarantees the greatest flexibility in the data analysis. As far as the memory requirement of the actual reweighting code is concerned, however, the method of choice is sometimes not so clear. Using directly histograms and lists, one typically has to store about $(6 - 8)V$ data, while working directly with the time series one needs $2N$ computer words. The cheaper solution (also in terms of CPU time) thus obviously depends on both, the system size V and the run length N. It is hence sometimes faster to generate from the time series first histograms and the required lists and then proceed with reweighting the latter quantities.

5.7.2 Multi-Histogram Technique

The basic idea of the multi-histogram technique [114] can be summarized as follows:

i)   Perform m MC simulations at $\beta_1, \beta_2, \ldots, \beta_m$ with N_i, $i = 1, \ldots, m$, measurements,
ii) reweight all runs to a common reference point β_0,
iii) combine at β_0 all information by computing error weighted averages,
iv) reweight the "combined histogram" to any other β.

Here we shall assume that the histograms $P_{\beta_i}(E)$ are "naturally" normalized, $\sum_E P_{\beta_i}(E) = N_i$, such that the statistical errors for each of the histograms $P_{\beta_i}(E)$ are approximately given by $\sqrt{P_{\beta_i}(E)}$. By choosing as reference point $\beta_0 = 0$ and working out the error weighted combined histogram one ends up with

$$\Omega(E) = \frac{\sum_{i=1}^{m} P_{\beta_i}(E)}{\sum_{i=1}^{m} N_i Z_i^{-1} e^{-\beta_i E}} \; , \tag{5.117}$$

where the unknown partition function values $Z_i \equiv Z(\beta_i)$ are determined self-consistently from

$$Z_i = \sum_{E} \Omega(E) e^{-\beta_i E} = \sum_{E} e^{-\beta_i E} \frac{\sum_{k=1}^{m} P_{\beta_k}(E)}{\sum_{k=1}^{m} N_k Z_k^{-1} e^{-\beta_k E}} \; , \tag{5.118}$$

up to an unimportant overall constant. A good starting point for the recursion is to fix, say, $Z_1 = 1$ and use single histogram reweighting to get an estimate of $Z_2/Z_1 = \exp[-(\hat{F}_2 - \hat{F}_1)]$, where $\hat{F}_i = \beta_i F(\beta_i)$. Once Z_2 is determined, the same procedure can be applied to estimate Z_3 and so on. In the limit of infinite statistics, this would already yield the solution of (5.118). In realistic simulations the statistics is of course limited and the (very few) remaining recursions average this uncertainty to get a self-consistent set of Z_i. In order to work in practice, the histograms at neighbouring β-values must have sufficient overlap, i.e., the spacings of the simulation points must be chosen according to the estimates (5.109)–(5.111).

Multiple-histogram reweighting has been widely applied in many different applications. Some problems of this method are that autocorrelations cannot properly be taken into account when computing the error weighted average (which is still correct but no longer optimized), the procedure for computing mixed quantities such as $\langle E^k M^l \rangle$ is difficult to justify (even though it does work as an "ad hoc" prescription quite well), and the statistical error analysis becomes quite cumbersome.

As an alternative one may compute by reweighting from each of the m simulations all quantities of interest as a function of β, including their statistical error bars which now also should take care of autocorrelations as discussed in Subsect. 5.5.3. In this way one obtains, at each β-value, m estimates, e.g. $e_1(\beta) \pm \Delta e_1, e_2(\beta) \pm \Delta e_2, \ldots, e_m(\beta) \pm \Delta e_m$, which may be optimally combined according to their error bars to give $e(\beta) \pm \Delta e$. If the relative error $\Delta e/e(\beta)$ is minimized, this leads to [109]

$$e(\beta) = \left(\frac{e_1(\beta)}{(\Delta e_1)^2} + \frac{e_2(\beta)}{(\Delta e_2)^2} + \cdots + \frac{e_m(\beta)}{(\Delta e_m)^2} \right) (\Delta e)^2 \; , \tag{5.119}$$

with

$$\frac{1}{(\Delta e)^2} = \frac{1}{(\Delta e_1)^2} + \frac{1}{(\Delta e_2)^2} + \cdots + \frac{1}{(\Delta e_m)^2} \; . \tag{5.120}$$

Notice that in this way the average for each quantity can be individually optimized.

5.8 Tempering Methods

Loosely speaking, tempering methods may be characterized as "dynamical multi-histogramming". Similarly to the static reweighting approach, in "simulated" as well as in "parallel" tempering one considers m simulation points $\beta_1 < \beta_2 < \cdots < \beta_m$ which here, however, are combined already during the simulation in a specific, dynamical way.

5.8.1 Simulated Tempering

In simulated tempering simulations [115,116] one starts from a joint partition function (expanded ensemble)

$$Z_{\mathrm{ST}} = \sum_{i=1}^{m} e^{g_i} \sum_{\{s\}} e^{-\beta_i H(\{s\})} , \tag{5.121}$$

where $g_i = \beta_i f(\beta_i)$ and the inverse temperature β is treated as an additional dynamical degree of freedom that can take the values $\beta_1, \ldots, \beta_m$. Employing a Metropolis algorithm, a proposed move from $\beta = \beta_i$ to β_j takes place with probability $\min[1, \exp[-(\beta_j - \beta_i)H(\{s\})] + g_j - g_i]$. Similar to multi-histogram reweighting (and also to multicanonical simulations), the free-energy parameters g_i are a priori unknown and have to be adjusted iteratively. To assure a reasonable acceptance rate for the β-update moves (usually between neighbouring β_i-values), the histograms at β_i and β_{i+1}, $i = 1, \ldots, m-1$, must overlap. An estimate for a suitable spacing $\delta\beta = \beta_{i+1} - \beta_i$ of the simulation points β_i is hence immediately given by the results (5.109)–(5.111) for the reweighting range,

$$\delta\beta \propto \begin{cases} L^{-D/2} & \text{off-critical ,} \\ L^{-1/\nu} & \text{critical ,} \\ L^{-D} & \text{first-order .} \end{cases} \tag{5.122}$$

Overall the simulated tempering method shows some similarities to the "avoiding rare events" variant of multicanonical simulations briefly discussed in the next section.

5.8.2 Parallel Tempering

In parallel tempering (exchange Monte Carlo, multiple Markov chain Monte Carlo) simulations [117, 118] the starting point is the product of partition functions (extended ensemble),

$$Z_{\mathrm{PT}} = \prod_{i=1}^{m} Z(\beta_i) = \prod_{i=1}^{m} \sum_{\{s\}_i} e^{-\beta_i H(\{s\}_i)} , \tag{5.123}$$

and all m systems at different simulation points $\beta_1 < \beta_2 < \cdots < \beta_m$ are simulated in parallel, using any legitimate update algorithm (Metropolis, cluster,...). This freedom in the choice of update algorithm is a big advantage of the parallel tempering method. After a certain number of sweeps, exchanges of the current configurations $\{s\}_i$ and $\{s\}_j$ are attempted (equivalently, the β_i may be exchanged, as is done in most implementations). Adapting the Metropolis criterion (5.34) to the present situation, the proposed exchange will be accepted with probability $W = \min(1, e^\Delta)$, where $\Delta = (\beta_j - \beta_i)[E(\{s\}_j) - E(\{s\}_i)]$. To assure a reasonable acceptance rate, usually only "nearest-neighbour" exchanges ($j = i \pm 1$) are attempted and the β_i should again be spaced with the $\delta\beta$ given in (5.122). In most applications, the smallest inverse temperature β_1 is chosen in the high-temperature phase where the autocorrelation time is expected to be very short and the system rapidly decorrelates. Conceptually this approach follows again the "avoiding rare events" strategy.

Notice that in parallel tempering no free-energy parameters must be adjusted. The method is thus very flexible and moreover can be almost trivially parallelized.

5.9 Multicanonical Ensembles

To conclude this introduction to simulation techniques, at least a very brief outline of multicanonical ensembles shall be given. For more details, in particular on practical implementations, see the recent reviews [4, 119–122]. Similar to the tempering methods of the last section, multicanonical simulations may also be interpreted as a dynamical multi-histogram reweighting method. This interpretation is stressed by the notation used in the original papers by Berg and Neuhaus [123, 124] and explains the name "*multi*canonical". At the same time, this method may also be viewed as a specific realization of non-Boltzmann sampling [125] which has been known since long to be a legitimate alternative to the more standard MC approaches [126]. The practical significance of non-Boltzmann sampling was first realized in the so-called "umbrella sampling" method [127], but it took many years before the introduction of the multicanonical ensemble [123, 124] turned non-Boltzmann sampling into a widely appreciated practical tool in computer simulation studies of phase transitions. Once the feasibility of such a generalized ensemble approach was realized, many related methods and further refinements were developed.

Conceptually the method can be divided into two main strategies. The first strategy can be best described as "avoiding rare events" which is close in spirit to the alternative tempering methods. In this variant one tries to connect the important parts of phase space by "easy paths" which go around suppressed rare-event regions which hence cannot be studied directly. The second approach is based on "enhancing the probability of rare event states", which is for example the typical strategy for dealing with the highly suppressed mixed-phase region of first-order phase transitions [16,122]. This allows a direct study of properties of the rare-event states such as, e.g., interface tensions or more generally free energy barriers, which would be very difficult (or practically impossible) with canonical simulations and also with the tempering methods discussed in Sect. 5.8.

In both multicanonical versions, the canonical Boltzmann distribution

$$\mathcal{P}_{\mathrm{can}}(\phi) \propto \exp(-\beta H(\phi)) \tag{5.124}$$

is replaced by an auxiliary distribution

$$\mathcal{P}_{\mathrm{muca}}(\phi) \propto W(\{Q_i\})\exp(-\beta H(\phi)) \equiv \exp(-\beta H(\phi) - f(\{Q_i(\phi)\})) , \tag{5.125}$$

where ϕ denotes generically the degrees of freedom and Q_i stands for a macroscopic observable such as the energy or magnetization. With a suitably chosen reweighting factor $W(\{Q_i\})$, the probability distribution $P_{\mathrm{muca}}(\{Q_i\})$ of the macroscopic variables $\{Q_i\}$ can be tuned to take any desired form. Canonical expectation values can always be recovered exactly by inverse reweighting,

$$\langle \mathcal{O} \rangle_{\mathrm{can}} = \langle \mathcal{O}W^{-1}(\{Q_i\}) \rangle_{\mathrm{muca}} / \langle W^{-1}(\{Q_i\}) \rangle_{\mathrm{muca}} , \tag{5.126}$$

similar to Eq. (5.116).

The Monte Carlo sampling of $\mathcal{P}_{\mathrm{muca}}(\phi)$ proceeds in the usual way by comparing $\beta H(\phi) + f(\{Q_i(\phi)\})$ before and after a proposed update move of ϕ. In most applications local update algorithms have been employed, but for certain classes of models also non-local multigrid methods are applicable [84,128]. A combination with non-local cluster update algorithms, on the other hand, is not straightforward. Only by making direct use of the random-cluster representation as a starting point, a multibondic variant [129–131] has been developed.

The performance of the simulation depends, however, in the first place on the choice of $\{Q_i\}$ and the reweighting factor $W(\{Q_i\})$, since for instance in the special case $W \equiv 1$ the troublesome canonical ensemble is recovered. The proper identification of the relevant set of Q_i's requires considerable physical intuition and insight into the specific system under study. While for disordered complex systems this may be a serious problem, in studies of first-order phase transitions the proper choice is clear since typically the energy E (temperature-driven transition) or magnetization M (field-driven transition) are the relevant variables. In both cases, the reweighting factor is usually chosen such that the multicanonical probability density $P_{\mathrm{muca}} = WP_{\mathrm{can}}$ is approximately flat between the two peaks of the canonical distribution. The most important technical point is the procedure for constructing the multicanonical weights, for which iterative procedures have been developed [4,119–122].

If P_{muca} was completely flat and the MC update moves would perform an ideal random walk, one would expect that after V^2 local updates the system has travelled on average a distance V in energy or magnetization. Since one lattice sweep consists of V local updates, the autocorrelation time should scale in this idealized picture as $\tau \propto V$. Numerical tests for various models with a first-order phase transition have shown that in practice the data are at best consistent with a behaviour $\tau \propto V^\alpha$, with $\alpha \geq 1$. While for the temperature-driven transitions of 2D Potts models the multibondic variant seems to saturate the bound [129–131], employing local update algorithms, typical fit results are $\alpha \approx 1.1-1.3$, and due to the limited accuracy of the data even a weak exponential growth cannot really be excluded. In fact, at least for the field-driven first-order transition of the 2D Ising model, it has been demonstrated recently [132, 133] that even for a perfectly flat multicanonical distribution a "hidden" free energy nucleation barrier leads to an exponential growth of τ, which is, however, much weaker than in the corresponding canonical simulation.

5.10 Concluding Remarks

The intention of these lecture notes was to give an elementary introduction to the basic concepts of modern Monte Carlo simulations and to illustrate their usefulness by applications to the very simple Ising lattice spin model. The basic Monte Carlo methods based on local update rules are straightforward to apply to all models with discrete degrees of freedom and with some extra care also to continuous variables and off-lattice models. Some generalizations of cluster update methods have already been indicated. Also other models may be efficiently simulated by this non-local method, but there is no guarantee that for a given model a cluster update procedure can be developed. The statistical error analysis part is obviously completely general, and also reweighting, tempering and multicanonical methods can be adapted to almost every problem at hand.

Acknowledgements

Many people have influenced these lecture notes with their advice, questions, discussions and active contributions. In particular I wish to thank Michael Bachmann, Bertrand Berche, Pierre-Emmanuel Berche, Bernd A. Berg, Alain Billoire, Kurt Binder, Elmar Bittner, Christophe Chatelain, Malte Henkel, Desmond A. Johnston, Ralph Kenna, David P. Landau, Eric Lorenz, Thomas Neuhaus, Andreas Nußbaumer, Michel Pleimling, Adriaan Schakel, and Martin Weigel for sharing their insight and knowledge with me.

This work was partially supported by the Deutsche Forschungsgemeinschaft (DFG) under grant Nos. JA 483/22-1 and JA 483/23-1, and the EU RTN-Network "ENRAGE": *Random Geometry and Random Matrices: From Quantum Gravity to Econophysics* under grant No. MRTN-CT-2004-005616.

References

1. M.E.J. Newman, G.T. Barkema: *Monte Carlo Methods in Statistical Physics* (Clarendon Press, Oxford, 1999)
2. D.P. Landau, K. Binder: *Monte Carlo Simulations in Statistical Physics* (Cambridge University Press, Cambridge, 2000)
3. K. Binder, D.W. Heermann: *Monte Carlo Simulations in Statistical Physics: An Introduction*, 4th edition (Springer, Berlin, 2002)
4. B.A. Berg: *Markov Chain Monte Carlo Simulations and Their Statistical Analysis* (World Scientific, Singapore, 2004)
5. W. Lenz: Phys. Z. **21**, 613 (1920); E. Ising: Z. Phys. **31**, 253 (1925)
6. L. Onsager: Phys. Rev. **65**, 117 (1944)
7. B.M. McCoy, T.T. Wu: *The Two-Dimensional Ising Model* (Harvard University Press, Cambridge, 1973)
8. R.J. Baxter: *Exactly Solved Models in Statistical Mechanics* (Academic Press, New York, 1982)
9. L. Onsager: Nuovo Cimento (Suppl.) **6**, 261 (1949); see also the historical remarks in Refs. [7,8]
10. C.N. Yang: Phys. Rev. **85**, 808 (1952)
11. C.H. Chang: Phys. Rev. **88**, 1422 (1952)
12. W.P. Orrick, B.G. Nickel, A.J. Guttmann, J.H. Perk: Phys. Rev. Lett. **86**, 4120 (2001); J. Stat. Phys. **102**, 795 (2001)
13. R.B. Potts: Proc. Camb. Phil. Soc. **48**, 106 (1952)
14. F.Y. Wu: Rev. Mod. Phys. **54**, 235 (1982)
15. F.Y. Wu: Rev. Mod. Phys. **55**, 315(E) (1983)
16. W. Janke: *First-Order Phase Transitions*, in: *Computer Simulations of Surfaces and Interfaces*, NATO Science Series, II. Mathematics, Physics and Chemistry – Vol. **114**, Proceedings of the NATO Advanced Study Institute, Albena, Bulgaria, 9–20 September 2002, eds. B. Dünweg, D.P. Landau, A.I. Milchev (Kluwer, Dordrecht, 2003), pp. 111–135
17. H.E. Stanley: *Introduction to Phase Transitions and Critical Phenomena* (Oxford Press, Oxford, 1979)
18. J.J. Binney, N.J. Dowrick, A.J. Fisher, M.E.J. Newman: *The Theory of Critical Phenomena* (Oxford University Press, Oxford, 1992)
19. D.A. Lavis, G.M. Bell: *Statistical Mechanics of Lattice Systems 2* (Springer, Berlin, 1999)
20. See the volumes of review articles edited by C. Domb, J.L. Lebowitz (eds.): *Phase Transitions and Critical Phenomena* (Academic Press, New York)
21. J.D. Gunton, M.S. Miguel, P.S. Sahni: in: *Phase Transitions and Critical Phenomena*, Vol. **8**, eds. C. Domb, J.L. Lebowitz (Academic Press, New York, 1983)
22. K. Binder: Rep. Prog. Phys. **50**, 783 (1987)
23. H.J. Herrmann, W. Janke, F. Karsch (eds.): *Dynamics of First Order Phase Transitions* (World Scientific, Singapore, 1992)
24. W. Janke: *Recent Developments in Monte Carlo Simulations of First-Order Phase Transitions*, in: *Computer Simulations in Condensed Matter Physics VII*, eds. D.P. Landau, K.K. Mon, H.-B. Schüttler (Springer, Berlin, 1994), p. 29
25. H. Kleinert: *Gauge Fields in Condensed Matter*, Vol. II (World Scientific, Singapore, 1989)

26. W. Janke, H. Kleinert: Phys. Rev. **B33**, 6346 (1986)
27. W. Janke, A.M.J. Schakel: Nucl. Phys. **B700**, 385 (2004); Comp. Phys. Comm. **169**, 222 (2005); Phys. Rev. **E71**, 036703 (2005); Phys. Rev. Lett. **95**, 135702 (2005); and e-print cond-mat/0508734. See also the extensive list of references to earlier work given therein
28. M. Weigel, W. Janke: Phys. Rev. **B62**, 6343 (2000)
29. K. Binder: in: *Monte Carlo Methods in Statistical Physics*, ed. K. Binder (Springer, Berlin, 1979)
30. M.E. Barber: in: *Phase Transitions and Critical Phenomena*, Vol. **8**, eds. C. Domb, J.L. Lebowitz (Academic Press, New York, 1983), p. 146
31. V. Privman (ed.): *Finite-Size Scaling and Numerical Simulations of Statistical Systems* (World Scientific, Singapore, 1990)
32. K. Binder: in: *Computational Methods in Field Theory*, Schladming Lecture Notes, eds. H. Gausterer, C.B. Lang (Springer, Berlin, 1992), p. 59
33. B. Kaufman: Phys. Rev. **76**, 1232 (1949)
34. A.E. Ferdinand, M.E. Fisher: Phys. Rev. **185**, 832 (1969)
35. M.E. Fisher, A.N. Berker: Phys. Rev. **B26**, 2507 (1982)
36. V. Privman, M.E. Fisher: J. Stat. Phys. **33**, 385 (1983)
37. K. Binder, D.P. Landau: Phys. Rev. **B30**, 1477 (1984)
38. M.S.S. Challa, D.P. Landau, K. Binder: Phys. Rev. **B34**, 1841 (1986)
39. V. Privman, J. Rudnik: J. Stat. Phys. **60**, 551 (1990)
40. C. Borgs, R. Kotecky: J. Stat. Phys. **61**, 79 (1990)
41. J. Lee, J.M. Kosterlitz: Phys. Rev. Lett. **65**, 137 (1990)
42. C. Borgs, R. Kotecky, S. Miracle-Solé: J. Stat. Phys. **62**, 529 (1991)
43. C. Borgs, W. Janke: Phys. Rev. Lett. **68**, 1738 (1992)
44. W. Janke: Phys. Rev. **B47**, 14757 (1993)
45. J.M. Hammersley, D.C. Handscomb: *Monte Carlo Methods* (London, 1965)
46. D.W. Heermann: *Computer Simulation Methods in Theoretical Physics*, 2nd ed., (Springer, Berlin, 1990)
47. K. Binder (ed.): *The Monte Carlo Method in Condensed Matter Physics* (Springer, Berlin, 1992)
48. N. Metropolis, A.W. Rosenbluth, M.N. Rosenbluth, A.H. Teller, E. Teller: J. Chem. Phys. **21**, 1087 (1953)
49. S. Kirkpatrick, C.D. Gelatt Jr., M.P. Vecchi: Science **220**, 671 (1983)
50. W. Janke: *Pseudo Random Numbers: Generation and Quality Checks*, invited lecture notes, in: Proceedings of the Euro Winter School *Quantum Simulations of Complex Many-Body Systems: From Theory to Algorithms*, eds. J. Grotendorst, D. Marx, A. Muramatsu, John von Neumann Institute for Computing, Jülich, NIC Series, Vol. **10**, pp. 447–458 (2002), and references therein
51. M. Creutz: Phys. Rev. **D36**, 515 (1987)
52. S.L. Adler: Phys. Rev. **D37**, 458 (1988)
53. H. Neuberger: Phys. Lett. **B207**, 461 (1988)
54. R. Gupta, J. DeLapp, G.G. Battrouni, G.C. Fox, C.F. Baillie, J. Apostolakis: Phys. Rev. Lett. **61**, 1996 (1988)
55. R.J. Glauber: J. Math. Phys. **4**, 294 (1963)
56. A.D. Rutenberg, A.J. Bray: Phys. Rev. **E51**, 5499 (1995)
57. P. Calabrese, A. Gambassi: J. Phys. **A38**, R133–R193 (2005)
58. M.P. Nightingale, H.W.J. Blöte: Phys. Rev. Lett. **76**, 4548 (1996); Phys. Rev. **B62**, 1089 (2000)

59. A. Barrat: Phys. Rev. **E57**, 3629 (1998)
60. C. Godrèche, J.-M. Luck: J. Phys.: Condens. Matter **14**, 1589 (2002)
61. L.F. Cugliandolo: in *Slow Relaxation and Non Equilibrium Dynamics in Condensed Matter*, Les Houches Lectures, eds. J.-L. Barrat, J. Dalibard, J. Kurchan, M.V. Feigel'man (Springer, Berlin, 2003)
62. F. Corberi, E. Lippiello, M. Zannetti: Phys. Rev. Lett. **90**, 099601 (2003) [Comment]; M. Henkel, M. Pleimling: Phys. Rev. Lett. **90**, 099602 (2003) [Reply]
63. L. Berthier, J.L. Barrat, J. Kurchan: Eur. Phys. J. **B11**, 635 (1999)
64. F. Corberi, E. Lippiello, M. Zannetti: Eur. Phys. J. **B24**, 359 (2001); Phys. Rev. **E65**, 046136 (2003)
65. M. Henkel: *Conformal Invariance and Critical Phenomena* (Springer, Berlin, 1999)
66. M. Henkel, M. Pleimling, C. Godrèche, J.-M. Luck: Phys. Rev. Lett. **87**, 265701 (2001)
67. M. Henkel: Nucl. Phys. **B641**, 405 (2002)
68. W. Zippold, R. Kühn, H. Horner: Eur. Phys. J. **B13**, 531 (2000)
69. M. Henkel, M. Paessens, M. Pleimling: Europhys. Lett. **62**, 664 (2003)
70. M. Henkel, M. Pleimling: Phys. Rev. **E68**, 065101 (R) (2003)
71. M. Henkel, A. Picone, M. Pleimling: Europhys. Lett. **68**, 191 (2004)
72. F. Bagnoli: J. Stat. Phys. **85**, 151 (1996)
73. T. Vojta: Phys. Rev. **E55**, 5157 (1997)
74. H. Hinrichsen, E. Domany: Phys. Rev. **E56**, 94 (1997)
75. E. Lorenz: Diploma thesis, Universität Leipzig (2005) [www.physik.uni-leipzig.de/~lorenz/diplom.pdf]
76. E. Lorenz, W. Janke: *Numerical Tests of Local Scale Invariance in Ageing q-State Potts Models*, Leipzig preprint (2006), EPL **77**, 10003 (2007)
77. M.B. Priestley: *Spectral Analysis and Time Series*, 2 vols. (Academic, London, 1981), Chaps. 5–7
78. T.W. Anderson: *The Statistical Analysis of Time Series* (Wiley, New York, 1971)
79. N. Madras, A.D. Sokal: J. Stat. Phys. **50**, 109 (1988)
80. A.D. Sokal, L.E. Thomas: J. Stat. Phys. **54**, 797 (1989)
81. A.M. Ferrenberg, D.P. Landau, K. Binder: J. Stat. Phys. **63**, 867 (1991)
82. A.D. Sokal: *Monte Carlo Methods in Statistical Mechanics: Foundations and New Algorithms*, Cours de Troisième Cycle de la Physique en Suisse Romande, Lausanne (1989)
83. A.D. Sokal: *Bosonic Algorithms*, in: *Quantum Fields on the Computer*, ed. M. Creutz (World Scientific, Singapore, 1992), p. 211
84. W. Janke, T. Sauer: J. Stat. Phys. **78**, 759 (1995)
85. B. Efron: *The Jackknife, the Bootstrap and Other Resampling Plans* (Society for Industrial and Applied Mathematics [SIAM], Philadelphia, 1982)
86. R.G. Miller: Biometrika **61**, 1 (1974)
87. W. Janke: *Statistical Analysis of Simulations: Data Correlations and Error Estimation*, invited lecture notes, in: Proceedings of the Euro Winter School *Quantum Simulations of Complex Many-Body Systems: From Theory to Algorithms*, eds. J. Grotendorst, D. Marx, A. Muramatsu, John von Neumann Institute for Computing, Jülich, NIC Series, Vol. **10**, pp. 423–445 (2002)
88. R.H. Swendsen, J.-S. Wang: Phys. Rev. Lett. **58**, 86 (1987)
89. U. Wolff: Phys. Rev. Lett. **62**, 361 (1989)

90. W. Janke: *Nonlocal Monte Carlo Algorithms for Statistical Physics Applications*, Mathematics and Computers in Simulations **47**, 329–346 (1998)

91. P.W. Kasteleyn, C.M. Fortuin: J. Phys. Soc. Japan **26** (Suppl.), 11 (1969)

92. C.M. Fortuin, P.W. Kasteleyn: Physica **57**, 536 (1972)

93. C.M. Fortuin: Physica **58**, 393 (1972)

94. C.M. Fortuin: Physica **59**, 545 (1972)

95. U. Wolff: Phys. Lett. **A228**, 379 (1989)

96. X.-L. Li, A.D. Sokal: Phys. Rev. Lett. **63**, 827 (1989); *ibid.* **67**, 1482 (1991)

97. P. Grassberger: Physica **A214**, 547 (1995); **A217**, 227 (1995) (erratum)

98. N. Ito, K. Hukushima, K. Ogawa, Y. Ozeki: J. Phys. Soc. Japan **69**, 1931 (2000)

99. D.W. Heermann, A.N. Burkitt: Physica **A162**, 210 (1990)

100. P. Tamayo: Physica **A201**, 543 (1993)

101. N. Ito, G.A. Kohring: Physica **A201**, 547 (1993)

102. U. Wolff: Nucl. Phys. **B322**, 759 (1989)

103. M. Hasenbusch: Nucl. Phys. **B333**, 581 (1990)

104. U. Wolff: Nucl. Phys. **B334**, 581 (1990)

105. C.F. Baillie: Int. J. Mod. Phys. **C1**, 91 (1990)

106. R.H. Swendsen, J.-S. Wang, A.M. Ferrenberg: in: *The Monte Carlo Method in Condensed Matter Physics*, ed. K. Binder (Springer, Berlin, 1992)

107. M. Hasenbusch, S. Meyer: Phys. Lett. **B241**, 238 (1990)

108. W. Janke: Phys. Lett. **A148**, 306 (1990)

109. C. Holm, W. Janke: Phys. Rev. **B48**, 936 (1993)

110. A.M. Ferrenberg, R.H. Swendsen: Phys. Rev. Lett. **61**, 2635 (1988)

111. A.M. Ferrenberg, R.H. Swendsen: Phys. Rev. Lett. **63**, 1658(E) (1989)

112. P.D. Beale: Phys. Rev. Lett. **76**, 78 (1996)

113. N. Wilding: *Computer Simulation of Continuous Phase Transitions*, in: *Computer Simulations of Surfaces and Interfaces*, NATO Science Series, II. Mathematics, Physics and Chemistry – Vol. **114**, Proceedings of the NATO Advanced Study Institute, Albena, Bulgaria, 9–20 September 2002, eds. B. Dünweg, D.P. Landau, A.I. Milchev (Kluwer, Dordrecht, 2003), pp. 161–171

114. A.M. Ferrenberg, R.H. Swendsen: Phys. Rev. Lett. **63**, 1195 (1989)

115. E. Marinari, G. Parisi: Europhys. Lett. **19**, 451 (1992)

116. A.P. Lyubartsev, A.A. Martsinovski, S.V. Shevkunov, P.N. Vorontsov-Velyaminov: J. Chem. Phys. **96**, 1776 (1992)

117. C.J. Geyer: in: *Computing Science and Statistics*, Proceedings of the 23rd Symposium on the Interface, ed. E.M. Keramidas (Interface Foundation, Fairfax, Virginia, 1991); pp. 156–163; C.J. Geyer, E.A. Thompson, J. Am. Stat. Assoc. **90**, 909 (1995)

118. K. Hukushima, K. Nemoto: J. Phys. Soc. Japan **65**, 1604 (1996)

119. B.A. Berg: Fields Inst. Comm. **26**, 1 (2000)

120. B.A. Berg: Comp. Phys. Comm. **104**, 52 (2002)

121. W. Janke: Physica **A254**, 164 (1998)

122. W. Janke: *Histograms and All That*, invited lectures, in: *Computer Simulations of Surfaces and Interfaces*, NATO Science Series, II. Mathematics, Physics and Chemistry – Vol. **114**, Proceedings of the NATO Advanced Study Institute, Albena, Bulgaria, 9–20 September 2002, eds. B. Dünweg, D.P. Landau, A.I. Milchev (Kluwer, Dordrecht, 2003), pp. 137–157

123. B.A. Berg, T. Neuhaus: Phys. Lett. **B267**, 249 (1991)

124. B.A. Berg, T. Neuhaus: Phys. Rev. Lett. **68**, 9 (1992)
125. W. Janke: Int. J. Mod. Phys. **C3**, 1137 (1992)
126. K. Binder: in *Phase Transitions and Critical Phenomena*, Vol. **5b**, eds. C. Domb, M.S. Green (Academic Press, New York, 1976), p. 1
127. G.M. Torrie, J.P. Valleau: Chem. Phys. Lett. **28**, 578 (1974); J. Comp. Phys. **23**, 187 (1977) 187; J. Chem. Phys. **66**, 1402 (1977)
128. W. Janke, T. Sauer: Phys. Rev. **E49**, 3475 (1994)
129. W. Janke, S. Kappler: Nucl. Phys. **B** (Proc. Suppl.) **42**, 876 (1995)
130. W. Janke, S. Kappler: Phys. Rev. Lett. **74**, 212 (1995)
131. M.S. Carroll, W. Janke, S. Kappler: J. Stat. Phys. **90**, 1277 (1998)
132. T. Neuhaus, J.S. Hager: J. Stat. Phys. **113** 47 (2003)
133. K. Leung, R.K.P. Zia: J. Phys. **A23**, 4593 (1990)

From Urn Models to Zero-Range Processes: Statics and Dynamics

C. Godrèche

Service de Physique Théorique, CEA Saclay, F-91191 Gif-sur-Yvette cedex, France
godreche@dsm-mail.saclay.cea.fr

Introduction

The aim of these notes is a description of the statics and dynamics of zero-range processes (ZRP) [1] and of related models. These models are simplified models of physical reality. Yet, besides the fact that they play an important role in the elucidation of conceptual problems of statistical mechanics and probability theory, they are instrumental in the understanding of a variety of complex physical situations. For instance the Ehrenfest model [2, 3] is the simplest example of a model belonging to the class of models described in the present text. A useful review of some of the applications of ZRP in physical situations can be founded in [4].

In these notes, we present a review of the subject, coming back on some of its conceptual aspects. We restrict all discussions to homogeneous models where all sites are equivalent. Before commencing, we summarise in a few words the main organisation of the text.

In Part I: Statics (Sects. 6.1–6.6), we first show that ZRP are special members of a class of stochastic processes which have the property that their stationary measures are known and have a product structure. The probability of a configuration of the system is given by the Boltzmann formula for an equilibrium urn model with independent sites. Reversibility (for symmetric dynamics) and pairwise balance (for asymmetric dynamics) are inherently related to the structure of the stationary measure. Generalisations to multiple-species ZRP are then addressed. The properties of the stationary measure of ZRP leading to a phase transition between a fluid phase and a condensed phase are finally briefly reviewed, as a preparation for the second part of these notes.

The stochastic nature of ZRP is fully revealed by the study of their dynamics. This is the subject of Part II (Sects. 6.7–6.9). We first address the non-stationary dynamical behaviour of the system when it evolves from a random initial disordered configuration to its stationary state. Then we investigate some aspects of its stationary dynamics, when the system fluctuates in its

C. Godrèche: *From Urn Models to Zero-Range Processes: Statics and Dynamics*, Lect. Notes Phys. **716**, 261–294 (2007)
DOI 10.1007/3-540-69684-9_6

stationary state. In both cases the model used is that giving rise to condensation.

Part I: Statics (Sects. 6.1–6.6)

6.1 Dynamical Urn Models and Zero-Range Processes

6.1.1 Dynamical Urn Models

We name *dynamical urn model* (DUM) the following stochastic process. Consider a finite connected graph, made of M sites (or urns), on which N particles are distributed. The occupation $N_i(t)$ of site i ($i = 1, \ldots, M$) is a random variable, and the total number of particles

$$\sum_{i=1}^{M} N_i(t) = N$$

is conserved in time. The model is defined by dynamical rules describing how particles hop from site to site. An elementary step of the dynamics consists in choosing a departure site d and an arrival site a connected to site d, and in transferring one of the particles present on site d to site a. This process takes place with rate $W_{k,l}$ per unit time, depending on the occupations both of the departure site, $k = N_d \neq 0$, and of the arrival site, $l = N_a$[1].

On the complete graph (i.e., in the mean-field geometry), all sites are connected, i.e., sites d and a are chosen independently at random. On finite-dimensional lattices, site a is chosen among the first neighbours of site d. In one dimension, site a is chosen to be the right neighbour of site d with probability p, or its left neighbour with probability $q = 1 - p$. In the following we consider the one-dimensional symmetric dynamics, corresponding to $p = 1/2$, and the general asymmetric one, corresponding to $p \neq 1/2$, both with periodic boundary conditions.

A configuration of the system is specified by the occupation numbers $N_i(t)$, i.e., a complete knowledge of its dynamics involves the determination of $\mathcal{P}(N_1, N_2, \ldots, N_M)$, the probability of finding the system in a given configuration at time t.

The process defined above can also be named a *Migration process*, and can be pictorially viewed in terms of colonies and migration. The sites are the colonies, or cities. An individual leaves its colony for another one, with a rate $W_{k,l}$ which depends on the number of members present in both the departure and the arrival colonies. Thus, for example, the philanthrope is characterized by a rate decreasing with k and increasing with l, the misanthrope by the converse.

---

[1] Throughout this text we use the notation N_i for the random occupation of site i, and k (an integer) for the value taken by this random variable.

6.1.2 Zero-Range Processes

Zero-range processes are just particular cases of DUM, with the additional restriction that the rate $W_{k,l}$ only depends on the occupation of the departure site:

$$W_{k,l} = u_k \ .$$

This simple restriction is enough to lead to a remarkable property of the stationary probability [1,5]. Indeed, the probability of a configuration of the system is equal to

$$\mathcal{P}(N_1, \dots, N_M) = \frac{1}{Z_{M,N}} \prod_{i=1}^{M} p_{N_i} \ , \tag{6.1}$$

where it is understood that $\sum N_i = N$, and where the factor $p_k = p_{N_i = k}$ satisfies the relation

$$u_k \, p_k = p_{k-1} \ , \tag{6.2}$$

which leads to the explicit form

$$p_0 = 1 \ , \qquad p_k = \frac{1}{u_1 \dots u_k} \ . \tag{6.3}$$

The normalisation factor, hereafter refered to as the partition function, reads

$$Z_{M,N} = \sum_{N_1} \cdots \sum_{N_M} p_{N_1} \cdots p_{N_M} \, \delta \left(\sum_i N_i, N \right) \ . \tag{6.4}$$

One important observation to make is that the stationary measure is insensitive to the bias.

These results can be proved by inspection. The master equation at stationarity reads

$$0 = \sum_{\mathcal{C}' \neq \mathcal{C}} M(\mathcal{C}|\mathcal{C}') \, \mathcal{P}(\mathcal{C}') - \sum_{\mathcal{C}'' \neq \mathcal{C}} M(\mathcal{C}''|\mathcal{C}) \, \mathcal{P}(\mathcal{C}) \ , \tag{6.5}$$

where $\mathcal{C} = \{N_1, \dots, N_M\}$, and $M(\mathcal{C}|\mathcal{C}')$ is the transition rate from $\mathcal{C}'$ to $\mathcal{C}$. Consider a system of $M = 3$ sites for simplicity. At stationarity the master equation reads explicitly

$$
\begin{aligned}
& p \left[\mathcal{P}(N_1 + 1, N_2 - 1, N_3) \, u_{N_1 + 1}(1 - \delta(N_2, 0)) + \text{c.p.} + \text{c.p.} \right] \\
+ \ & q \left[\mathcal{P}(N_1 + 1, N_2, N_3 - 1) \, u_{N_1 + 1}(1 - \delta(N_3, 0)) + \text{c.p.} + \text{c.p.} \right] \\
= \ & \mathcal{P}(N_1, N_2, N_3) [\, u_{N_1}(1 - \delta(N_1, 0)) + \text{c.p.} + \text{c.p.}] \ ,
\end{aligned}
\tag{6.6}
$$

where c.p. stands for circular permutation. Carrying the product form (6.1) into the equation, and using (6.2), satisfies the master equation. This is the *unique* solution of the problem.

The cancelation of terms in the equation occurs by pair. Pairs correspond to terms bearing the same p (respectively q) factor, and the same $1 - \delta(N_i, 0)$ factor. Hence we have for example, after cancelling terms $1 - \delta(N_2, 0)$ on both sides,

$$p\mathcal{P}(N_1 + 1, N_2 - 1, N_3)\, u_{N_1+1} = p\mathcal{P}(N_1, N_2, N_3)\, u_{N_2} , \qquad (6.7)$$

i.e., using the product form (6.1), with $N_1 = k$ and $N_2 = l$,

$$p_{k+1}\, p_{l-1}\, u_{k+1} = p_k\, p_l\, u_l , \qquad (6.8)$$

which is precisely the relation that leads to (6.2).

It is interesting to emphasize the interpretation of (6.7), or (6.8). Consider the following configurations:

$$\mathcal{C} = (N_1, N_2, N_3) ,$$
$$\mathcal{C}' = (N_1 + 1, N_2 - 1, N_3) ,$$
$$\mathcal{C}'' = (N_1, N_2 - 1, N_3 + 1) ,$$

and the corresponding rates

$$M(\mathcal{C}|\mathcal{C}') = p\, u_{N_1+1} ,$$
$$M(\mathcal{C}'|\mathcal{C}) = q\, u_{N_2} ,$$
$$M(\mathcal{C}''|\mathcal{C}) = p\, u_{N_2} .$$

In general, i.e. for a general value of $0 < p < 1$, (6.7) reads

$$M(\mathcal{C}|\mathcal{C}')\mathcal{P}(\mathcal{C}') = M(\mathcal{C}''|\mathcal{C})\mathcal{P}(\mathcal{C}) .$$

This is a condition for *pairwise balance* [6]. It expresses the equality between the probability fluxes flowing from $\mathcal{C}'$ to $\mathcal{C}$, and from $\mathcal{C}$ to $\mathcal{C}''$. In the particular case where $p = 1/2$, or more generally when the dynamics is symmetric, then (6.7) becomes the condition for detailed balance

$$M(\mathcal{C}|\mathcal{C}')\mathcal{P}(\mathcal{C}') = M(\mathcal{C}'|\mathcal{C})\mathcal{P}(\mathcal{C}) .$$

6.1.3 Equilibrium Urn Models with Independent Sites

We now adopt a completely different point of view. We consider equilibrium urn models with independent sites, on which a dynamics is then defined, in such a way that equilibrium is recovered at long times.

As above, we consider a finite connected graph, made of M sites (or urns), on which N particles are distributed. The number of particles on site i is the random variable N_i, with $\sum N_i = N$. The total energy of the system is defined as the sum

$$E(N_1, \ldots, N_M) = \sum_{i=1}^{M} E(N_i) .$$

Let

$$p_{N_i} = e^{-\beta E(N_i)} \tag{6.9}$$

be the unnormalized Boltzmann weight attached to site i. Then, clearly, the probability of a configuration of the system is given by the product form (6.1), and $Z_{M,N}$ appears as the usual partition function for this statistical mechanical system.

We now define a dynamics for this model, such that equilibrium is attained in the limit of long times. We therefore choose a rule obeying detailed balance for the move of a particle. This implies that the dynamics should be symmetric. Restricting to the one-dimensional case, $(p = 1/2)$, if $N_i = k$ and $N_{i\pm1} = l$, we have

$$p_k p_l \, W_{k,l} = p_{k-1} p_{l+1} \, W_{l+1,k-1} \,, \tag{6.10}$$

which expresses the probability balance between the configurations $\{N_d = k, N_a = l\}$ where $(d = i, a = i \pm 1)$, and $\{N_d = l + 1, N_a = k - 1\}$ where $(d = i \pm 1, a = i)$. It applies as well to the case of the complete graph. For example, with the Metropolis rule, the move is allowed with probability $\min(1, \exp(-\beta \Delta E))$, where ΔE is the change in energy due to the move.

Let us mention two well-studied models in this class: the backgammon model and the zeta-urn model, that we briefly describe. The backgammon model is a simple example of a system which exhibits slow relaxation due to entropy barriers [7, 8]. The following choice of an energy function is done:

$$E(N_i) = -\delta(N_i, 0) \,.$$

The statics of this model is trivial. Its interest lies in its dynamical behaviour. The dynamics of the model has been thoroughly studied in the mean-field geometry, with Metropolis dynamics, and with the additional rule that a particle (instead of a site) is chosen at random. The rate for the Metropolis rule reads

$$W_{k,l} = \min\left(1, \frac{p_{k-1} p_{l+1}}{p_k p_l}\right) \,.$$

From (6.9), we have

$$p_0 = e^\beta \,, \qquad p_k = 1 \,, \quad (k > 1) \,,$$

and therefore $W_{k,0} = e^{-\beta}$ for any $k > 1$, and $W_{k,l} = 1$ otherwise, or in compact form:

$$W_{k,l} = 1 + (e^{-\beta} - 1)\delta_{l,0}(1 - \delta_{k,1}) \qquad (k > 0) \,.$$

As can be read on this expression, at low temperature increasing the number of empty sites is not favoured. The total energy is indeed equal to minus the number of empty sites, so that particles tend to condensate in fewer and fewer sites as times passes, at least at low temperature.

The static zeta urn model has energy function

$$E(N_i) = \ln(N_i + 1) , \tag{6.11}$$

hence

$$p_k = \frac{1}{(1 + k)^\beta} .$$

The model was initially introduced as a mean-field model of discretized quantum gravity [9]. Its dynamics was subsequently defined and investigated in the mean-field geometry with heat-bath dynamics [10, 11].

If instead, the transfer rate is taken to be that of a ZRP, with $W_{k,l} = u_k$, where $u_k = p_{k-1}/p_k$, the universal properties of the dynamics of the zeta urn model are not changed [12]. We thus get

$$u_k = \left(1 + \frac{1}{k}\right)^\beta \approx 1 + \frac{\beta}{k} .$$

The model is therefore in the same universality class as the ZRP with condensation studied in the rest of this text, and defined with the rate $u_k = 1 + b/k$. The parameter b for this model can therefore be identified with the inverse temperature.

To summarise at this point, we have so far encountered two classes of dynamical urn models with stationary product measures. On the one hand, ZRP are defined for any value of the drive, and are such that the transfer rate $W_{k,l}$ only depends on k. On the other hand, equilibrium urn models with independent sites are defined from the start without drive, but the transfer rate has the full dependence in both k and l. A natural question to ask is whether there exist models possessing both features, namely models with stationary product measure, even when submitted to a drive, and with transfer rate $W_{k,l}$ not restricted to depend only on k.

6.1.4 Dynamical Urn Models with Stationary Product Measure

We address the question just posed. Given a DUM, what choice of rate $W_{k,l}$ is compatible with a stationary measure of the form (6.1), *even if the dynamics is not symmetric?*

Let us restrict to the case of the one-dimensional geometry with asymmetric hops. The results are as follows:

- In the general case, $0 < p < 1$, two conditions are imposed on the rate $W_{k,l}$. The first condition is

$$p_k p_l \, W_{k,l} = p_{k-1} p_{l+1} \, W_{l+1,k-1} . \tag{6.12}$$

  The second condition reads

$$W_{k,l} - W_{l,k} = W_{k,0} - W_{l,0} . \tag{6.13}$$

  Equation (6.12) expresses the condition of pairwise balance.

- In the symmetric case ($p = 1/2$), the only condition imposed on the transfer rate is (6.12), or equivalently (6.10). It expresses the condition of detailed balance. In other words, if the stationary measure is a product, it is necessarily an equilibrium measure and we are taken back to the situation of Sect. 6.1.3.

Let us give the proof. By hypothesis, the stationary probability is given and has the product form (6.1), with given p_{N_i}. We rewrite the master equation (6.5) as an equality between gain and loss terms, after dividing both hand sides by $\mathcal{P}(\mathcal{C})$,

$$pG_R + qG_L = pL_R + qL_L ,$$

with right and left contributions

$$L_R = \sum_i W_{N_i,N_{i+1}} , \qquad L_L = \sum_i W_{N_{i+1},N_i} ,$$

$$G_R = \sum_i W_{N_i+1,N_{i+1}-1} \frac{p_{N_i+1}p_{N_{i+1}-1}}{p_{N_i}p_{N_{i+1}}} ,$$

$$G_L = \sum_i W_{N_{i+1}+1,N_i-1} \frac{p_{N_{i+1}+1}p_{N_i-1}}{p_{N_i}p_{N_{i+1}}} .$$

We now specialize to the configuration where all sites are empty except for sites i and $i+1$:

$$\mathcal{C} = \{N_1 = 0,\dots, N_{i-1} = 0, N_i = k, N_{i+1} = l, N_{i+2} = 0,\dots, N_M = 0\} .$$

We obtain

$$p\left(W_{1,k-1} \frac{p_1 p_{k-1}}{p_0 p_k} + W_{k+1,l-1} \frac{p_{k+1}p_{l-1}}{p_k p_l} \right)$$

$$+q\left(W_{1,l-1} \frac{p_1 p_{l-1}}{p_0 p_l} + W_{l+1,k-1} \frac{p_{k-1}p_{l+1}}{p_k p_l} \right)$$

$$= p(W_{k,l} + W_{l,0}) + q(W_{l,k} + W_{k,0}) . \tag{6.14}$$

Taking $k = 0$, (6.14) reduces to

$$p_1 p_{l-1} W_{1,l-1} = p_0 p_l W_{l,0} , \tag{6.15}$$

which expresses the probability balance between the configurations $\{N_d = 1, N_a = l - 1\}$ and $\{N_d = l, N_a = 0\}$. This equality is then used in (6.14) to yield the fundamental equation

$$p(W_{k,l} - W_{k,0}) + q(W_{l,k} - W_{l,0})$$

$$= p\left(W_{k+1,l-1} \frac{p_{k+1}p_{l-1}}{p_k p_l} - W_{l,0} \right) + q\left(W_{l+1,k-1} \frac{p_{k-1}p_{l+1}}{p_k p_l} - W_{k,0} \right). \tag{6.16}$$

From this equation, the two conditions (6.12) and (6.13) are obtained, as shown in the appendix. The conditions thus found are necessary. They are also sufficient as one can convince oneself by redoing the reasoning for a generic configuration. The analysis done here applies as well to the complete graph, for which the dynamics is symmetric.

Coming back to the case of a ZRP, condition (6.13) is trivially satisfied, while the pairwise balance condition (6.12) yields (6.8), rewritten here for convenience,

$$p_k p_l \, u_k = p_{k-1} p_{l+1} \, u_{l+1} \, .$$

The ZRP appears as the minimal model of the class of DUM leading to a product measure in the stationary state independent of the asymmetry. It is important to realize that this measure is that of an equilibrium urn model with independent sites (see Subsect. 6.1.3) and therefore any result on the statics of a ZRP pertains to the field of equilibrium statistical mechanics.

The original work on the question posed in the present section is due to [13]. The dynamical urn model described in the present notes is named a *misanthrope process* in [13] because the rates $W_{k,l}$ considered in this reference are increasing functions of k. Yet another presentation, restricted to the 1D totally asymmetric case ($p = 1$) can be found in [4].

6.2 A Counterexample

Let us now examine the case where the transfer rate only depends on the occupation of the arrival site,

$$W_{k,l} = v_l (1 - \delta_{k,0}) \, . \tag{6.17}$$

If the dynamics is symmetric, the only constraint to take into account in order to have product probability in the stationary state is the detailed balance condition (6.12), which reads here $p_{k+1} p_{l-1} \, v_{l-1} = p_k p_l \, v_k$. The relation $p_{l-1} v_{l-1} = p_l$ follows, which determines the measure fully. However, if the dynamics is not symmetric, (6.13) is violated by (6.17), which rules out the possibility of stationary product measure for this case.

Let us illustrate the difficulty on the simple case of a system of $M = 3$ sites. The stationary master equation reads

$$\begin{aligned}
p \, &[\mathcal{P}(N_1 + 1, N_2 - 1, N_3) \, v_{N_2-1}(1 - \delta(N_2, 0)) + \text{c.p.} + \text{c.p.}] \\
&+ q \, [\mathcal{P}(N_1 + 1, N_2, N_3 - 1) \, v_{N_3-1}(1 - \delta(N_3, 0)) + \text{c.p.} + \text{c.p.}] \\
&= p \mathcal{P}(N_1, N_2, N_3)[v_{N_1}(1 - \delta(N_3, 0)) + \text{c.p.} + \text{c.p.}] \\
&+ q \mathcal{P}(N_1, N_2, N_3)[v_{N_1}(1 - \delta(N_2, 0)) + \text{c.p.} + \text{c.p.}] \, .
\end{aligned}$$

There is no way of pairing the terms in the master equation to obtain their mutual cancellation if $p \neq 1/2$, while this is possible for $p = 1/2$. More generally, the stationary probability is unknown for the asymmetric process (for arbitrary system size M) [14].

6.3 Two-Species ZRP: Conditions for Product Measure

A simple generalization of the ZRP defined so far consists in considering two (or more generally n) coexisting species on each site [15, 16], named particles of type A and B respectively. The hopping rates for A and B particles only depend on the occupations of the departure site: $N_i^A = k$, $N_i^B = l$. They are respectively denoted by $u_{k,l}$ and $v_{k,l}$. The new fact is that the condition for product stationary measure imposes a constraint on the rates $u_{k,l}$ and $v_{k,l}$ [15, 16], given by equation (6.21).

We revisit this problem, keeping the line of thought followed for the (single-species) ZRP in Sect. 6.1. We want to show that, as was the case for the single-species ZRP, for the two-species ZRP satisfying (6.21) the following properties come together:

- The stationary probability is a product and is insensitive to the bias. It is the stationary probability of an equilibrium urn model with independent sites.
- If the dynamics is symmetric, the process is reversible, i.e. satisfies detailed balance, otherwise, in the presence of a bias, pairwise balance holds.

6.3.1 Equilibrium Urn Models with Independent Sites

Let us first consider an equilibrium urn model for two species with independent sites. A configuration of the system is denoted by $\mathcal{C} = \{\boldsymbol{N}_1, \dots, \boldsymbol{N}_M\}$, where $\boldsymbol{N}_i = (N_i^A, N_i^B)$. The energy is given by the sum

$$E(\mathcal{C}) = \sum_i{}' E(\boldsymbol{N}_i) .$$

The Boltzmann weight reads

$$\mathcal{P}(\mathcal{C}) = \frac{1}{Z_{M,N^A,N^B}} \prod_i p_{\boldsymbol{N}_i} , \tag{6.18}$$

where N^A and N^B are respectively the total number of A and B particles, and Z the partition function. A dynamics yielding this equilibrium measure should fulfill detailed balance. We restrict the rates to depend only on the departure site. With for the departure site: $N_d^A = k, N_d^B = l$, and the arrival site: $N_a^A = m, N_a^B = n$, we must impose

$$p_{k,l}\, p_{m,n}\, u_{k,l} = p_{k-1,l}\, p_{m+1,n}\, u_{m+1,n}$$

$$p_{k,l}\, p_{m,n}\, v_{k,l} = p_{k,l-1}\, p_{m,n+1}\, v_{m,n+1} , \tag{6.19}$$

hence,

$$u_{k,l}\, p_{k,l} = p_{k-1,l} , \qquad v_{k,l}\, p_{k,l} = p_{k,l-1} . \tag{6.20}$$

These relations generalise (6.2). Consideration of the two possible paths leading from $p_{k,l}$ to $p_{k-1,l-1}$, using (6.21), imposes a "gauge" condition on the rates:

$$u_{k,l}\, v_{k-1,l} = v_{k,l}\, u_{k,l-1} . \tag{6.21}$$

6.3.2 Product Measure

We can now proceed as for the single-species ZRP. We claim that, even in the presence of a bias, (6.18), (6.20), hold. The proof is by inspection: (6.18) carried into the master equation of the process is seen to be a solution if (6.19), or (6.20), hold. The constraint (6.21) follows. See [15, 16].

6.3.3 Reversibility Implies Stationary Product Measure

Finally we show by a direct route that (6.21) is a consequence of reversibility, when the dynamics is symmetric. We use the Kolmogorov condition, a necessary and sufficient condition for the reversibility of a Markov process (i.e., for detailed balance to hold), which states that the product of rates along any cycle in the state space of the process and for the reverse cycle should be equal [17, 18]. Consider the configuration

$$(\boldsymbol{N}_1 = (k, l), \boldsymbol{N}_2 = (m, n), \ldots) .$$

We consider the following cycle in the space of states of the process

$$(k, l; m, n) \rightarrow (k, l-1; m, n+1) \rightarrow (k+1, l-1; m-1, n+1)$$
$$\rightarrow (k+1, l; m-1, n) \rightarrow (k, l; m, n) .$$

For the cycle considered above, the Kolmogorov condition yields

$$\frac{v_{k,l}\, u_{k+1,l}}{v_{k+1,l}\, u_{k+1,l-1}} = \frac{u_{m,n}\, v_{m,n+1}}{u_{m,n+1}\, v_{m-1,n+1}} .$$

This condition is satisfied if and only if (6.21) holds.

6.3.4 An Example of a Two-Species ZRP with Non Product Stationary Measure

Consider the ZRP defined by the following rates [19]

$$u_{k,l} = 1 + \frac{b}{l} , \qquad v_{k,l} = 1 + \frac{b}{k} .$$

These rates violate (6.21), and therefore, as explained above, this process violates time reversal symmetry *even in the absence of a bias*. The study of the stationary properties of the model is addressed in [20].

6.4 Two Extreme Cases

6.4.1 The Case of Two Sites

We come back to the case of a general dynamical urn model, with one species, where now the number of sites is $M = 2$. This case is interesting for several reasons. Firstly the model stands by itself, for instance the Ehrenfest urn model belongs to this class, as shown below. Secondly, it illustrates some aspects of the general theory for a system of arbitrary size M. Finally, it relates to the other case considered in this section, a thermodynamic system on the complete graph, the master equation of which is formally that of a two-site system.

Since $N_2 = N - N_1$, a configuration of the system is entirely defined by the occupation of site 1, N_1, and the hopping rate only depends on one variable: $W_{k,l} = u_k$. Let us denote the occupation probability of site 1, i.e., the probability of a configuration of the system, by

$$f_k(t) = \mathcal{P}(N_1(t) = k) .$$

It obeys the master equation

$$\frac{\mathrm{d}f_k(t)}{\mathrm{d}t} = \mu_{k+1} f_{k+1} + \lambda_{k-1} f_{k-1} - (\mu_k + \lambda_k)f_k \qquad (1 \le k \le N - 1) ,$$

$$\frac{\mathrm{d}f_0(t)}{\mathrm{d}t} = \mu_1 f_1 - \lambda_0 f_0 , \qquad\qquad\qquad (6.22)$$

$$\frac{\mathrm{d}f_N(t)}{\mathrm{d}t} = \lambda_{N-1} f_{N-1} - \mu_N f_N ,$$

where λ_k and μ_k are respectively the rate at which a particle enters site 1, coming from site 2, or leaves site 1 for site 2:

$$\lambda_k = u_{N-k} , \qquad \mu_k = u_k .$$

The equations for $k = 0$ or $k = N$ are special, since $u_0 = 0$. The above equations describe a biased random walk on the interval $(0, N)$, with reflecting boundaries at 0 and N, the position of the walker being the random variable $N_1(t)$, i.e., the number of particles on site 1.

The time-independent solution to (6.22) satisfies

$$\mu_{k+1} f_{k+1,\mathrm{eq}} - \lambda_k f_{k,\mathrm{eq}} = \ldots = \mu_1 f_{1,\mathrm{eq}} - \lambda_0 f_{0,\mathrm{eq}} = 0 ,$$

which yields the detailed balance condition at equilibrium

$$\mu_{k+1} f_{k+1,\mathrm{eq}} = \lambda_k f_{k,\mathrm{eq}} . \qquad\qquad (6.23)$$

From this equation it is easy to obtain

$$f_{k,\mathrm{eq}} = \frac{p_k\, p_{N-k}}{Z_{2,N}} , \qquad Z_{2,N} = \sum_{k=0}^{N} p_k p_{N-k} ,$$

where the p_k are given by (6.3). These expressions are special instances of Eqs. (6.1) and (6.4) which hold for the general case. Elements on the dynamics of the two-site model can be found in [21].

Remark. This historical Ehrenfest model [2,3] is a special instance of a 2-site dynamical urn model. Consider N particles, labeled from 1 to N, which are distributed in two urns (sites). At random times, given by a Poisson process with unit rate, a particle is chosen at random (i.e., an integer between 1 and N is chosen at random), and moved from the site on which it is to the other site. The master equation reads

$$\frac{df_k(t)}{dt} = \frac{k+1}{N} f_{k+1}(t) + \frac{N+1-k}{N} f_{k-1}(t) - f_k(t) \,. \tag{6.24}$$

Indeed, a move of a particle from site number 1 to site number 2 (resp. from site number 2 to site number 1) occurs with a rate k/N (resp. $(N-k)/N$) per unit time.

Note that the rule of choosing a labeled particle is different from the rule adopted above for dynamical urn models (as was already the case for the backgammon model). Yet we can describe this model as a 2-site dynamical urn model, by taking $u_k = k$ (dropping the factor N which enters the scale of time). Then $p_k = 1/k!$, and the distribution of particles amongst the two sites is binomial,

$$f_{k,\text{eq}} = 2^{-N} \binom{N}{k} \quad (k = 0, \ldots, N) \,, \tag{6.25}$$

as is well known for the Ehrenfest model.

6.4.2 A Thermodynamic System on the Complete Graph

In the mean-field geometry, for a thermodynamic system, the temporal evolution of the occupation probability $f_k(t)$ is given by the master equation

$$\frac{df_k(t)}{dt} = \mu_{k+1} f_{k+1} + \lambda_{k-1} f_{k-1} - (\mu_k + \lambda_k) f_k \quad (k \geq 1) \,,$$
$$\frac{df_0(t)}{dt} = \mu_1 f_1 - \lambda_0 f_0 \,, \tag{6.26}$$

where

$$\mu_k = u_k \,, \quad (k > 0) \,, \qquad \lambda_k = \sum_{l=1}^{\infty} u_l f_l \equiv \bar{u}_t \,, \quad (k \geq 0) \,. \tag{6.27}$$

These are respectively the rates at which a particle leaves site 1, or arrives on this site. In other words, on the complete graph, all sites other than site 1 play the role of a single site from which particles are emitted with rate $\bar{u}_t$, and therefore (6.26) is formally similar to the master equation (6.22) for a system of two sites. In the present case this set of equations is non linear because $\bar{u}_t$ is itself a function of the $f_k(t)$.

In the stationary state the detailed balance condition (6.23) reads

$$\frac{f_{k+1,\mathrm{eq}}}{f_{k,\mathrm{eq}}} = \frac{\lambda_k}{\mu_{k+1}} = \frac{\bar{u}_{\mathrm{eq}}}{u_{k+1}} \; ,$$

yielding

$$f_{k,\mathrm{eq}} = \frac{\lambda_0 \dots \lambda_{k-1}}{\mu_1 \dots \mu_k} f_{0,\mathrm{eq}} \; ,$$

where $f_{0,\mathrm{eq}}$ is fixed by normalisation. Hence

$$f_{k,\mathrm{eq}} = \frac{p_k \bar{u}_{\mathrm{eq}}^k}{\sum_{k=0}^{\infty} p_k \bar{u}_{\mathrm{eq}}^k} \; , \tag{6.28}$$

with the p_k given by (6.3). This expression is a particular instance of the general case (6.37).

6.5 Statics of ZRP: Fundamental Properties

We collect here the results found so far concerning single-species ZRP's. A ZRP is a dynamical urn model, for which the rate of transfer of a particle, u_k, only depends on the occupation of the departure site, k. The stationary state of a ZRP is that of an equilibrium urn model with independent sites: the probability of a configuration of the system is (independently of the asymmetry)

$$\mathcal{P}(N_1, \dots, N_M) = \frac{1}{Z_{M,N}} \prod_{i=1}^{M} p_{N_i} \; , \tag{6.29}$$

with partition function

$$Z_{M,N} = \sum_{N_1} \cdots \sum_{N_M} p_{N_1} \cdots p_{N_M} \, \delta\left(\sum_i N_i, N \right) . \tag{6.30}$$

The factor p_{N_i} obeys the pairwise balance condition (6.12), i.e., $p_k p_l \, u_k = p_{k-1} p_{l+1} \, u_{l+1}$, and hence

$$p_k \, u_k = p_{k-1} \; ,$$

which gives the explicit form of p_k (for u_k given)

$$p_0 = 1 \; , \qquad p_k = \frac{1}{u_1 \dots u_k} \; . \tag{6.31}$$

The value given to p_0 is arbitrary. The energy function associated to the underlying equilibrium urn model mentioned above is defined using Eq. (6.9). The partition function $Z_{M,N}$ obeys the recursion formula

$$Z_{M,N} = \sum_{k=0}^{N} p_k \, Z_{M-1,N-k} \; . \tag{6.32}$$

This ensures that the stationary single-site occupation probability

$$f_{k,\text{st}} = \mathcal{P}(N_1 = k) = \frac{p_k \, Z_{M-1,N-k}}{Z_{M,N}} \qquad (6.33)$$

is normalised. We have

$$Z_{0,N} = \delta_{N,0} , \qquad Z_{1,N} = p_N , \qquad Z_{2,N} = \sum_{k=0}^{N} p_k p_{N-k} , \qquad (6.34)$$

and so on. Using an integral representation of the Kronecker delta function,

$$\delta(m,n) = \oint \frac{dz}{2\pi i z^{n+1}} z^m ,$$

we obtain

$$Z_{M,N} = \oint \frac{dz}{2\pi i z^{N+1}} P(z)^M , \qquad (6.35)$$

where the generating series of the weights p_k reads

$$P(z) = \sum_{k\geq 0} p_k z^k .$$

In other words, $Z_{M,N}$ is the coefficient of z^N in $P(z)^M$. Static properties of the ZRP are therefore entirely encoded in this series.

In the thermodynamic limit ($M \to \infty$ at fixed density $N/M = \rho$), the free energy per site,

$$\mathcal{F} = - \lim_{M\to\infty} \frac{1}{M} \ln Z_{M,N} ,$$

can be obtained by evaluating the contour integral in (6.35) by the saddle-point method. The saddle-point value z_0 depends on the density ρ through the equation

$$\frac{z_0 P'(z_0)}{P(z_0)} = \rho . \qquad (6.36)$$

The free energy per site is $\mathcal{F} = \rho \ln z_0 - \ln P(z_0)$, and the stationary occupation probability reads

$$f_{k,\text{st}} = \frac{p_k \, z_0^k}{P(z_0)} . \qquad (6.37)$$

Equation (6.36) can be rewritten as

$$\langle N_1 \rangle = \sum_k k f_{k,\text{st}} = \rho . \qquad (6.38)$$

Note that the function

$$\rho(z_0) = z_0 \frac{P'(z_0)}{P(z_0)}$$

is increasing with z_0 because

$$z_0 \frac{d\rho(z_0)}{dz_0} = \mathrm{Var}\, N_1 \,.$$

Finally the stationary average rate reads

$$\bar{u}_{\mathrm{st}}(M,N) = \langle u_{N_1} \rangle = \sum_k u_k \, f_{k,\mathrm{st}} = \sum_k u_k \frac{p_k \, Z_{M-1,N-k}}{Z_{M,N}} = \frac{Z_{M,N-1}}{Z_{M,N}} \,. \quad (6.39)$$

In the thermodynamic limit, we have $\bar{u}_{\mathrm{st}} = z_0$ (defined in (6.36) above). The expression (6.28) found for the case of the complete graph in the thermodynamic limit is a particular example of (6.37).

6.6 Statics of ZRP: Examples and the Phenomenon of Condensation

We illustrate through examples the considerations of the previous section. In particular we discuss the possible solutions of Eq. (6.36) (or (6.38)). Two possible situations can arise. Either $\rho(z_0)$ is allowed to increase without bounds, in which case the equation has a solution in z_0 for any value of ρ. Or $\rho(z_0)$ reaches a maximal value, ρ_c, in which case the equation has no solution if $\rho > \rho_c$.

6.6.1 Two Simple Examples

Let $u_k = k$. This model can be seen as a multi-urn generalisation of the Ehrenfest model. We have $P(z) = e^z$. The radius of convergence of this series is infinite. Hence Eq. (6.36) has a solution for any value of ρ: $\rho(z_0) = z_0$, hence $z_0 = \rho$, and

$$f_{k,\mathrm{st}} = e^{-\rho} \frac{\rho^k}{k!} \,,$$

which is a Poisson distribution. The fast decay of the distribution is characteristic of an homogeneous fluid phase.

As a second example let $u_k = 1$. Then $P(z) = 1/(1-z)$. The partition function of a finite system is

$$Z_{M,N} = \binom{M+N-1}{N} \,.$$

The radius of convergence of $P(z)$ is equal to 1. At this maximal allowed value of z, $\rho(z) = z/(1-z)$ is infinite. Therefore (6.36) has a solution for any value of ρ: $z_0 = \rho/(1+\rho)$, and finally

$$f_{k,\mathrm{st}} = \frac{1}{1+\rho} \left(\frac{\rho}{1+\rho} \right)^k \,.$$

The system is again in a fluid phase.

6.6.2 The Canonical Example for the Phenomenon of Condensation

We consider the ZRP with transfer rate

$$u_k = 1 + \frac{b}{k} \ .$$

This case, and closely related models, have been studied in various references [9–12, 22–27]. We follow here the approach and notations of [21]. For this choice of rate,

$$p_k = \frac{\Gamma(b+1)\,k!}{\Gamma(k+b+1)} = \int_0^1 du\, u^k\, b(1-u)^{b-1} \approx \frac{\Gamma(b+1)}{k^b} \ ,$$

$$P(z) = \int_0^1 du\, \frac{b(1-u)^{b-1}}{1-zu} = {}_2F_1(1,1;b+1;z) \ , \tag{6.40}$$

where $_2F_1$ is the hypergeometric function. The function $P(z)$ has a branch cut at $z = 1$, with a singular part of the form[2]

$$P_{\mathrm{sg}}(z) \approx A\,P(1)(1-z)^{b-1} \ ,$$

so that $P(z)$ is only differentiable $n \equiv \mathrm{Int}(b) - 1$ many times at $z = 1$:

$$P(z) \approx P(1) + (1-z)\,P'(1) + \cdots + \frac{(1-z)^n}{n!}P^{(n)}(1) + P_{\mathrm{sg}}(z) \ .$$

The following values are of interest:

$$P(1) = \frac{b}{b-1} \ , \qquad\qquad A = \frac{(b-1)\pi}{\sin \pi b} \ ,$$

$$P'(1) = \frac{b}{(b-1)(b-2)} \ , \qquad P''(1) = \frac{4b}{(b-1)(b-2)(b-3)} \ . \tag{6.41}$$

For $b \leq 2$, $\rho(1)$ is infinite. The system is in a fluid phase:

$$f_{k,\mathrm{st}} \sim k^{-b}\, e^{-k|\ln z_0|} \tag{6.42}$$

For $b > 2$, $\rho(1)$ is finite. The system has a continuous phase transition at a finite critical density

$$\rho_c = \frac{P'(1)}{P(1)} = \frac{\sum_k k\, p_k}{\sum_k p_k} = \frac{1}{b-2} \ ,$$

such that the saddle point z_0 reaches the singular point $z = 1$. This critical density separates a fluid phase ($\rho < \rho_c$) and a condensed phase ($\rho > \rho_c$).

---

[2] Whenever $b = n \geq 2$ is an integer, the amplitude A diverges. The singular part of the generating series is of the form $P_{\mathrm{sg}}(z) \approx n(-1)^n(1-z)^{n-1}\ln(1-z)$.

Fluid Phase ($\rho < \rho_c$)

The equation (6.36) has a solution for any $\rho < \rho_c$. The single site probability has the form (6.42).

Critical Density ($\rho = \rho_c$)

The occupation probability

$$f_{k,\mathrm{st}} = \frac{p_k}{P(1)} \approx \frac{(b-1)\Gamma(b)}{k^b} \tag{6.43}$$

falls off as a power-law in the thermodynamic limit. The critical free energy reads

$$\mathcal{F}_c = -\ln P(1) = -\ln \frac{b}{b-1} \ .$$

The second moment of the occupation probability,

$$\mu_c = \langle N_1^2 \rangle = \sum_{k \geq 0} k^2 f_{k,\mathrm{st}} = \frac{P'(1) + P''(1)}{P(1)} = \frac{b+1}{(b-2)(b-3)} \ , \tag{6.44}$$

is convergent for $b > 3$ (regime of normal fluctuations), and divergent for $2 < b < 3$ (regime of anomalous fluctuations).

Condensed Phase ($\rho > \rho_c$)

A large and finite system in the condensed phase essentially consists of a uniform critical background, containing on average $N_c = M \rho_c$ particles, and of a macroscopic condensate, containing on average $\Delta = N - N_c = M(\rho - \rho_c)$ excess particles with respect to the critical state.

The occupation probability $f_{k,\mathrm{st}}$ accordingly splits into two main contributions [24]. The contribution of the critical background, corresponding to small values of the occupation ($k \ll M$), is approximately given by (6.43). The contribution of the condensate shows up as a hump located around $k = \Delta$. The hump is a Gaussian whose width scales as $M^{1/2}$ whenever μ_c is finite, i.e., for $b > 3$, whereas it has power-law tails and a larger width, scaling as $M^{1/(b-1)}$, in the regime of anomalous fluctuations ($2 < b < 3$). The weight of the condensate probability hump is approximately $1/M$, in accord with the picture that the system typically contains a well-defined condensate located on a single site at any given time.

6.6.3 Rate $u_k = 1 + a/k^\sigma$: Stretched-Exponential Critical Behaviour

Consider the ZRP with hopping rate [21, 22, 26]

$$u_k = 1 + \frac{a}{k^\sigma} \, , \tag{6.45}$$

where σ is an arbitrary exponent. The situation of interest corresponds to $0 < \sigma < 1$. Equation (6.3) leads to the estimate

$$p_k \sim \exp\left(-a\sum_{\ell=1}^{k}\frac{1}{\ell^\sigma}\right) \sim \exp\left(-\frac{a}{1-\sigma}k^{1-\sigma}\right) \, . \tag{6.46}$$

The generating series $P(z)$ has an essential singularity at $z = 1$ with an exponentially small discontinuity. The critical density

$$\rho_c = \frac{P'(1)}{P(1)} = \frac{\sum_k k\, p_k}{\sum_k p_k}$$

is finite. The occupation probability at the critical density, $f_{k,\mathrm{st}} = p_k/P(1)$, decays as a stretched exponential law.

Part II: Dynamics (Sects. 6.7–6.9)

6.7 Zero-Range Processes: Nonstationary Dynamics (I)

The question is to determine the temporal evolution of the system starting from a random disordered initial condition. Here we study the dynamics of the class of ZRP giving rise to a condensation transition in their stationary state. For simplicity we will choose the hopping rate

$$u_k = 1 + \frac{b}{k} \, .$$

We address the question first in the fully connected geometry.

The same question can be addressed for dynamical urn models (see e.g. [8]). The analysis that follows [12], as well as that contained in the next section, are essentially the same as that performed for the zeta-urn model [10, 11].

6.7.1 Dynamics on the Complete Graph

We wish to determine the temporal evolution of the occupation probability $f_k(t)$. Conservation of probability and of density yields

$$\sum_{k=0}^{\infty} f_k(t) = 1 \, , \tag{6.47}$$

$$\sum_{k=1}^{\infty} k\, f_k(t) = \rho \, , \tag{6.48}$$

where we have taken the thermodynamic limit $N \to \infty, M \to \infty$, with fixed density $\rho = N/M$. We consider a system with Poissonian initial distribution of occupation probabilities,

$$f_k(0) = e^{-\rho} \frac{\rho^k}{k!} ,$$

i.e., such that initially particles are distributed at random amongst sites.

Since the Eq. (6.26) are non-linear they have no explicit solution in closed form. Yet one can extract from them an analytical description of the dynamics of the system at long times, both in the condensed phase, and at criticality. The structure of the reasoning borrows to former studies on urn models [10, 11]. (For a review, see [8].)

As we show below, there exists two different regimes in the evolution of the system, both in the condensed phase or at criticality, which we study successively.

(a) Nonequilibrium Dynamics of Condensation ($\rho > \rho_c$)

Since $\bar{u}_{eq} = 1$, we set, for large times,

$$\bar{u}_t \approx 1 + A \, \varepsilon_t , \tag{6.49}$$

where the small time scale ε_t is to be determined, and A is an unknown amplitude.

Regime I: k fixed, t large.

For t large enough, sites empty (u_k) faster than they fill ($\bar{u}_t$). In this regime there is convergence to equilibrium, hence we set

$$f_k(t) \approx f_{k,eq}(1 + v_k \, \varepsilon_t) , \tag{6.50}$$

with $f_{k,eq}$ given by (6.43), and where the v_k are unknown. This expression carried into (6.26) yields the stationary equation $\dot{f}_k = 0$, because the derivative $\dot{f}_k$, proportional to $\dot{\varepsilon}_t$, is negligible compared to the right-hand side. We thus obtain an equation similar to the detailed balance condition:

$$\frac{f_{k+1,eq}}{f_{k,eq}} \frac{1 + v_{k+1} \, \varepsilon_t}{1 + v_k \, \varepsilon_t} = \frac{1 + A \varepsilon_t}{u_{k+1}} .$$

Using (6.43) and (6.3), we obtain, at leading order in ε_t, $v_{k+1} - v_k = A$, and finally

$$v_k = v_0 + k A . \tag{6.51}$$

At this stage, v_0 and the amplitude A are still to be determined.

Regime II: k and t are simultaneously large.

This is the scaling regime, with scaling variable $x = k\,\varepsilon_t$. Following the treatment of [10, 11], we look for a similarity solution of (6.26) of the form

$$f_k(t) \approx (\rho - \rho_c)\,\varepsilon_t^2\,g(x) . \tag{6.52}$$

We thus obtain for $g(x)$ the linear differential equation

$$g''(x) + \left(\frac{x}{2} - A + \frac{b}{x}\right) g'(x) + \left(1 - \frac{b}{x^2}\right) g(x) = 0 ,$$

with $\varepsilon_t \approx t^{-\frac{1}{2}}$. This is precisely the differential equation found in [10, 11], for the zeta-urn model. The amplitude A can be determined by the fact that the equation has an acceptable solution $g(x)$ vanishing as $x \to 0$ and $x \to \infty$ [10]. The amplitude A and the scaling function $g(x)$ are universal quantities, only depending on the value of b. The sum rules (6.47) and (6.48) yield respectively

$$\int_0^\infty \mathrm{d}x\, g(x) = \frac{v_0 + A\rho_c}{\rho_c - \rho} ,$$

$$\int_0^\infty \mathrm{d}x\, x g(x) = 1 .$$

The differential equation above has no closed form solution. However further information on the form of the solution $g(x)$ can be found in [10, 11].

An intuitive description of the dynamics of condensation in the scaling regime is as follows. The typical occupancy k_{cond} of the sites making the condensate scales as $t^{\frac{1}{2}}$. The total number of particles in the condensate is equal to $M(\rho - \rho_c)$, the remaining $M\rho_c$ lying in the fluid. Therefore the number of sites belonging to the condensate scales as $M(\rho - \rho_c)t^{-\frac{1}{2}}$.

(b) Nonequilibrium Critical Dynamics ($\rho = \rho_c$)

The analysis follows closely that done in [11]. We set

$$\bar{u}_t \approx 1 + A\,\varepsilon_t ,$$

with $\varepsilon_t = t^{-\omega}$, where the exponent ω is to be determined, and we consider the same two regimes as above. In regime I, we still set (6.50) for $f_k(t)$. The reasoning leading to the relationship $v_k = v_0 + k\,A$ (see (6.51)) is still valid here. In regime II, we look for a similarity solution to (6.26) of the form

$$f_k(t) \approx f_{k,\mathrm{eq}}\, g_c(x) \qquad x = k\, t^{-\frac{1}{2}} . \tag{6.53}$$

Indeed, for any large but finite time t, the system looks critical, i.e., the occupation probabilities $f_k(t)$ have essentially converged toward their equilibrium values (6.43), for $k \ll t^{1/2}$, while for $k \gg t^{1/2}$ the system still looks disordered. The $f_k(t)$ are expected to fall off very fast, which is confirmed by the following analysis.

The sum rules (6.47) and (6.48) lead respectively to the following equations, provided that $b > 3$,

$$v_0 + A\rho_c = 0 \, , \tag{6.54}$$

$$t^{-\omega}(v_0\rho_c + A\mu_c) = t^{-(b-2)/2}(b-1)\Gamma(b)\int_0^\infty du\, u^{1-b}(1 - g_c(u)) \, , \tag{6.55}$$

where $\mu_c = \sum k^2 f_{k,\text{eq}}$ is given in Eq. (6.44). Equation (6.55) fixes the value of ω:

$$\omega = (b-2)/2 \, . \tag{6.56}$$

The differential equation obeyed by $g_c(x)$ is obtained by carrying (6.53) into (6.26). It reads

$$g_c''(x) + \left(\frac{x}{2} - \frac{b}{x}\right)g_c'(x) = 0 \, ,$$

the solution of which is, with $g_c(0) = 1$,

$$g_c(x) = \frac{2^{-b}}{\Gamma(\frac{b+1}{2})}\int_x^\infty dy\, y^b e^{-y^2/4} \, . \tag{6.57}$$

The fall-off of $g_c(x)$ for $x \gg 1$ is very fast: $g_c(x) \sim \exp(-x^2/4)$, hence $f_k(t) \sim \exp(-k^2/4t)$. We finally obtain

$$A = \frac{(b-1)\Gamma(b)}{\mu_c - \rho_c^2}\int_0^\infty du\, u^{1-b}(1 - g_c(u)) = \frac{(b-2)(b-3)}{b-1}\Gamma\left(\frac{b}{2}\right) \, .$$

Let us mention that for any hopping rate of the form $u_k \approx 1 + b/k$, the scaling functions, $g(x)$ in the condensed phase (more precisely: $g(x)/(\rho - \rho_c)$), and $g_c(x)$ at criticality, are universal. In both cases the scaling variable is $x = kt^{-1/2}$. The critical density ρ_c, and, as a consequence, any quantity depending on ρ_c, such as the amplitude v_0, are non universal, with values depending on the precise definition of u_k. As noted above, the amplitude A is a universal quantity in the condensed phase.

6.7.2 Late Stages of the Dynamics and the Case of One Dimension

As mentioned above, in the first stage of the dynamics, in the MF geometry, the number of most populated sites decays as $M/t^{1/2}$. Hence, after a time of order M^2, the system contains a finite number of highly populated sites, i.e., condensate precursors.

The late stage of the non-stationary dynamics, where all but one of the precursors die out, is thus expected to also last a length of time of the order of the diffusive timescale M^2. This is substantiated by numerical simulations in [25]. Another argument is presented in Subsect. 6.9.4. The whole non-stationary process of the formation of the condensate is therefore characterised by a single timescale

$$\tau_{\text{non}-\text{st}} \sim M^2 \ .$$

The same results hold for the 1DAS case. The analysis relies upon numerical work or heuristic and scaling arguments [12, 25].

A similar scenario holds in the 1DS geometry, the only difference being that $\tau_{\text{non}-\text{st}}$ now scales as M^3. The shift of the dynamical exponent by one unit in the 1DS geometry has a common origin [12, 25]: it stems from the Gambler's ruin problem [28]. An analogous phenomenon is encountered for example in the coarsening law for the domain growth, and in the motion of a tagged particle, in 1D Kawasaki dynamics [29].

We refer to the original references for further results (scaling functions, critical case, etc.).

6.8 Zero-Range Processes: Nonequilibrium Dynamics (II)

So far we considered the dynamics of one-time quantities, related to the random variable $N_1(t)$. We now explore another facet of the nonequilibrium dynamics of the ZRP with hopping rate $u_k = 1 + b/k$, namely the two-time nonstationary aspects of its dynamics. This essentially means that any function of the two times depends on both times, instead of depending on their difference, which would be the case at stationarity. The situation here is analogous to that encountered when a ferromagnetic spin system is quenched from a high temperature, corresponding to an initial disordered configuration, to a lower temperature, $T \leq T_c$ [30, 31].

We consider the same ZRP as in the previous section, on the complete graph, in the thermodynamic limit. The system relaxes from a nonequilibrium initial condition towards equilibrium. In order to characterize the fluctuations of the local density of particles, $N_1(t)$, around its mean $\langle N_1(t) \rangle = \rho$, we study its associated two-time correlation and response functions, and fluctuation-dissipation ratio.

6.8.1 General Framework

The connected two-time correlation function of the density between time s (waiting time) and time t (observation time), with $s \leq t$, is defined as

$$C(t, s) = \langle N_1(s) N_1(t) \rangle - \rho^2 \ .$$

It can be rewritten as

$$C(t, s) = \sum_{k \geq 1} k \, \gamma_k(t, s) - \rho^2 \ ,$$

where the function $\gamma_k(t, s)$ is defined by

$$\gamma_k(t, s) = \sum_{j \geq 1} j \, f_j(s) \, \mathcal{P}\{N_1(t) = k \mid N_1(s) = j\}$$

with the initial value at $t = s$

$$\gamma_k(s, s) = k \, f_k(s) \, .$$

Its temporal evolution for $t \geq s$ is given by the master equation (6.26):

$$\frac{\partial \gamma_k(t, s)}{\partial t} = \mu_{k+1} \gamma_{k+1} + \lambda_{k-1} \gamma_{k-1} - (\mu_k + \lambda_k) \gamma_k \qquad (k \geq 1) \, ,$$

$$\frac{\partial \gamma_0(t, s)}{\partial t} = \mu_1 \gamma_1 - \lambda_0 \gamma_0 \, . \tag{6.58}$$

The rates λ_k and μ_k are defined in (6.27). The rate λ_k only depends on the $f_k(t)$, hence (6.58) are linear equations for the $\gamma_k(t, s)$.

The local response function measures the influence on the mean density on site number 1 of a perturbation in the canonically conjugate variable, i.e., the local chemical potential acting on the same site. Suppose that site number 1 is subjected to a small time-dependent chemical potential $\alpha_1(t)$, so that the total reduced energy of the system (see Sect. 6.5) is now

$$\beta E(\{N_i\}) = \sum_{i=1}^{M} \beta E(N_i) + \alpha_1(t) N_1 \, .$$

The mean density on site number 1 reads

$$\langle N_1(t) \rangle = \rho + \int_0^t ds \, R(t, s) \, \alpha_1(s) + \cdots \, ,$$

where only the term linear in $\alpha(s)$ is written explicitly. The kernel of the linear response is the two-time response function

$$R(t, s) = \frac{\delta \langle N_1(t) \rangle}{\delta \alpha_1(s)} \, .$$

The temporal evolution of this function is given by a master equation similar to (6.26) [11].

The zero-range processes that we consider here have a fast convergence towards equilibrium, with a finite relaxation time τ_{relax} in their fluid phase, as is the case for a generic statistical-mechanical model in its high-temperature disordered phase. If the earlier time exceeds the relaxation time ($s \gg \tau_{\text{relax}}$), the system is at equilibrium. One-time quantities take their equilibrium values. Two-time quantities, such as the correlation and response functions, are invariant under time translations:

$$C(t,s) = C_{eq}(\tau), \quad R(t,s) = R_{eq}(\tau), \tag{6.59}$$

where $\tau = t - s \geq 0$. They are related by the fluctuation-dissipation theorem

$$R_{eq}(\tau) = -\frac{dC_{eq}(\tau)}{d\tau}. \tag{6.60}$$

In the condensed phase and at criticality the relaxation time τ_{relax} becomes infinite. If the waiting time s and the observation time t are much smaller than τ_{relax}, both time-translation invariance (6.59) and the fluctuation-dissipation theorem (6.60) are violated. It is convenient [32] to characterize departure from equilibrium by the fluctuation-dissipation ratio

$$X(t,s) = \frac{R(t,s)}{\dfrac{\partial C(t,s)}{\partial s}}. \tag{6.61}$$

In general, this dimensionless quantity depends on both times s and t and on the observable under consideration. It may also exhibit a non-trivial scaling behavior in the two-time plane. In all known cases it is observed that

$$0 \leq X(t,s) \leq 1. $$

6.8.2 Application: ZRP with Condensation ($u_k = 1 + b/k$)

Nonequilibrium Critical Dynamics ($\rho = \rho_c$)

Let us first note that the variance of the population of site number 1 converges to its equilibrium value $C_{eq} = \mu_c - \rho_c^2$ as a power law:

$$C(t,t) = \langle N_1(t)^2 \rangle - \rho_c^2 \approx C_{eq} - \frac{2^{3-b} t^{-(b-3)/2}}{(b-3)\,\Gamma\left((b+1)/2\right)}. \tag{6.62}$$

The derivation of the behaviour of the two-time density correlation and response functions is the same as in [11]. In the nonequilibrium scaling regime ($s, t \gg 1$), one finds

$$C(t,s) \approx s^{-(b-3)/2}\,\Phi(x),$$
$$\frac{\partial C(t,s)}{\partial s} \approx s^{-(b-1)/2}\,\Phi_1(x), \tag{6.63}$$
$$R(t,s) \approx s^{-(b-1)/2}\,\Phi_2(x),$$

where

$$x = t/s \geq 1.$$

As a consequence, in the scaling regime, the fluctuation-dissipation ratio $X(t,s)$ only depends on x:

$$X(t,s) \approx \mathcal{X}(x) = \frac{\Phi_2(x)}{\Phi_1(x)}.$$

The dimensionless scaling function $\mathcal{X}(x)$ is universal, and it admits a non-trivial limit value in the regime where the two time variables s and t are well separated in the scaling regime [33]:

$$X_\infty = \lim_{s\to\infty} \lim_{t\to\infty} X(t,s) = \mathcal{X}(\infty) .$$

Explicit expressions for the above scaling functions can be derived, using a spectral decomposition in Laguerre polynomials [11]. The limit fluctuation-dissipation ratio thus obtained

$$X_\infty = \frac{b+1}{b+2} \quad (b > 3) ,$$

lies in an unusually high range $(4/5 < X_\infty < 1)$ for a critical system. Indeed, statistical-mechanical models such as ferromagnets are observed to have $0 < X_\infty \leq 1/2$ at their critical point. The upper bound $X_\infty = 1/2$, corresponding to the mean-field situation [33], is also observed in a range of simpler models [32, 34].

The above results illustrate general predictions on nonequilibrium critical dynamics [31, 33–35]. The exponent of the waiting time s in the first line of (6.63) already appears in (6.62). It is related to the anomalous dimension of the observable under consideration, and would read $(d-2+\eta)/z_c$ for a d-dimensional ferromagnet, where η is the equilibrium correlation exponent and z_c the dynamical critical exponent. The scaling functions $\Phi(x)$, $\Phi_{1,2}(x)$ are universal up to an overall multiplicative constant, and they obey a common power-law fall-off in $x^{-b/2}$. The latter exponent is not related to exponents pertaining to usual equilibrium critical dynamics. It reads $-\lambda_c/z_c = \Theta_c - d/z_c$ for a ferromagnet, where λ_c is the critical autocorrelation exponent [36] and Θ_c is the critical initial-slip exponent [35].

Nonequilibrium Dynamics of Condensation $(\rho > \rho_c)$

In the scaling regime, two-time quantities are found to scale as [11]

$$\begin{aligned}
C(t,s) &\approx (\rho - \rho_c)s^{1/2}\,\Phi(x) , \quad (x = t/s) , \\
\frac{\partial C(t,s)}{\partial s} &\approx (\rho - \rho_c)s^{-1/2}\,\Phi_1(x) , \\
R(t,s) &\approx (\rho - \rho_c)s^{-1/2}\,\Phi_2(x) , \\
X(t,s) &\approx \mathcal{X}(x) = \frac{\Phi_2(x)}{\Phi_1(x)} .
\end{aligned} \qquad (6.64)$$

The scaling functions $\Phi(x)$, $\Phi_{1,2}(x)$ have finite values, both at coinciding times $(x = 1)$ and in the limit of large time separations $(x = \infty)$. The limit fluctuation-dissipation ratio $X_\infty = \mathcal{X}(\infty)$ depends continuously on b throughout the condensed phase $(b > 2)$, and vanishes only as

$$X_\infty = b^{-1/2} - \frac{b^{-3/4}}{4} + \cdots$$

for b large, which corresponds formally to low temperature, while coarsening systems are known [37] to have identically $X_\infty = 0$ throughout their low-temperature phase. In Fig. 6.1 a summary of the values of X_∞ is presented.

This dynamics is different from the usual phase-ordering dynamics [30]. Indeed, when a ferromagnet is quenched below its critical temperature, domain growth and phase separation take place in a statistically homogeneous way, at least for an infinite system. In the present situation, condensation takes place in a very inhomogeneous fashion, since fewer and fewer sites are involved in the process.

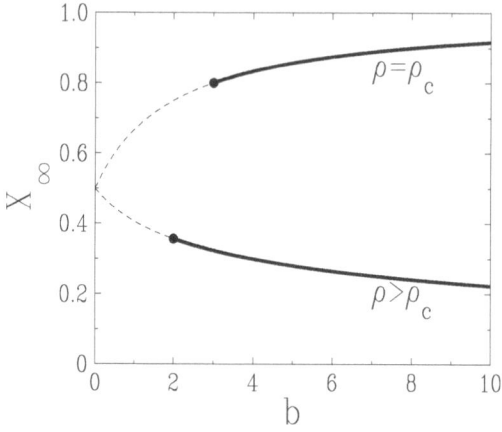

Fig. 6.1. Plot of the limit fluctuation-dissipation ratio X_∞ against b. *Upper curve*: critical point ($b > 3$, $\rho = \rho_c$). *Lower curve*: condensed phase ($b > 2$, $\rho > \rho_c$). *Thin dashed lines*: continuation of the results to lower values of b

6.8.3 One Dimension

For both the symmetric and asymmetric cases the response can be defined in the same fashion as above. There is no analytical tools at our disposal to compute these functions, even in the scaling regime. However, for the symmetric case, the fluctuation-dissipation still holds at equilibrium, while it should be violated in the stationary state of the asymmetric case.

6.9 Stationary Dynamics of the Condensate

6.9.1 The Question Posed

Consider a ferromagnetic system, an Ising spin system for instance. At equilibrium in the low temperature phase, the spin symmetry is spontaneously

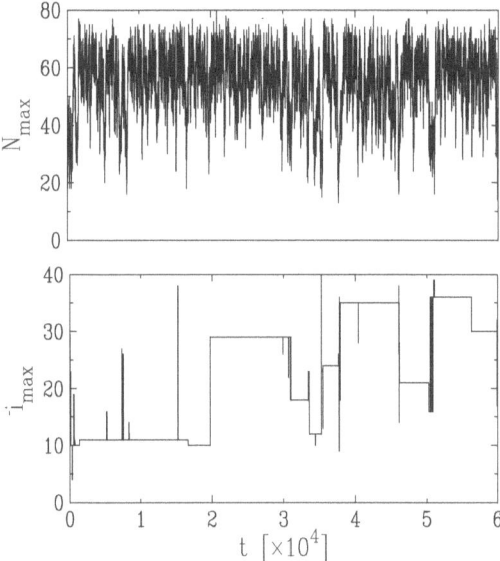

Fig. 6.2. Dynamics of the condensate (1DS geometry with $b = 4$, $M = 40$, $N = 80$). *Upper panel*: instantaneous number of particles $N_{max}(t)$ on the most populated site. *Lower panel*: location $i_{max}(t)$ of that site

broken. There are two possible equilibrium states, one with positive magneti-sation, the other one with negative magnetisation. However, if one observes a large but finite system, then as time passes, the magnetisation keeps chang-ing sign, the system flipping between the two possible equilibrium states. Ergodicity is restored for a finite system. The typical time between two flips is exponential in L^{d-1}, where L is the linear size of the system, and d the dimension of space.

A similar situation occurs for in the condensed phase of a ZRP. Here the spontaneously broken symmetry is translational invariance. For a large but finite system in the stationary state, as time passes, the condensate keeps moving across the system. It spends long lengths of time on a given site, before suddenly disappearing and reappearing on another site. The typical value of these lengths of time defines the characteristic time τ of the dynamics of the condensate. The aim of this section is to analyse the nature of this motion and in particular to characterise how τ scales with the system size M.

6.9.2 Numerical Observations

An intuitive understanding of the phenomenon is easily gained by perform-ing Monte-Carlo simulations. These simulations, done in the three geome-tries: mean-field (MF), one-dimensional asymmetric (1DAS) ($p = 1$), and one-dimensional symmetric (1DS) ($p = q = 1/2$), lead to a common picture.

The condensate is immobile for rather long lapses of time; it then performs sudden random non-local jumps all over the system, at Poissonian times whose characteristic scale grows rapidly with the system size M. Figure 6.2 illustrates this process for the 1DS case, for a system of size $M = 40$, with $N = 80$ particles, i.e., $\rho = 2$, and $b = 4$, hence $\rho_c = 1/2$. The upper panel shows the track of the instantaneous number of particles $N_{\max}(t)$ on the most populated site. The signal for $N_{\max}(t)$ fluctuates around $\Delta \equiv M(\rho - \rho_c) = 60$, the mean size of the condensate. The lower panel shows the label $i_{\max}(t)$ of that site, i.e., the location of the condensate. The non-local character of the motion of the condensate is clearly visible, whereas the longest lapses of time where the condensate stays still give a heuristic measure of the characteristic time τ.

We show in what follows that $\tau \sim M^b$ for the fully connected geometry and the directed case, while $\tau \sim M^{b+1}$ for the symmetric case. Moving the condensate is therefore slower than forming the condensate ($\tau_{\text{non-st}} \sim M^2$, $\tau_{\text{non-st}} \sim M^3$ respectively, see Subsect. 6.7.2).

6.9.3 Theoretical Analysis

All the idea relies on a problem of barrier crossing. Defining the potential as $V_k = -\ln f_{k,\text{st}}$, then a dip in the probability $f_{k,\text{st}}$ corresponds to a barrier in the potential. The flipping time τ is the time to cross the barrier, or the first-passage time from right to left. Let us explain these ideas in more detail.

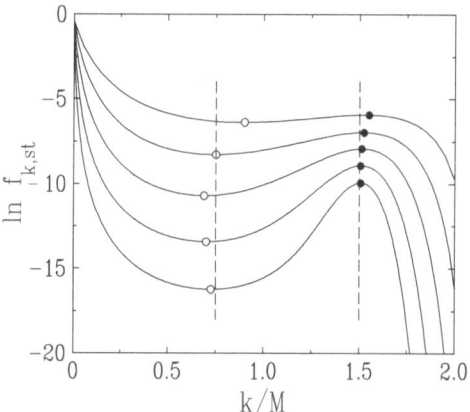

Fig. 6.3. Logarithmic plot of the occupation probability $f_{k,\text{st}}$ in the condensed phase ($b = 4, \rho = 4, \rho_c = 2$), against the ratio k/M. *Top* to *bottom*: $M = 20, 40, 80, 160$, and 320. *Full lines*: $f_{k,\text{st}}$ obtained by (6.32) and (6.33). Full (*empty*) symbols: maxima (minima) of occupation probability. *Dashed vertical lines*: asymptotic locations of the minima: $k/M = \Delta/(2M) = (\rho - \rho_c)/2 = 3/4$, and of the maxima: $k/M = \Delta/M = \rho - \rho_c = 3/2$

We first analyse the behaviour of the occupation probability $f_{k,\text{st}}$ in the condensed phase. Figure 6.3 shows a logarithmic plot of $f_{k,\text{st}}$, computed using equations (6.32) and (6.33), against the ratio k/M, for $b = 4$, $\rho = 2$, and several values of M. This plot exhibits the following features.

- For $k \ll M$, the distribution $f_{k,\text{st}}$ is approximately given by the power law (6.43) of an infinite critical system.
- The contribution of the condensate slowly builds up as a probability hump around $k = \Delta = M(\rho - \rho_c)$, the mean number of excess particles.
- One observes a broad and shallow probability "dip" in the region located between the critical background and the condensate hump, i.e., in the region $k \gg 1$ and $\Delta - k \gg 1$.

The region of the dip is dominated by configurations where the excess particles are shared by *two* sites. Indeed, one has (see [21] for a proof)

$$f_{k,\text{st}} \approx (b-1)\Gamma(b) \frac{\Delta^b}{k^b(\Delta - k)^b} \qquad (k \gg 1, \Delta - k \gg 1) . \tag{6.65}$$

The observed locations of the maxima ($k \approx \Delta$) and minima ($k \approx \Delta/2$) of the occupation probability corroborate this picture, as explained in the caption of Fig. 6.3.

These observations lead to the following crude estimate for the characteristic time:

$$\tau \sim \frac{1}{f_{\text{min}}} , \tag{6.66}$$

since the minimum f_{min} of $f_{k,\text{st}}$ corresponds to a barrier to cross, in the spirit of the Arrhenius law. The limiting scale of time is that required for the passage of this potential barrier. Equation (6.65) implies that f_{min} is reached near the middle of the dip region ($k \approx N/2$), and therefore (6.66) yields

$$\tau \sim \Delta^b . \tag{6.67}$$

We now present a more precise treatment. Assume that the condensate is on site number 1 at the initial observation time ($t = 0$). The number $N_1(t)$ of particles on that site is initially very large, $N_1(0) \approx \Delta$, and therefore evolves slowly, until the condensate dissolves into the critical background. Thus

- We single out $N_1(t)$ as the collective co-ordinate of the system, that is the appropriate slow variable describing the dynamics of the condensate.
- We model the dynamics of $N_1(t)$ by (6.22), i.e., by a biased diffusive motion on the interval $k = 0, \ldots, N$. The left hopping rate is taken equal to the microscopic rate: $\mu_k = u_k$. The right hopping rate λ_k is chosen such that, in the stationary state, the probability $f_{k,\text{st}}$ of the effective model coincide with the occupation probability (6.33) of the original ZRP. The detailed balance condition (6.23) yields

$$\lambda_k = \frac{\mu_{k+1} f_{k+1,\mathrm{st}}}{f_{k,\mathrm{st}}} = \frac{Z_{M-1,N-k-1}}{Z_{M-1,N-k}} = \bar{u}_{\mathrm{st}}(M-1,N-k) ,$$

where the right side of the equation is the average rate coming from $M-1$ sites, containing $N-k$ particles (see (6.39)). The rate λ_k is thus a function of k, M, and N. For $M = 2$, this formula gives $\lambda_k = u_{N-k}$, as expected. In the fluid phase, in the thermodynamic limit, the rates λ_k converge to z_0, defined in (6.36). Finally, for the condensed phase, in the dip region, we obtain

$$\lambda_k \approx 1 + \frac{b}{\Delta - k} \equiv u_{\Delta-k} .$$

This effective description reduces the full model to a Markovian model for one degree of freedom in an *asymmetric* potential. The two valleys of the potential are separated by a high (power-law) barrier. The left potential valley, corresponding to the critical background, has a weight $P_{\mathrm{L}} \approx 1$, whereas the right potential valley, corresponding to the hump of the condensate, has a weight $P_{\mathrm{R}} \approx 1/M \ll 1$ (see Subsect. 6.6.2).

In this framework the stationary dynamics of the condensate is characterised by a single diverging timescale. We choose to define this timescale, denoted by τ_{Markov}, to be the crossing time T_{L} from the right valley to the left one in the effective Markovian problem. The characteristic time is thus expressed by

$$\tau_{\mathrm{Markov}} \equiv T_{\mathrm{L}} = \sum_{\ell=1}^{N} \frac{1}{\mu_\ell f_{\ell,\mathrm{st}}} \sum_{m=\ell}^{N} f_{m,\mathrm{st}} , \qquad (6.68)$$

in terms of known quantities, the rates μ_k and the stationary probabilities $f_{k,\mathrm{st}}$. Its asymptotic growth is easily determined by noting that (6.68) is dominated by the behaviour of the probability $f_{k,\mathrm{st}}$ in the region of the dip. Hence, inserting the expression (6.65) into (6.68), and evaluating the sum as an integral, we obtain

$$\tau_{\mathrm{Markov}} \approx \frac{b\Gamma(b+1)}{(b-1)\Gamma(2b+2)} \frac{\Delta^{b+1}}{M} = \frac{b\Gamma(b+1)}{(b-1)\Gamma(2b+2)} (\rho - \rho_c)^{b+1} M^b . \quad (6.69)$$

In order to compare the above theoretical predictions to the measured flipping time τ, we compute the two-time stationary correlation function

$$C(t,0) = \langle N_1(t) N_1(0) \rangle - \rho^2 .$$

This quantity decays exponentially with a relaxation constant which gives a natural measure of τ. It is found that $\tau \sim \tau_{\mathrm{Markov}} \sim M^b$ in the MF and 1DAS geometries, and that $\tau \sim M\tau_{\mathrm{Markov}} \sim M^{b+1}$ in the 1DS geometry. For the latter case the occurrence of one supplementary power in the system size has the same origin as for nonequilibrium dynamics.

6.9.4 Last Remarks

Table 6.1 summarises the values of the dynamical exponents z and Z, such that $\tau_{\mathrm{non-st}} \sim M^z$ and $\tau \sim M^Z$, where τ is the characteristic timescale for the stationary motion of the condensate.

As recalled above, the non-stationary dynamical exponents are insensitive to the exponent b, and more generally to the statics, provided the system is in its condensed phase. This feature is easily understood in the context of the Markovian Ansatz proposed in the present work. Indeed the last stage of the formation of the condensate, i.e., the disappearance of the smaller of the last two precursors, implies no barrier crossing. In terms of the occupation of the condensate, it corresponds to the transition from N_1 to Δ, where the initial occupation N_1 of the larger precursor was already larger than $\Delta/2$, corresponding to the top of the potential barrier. This explains why $\tau_{\mathrm{non-st}}$ is given by the diffusive timescale, both in the framework of the Markovian Ansatz and in the MF and 1DAS geometries.

Table 6.1. Non-stationary and stationary dynamical exponents of the ZRP with static exponent $b > 2$

Geometry	z	Z
MF, 1DAS	2	b
1DS	3	$b + 1$

6.10 Further References

Complementary aspects to the present notes can be found in [4, 38–40].

Acknowledgements

It is a pleasure to thank J-M Drouffe and J-M Luck, with whom I have been collaborating through the years on the subject of these notes, as well as R Blythe, M Evans, S Grosskinsky, T Hanney, E Levine, S Majumdar, D Mukamel, G Schütz, and H Spohn for interesting discussions.

Appendix: Proof of Eqs. (6.12) and (6.13)

We recall the fundamental equation (6.16)

$$p(W_{k,l} - W_{k,0}) + q(W_{l,k} - W_{l,0})$$
$$= p\left(W_{k+1,l-1}\frac{p_{k+1}p_{l-1}}{p_k p_l} - W_{l,0}\right) + q\left(W_{l+1,k-1}\frac{p_{k-1}p_{l+1}}{p_k p_l} - W_{k,0}\right). \text{ (A.1)}$$

We first prove (6.12) for the symmetric case, $p = 1/2$. Setting $x_k = W_{k,l}\, p_k p_l$, where $k + l = n$, (A.1) can be rewritten as

$$x_k - x_{n-(k-1)} = x_{k+1} - x_{n-k}\ .$$

This expression is therefore a constant independent of k, which is equal to zero, as can be seen by taking $k = n$. We thus obtain

$$x_{k+1} = x_{n-k}$$

which is the detailed balance condition

$$p_k p_l\, W_{k,l} = p_{k-1}p_{l+1}\, W_{l+1,k-1}\ . \tag{A.2}$$

We now show that in the general case, $p \neq 1/2$, Eq. (A.1) yields two constraints on the rate: Eq. (A.2) to be interpreted as the pairwise balance condition, and Eq. (6.13)

$$W_{l,k} - W_{k,l} = W_{l,0} - W_{k,0}\ . \tag{A.3}$$

Set $a_k = p_k p_l\, W_{k,l}$. Equation (A.1) can be rewritten as

$$y_{k+1} - y_k = (p - q)(a_{n-k} - a_k) \tag{A.4}$$

where

$$y_k = p\, x_k - q\, x_{n-(k-1)}, \qquad y_0 = 0\ . \tag{A.5}$$

If

$$y_{k+1} = y_{n-k} \tag{A.6}$$

then it follows immediately that $x_{k+1} = x_{n-k}$, which is the condition for pairwise balance seen above. This relation itself plugged into (A.5), yields

$$x_{k+1} - x_k = a_{n-k} - a_k$$

which is (A.3). In order to prove (A.6) we set

$$A_k = a_1 + \cdots + a_k$$

i.e. $a_k = A_k - A_{k-1}$. We thus have

$$y_k + (p - q)(A_{k-1} + A_{n-k}) = y_{k+1} + (p - q)(A_k + A_{n-k-1})$$

which is equal to $(p - q)A_n$, hence

$$y_k = (p - q)(A_n - A_{n-k} - A_k + a_k)\ .$$

Therefore $y_k - (p - q)a_k$ is symmetric in the change $k \to n - k$, and finally

$$y_k - (p - q)a_k = y_{n-k} - (p - q)a_{n-k}$$

which, using (A.4), yields (A.6).

References

1. Spitzer F. 1970: Advances in Math. **5**, 246
2. Ehrenfest P and T 1907 Phys. Zeit. **8** 311
3. Kac M 1947 Amer. Math. Monthly **54** 369; Kac M 1959 *Probability and Related Topics in Physical Sciences* Lectures in Applied Mathematics vol **1 A** (American Mathematical Society)
4. Evans M R and Hanney T 2005 J. Phys. A **38** R195
5. Andjel E D 1982 Ann. Prob. **10** 525
6. Schütz G M, Ramaswamy R and Barma M 1996 J. Phys. A **29** 837
7. Ritort F 1995 Phys. Rev. Lett. **75** 1190
8. For a review, see: Godrèche C and Luck J M 2002 J. Phys. Cond. Matt. **14** 1601
9. Bialas P, Burda Z and Johnston D 1997 Nucl. Phys. B **493** 505; Bialas P, Burda Z and Johnston D 1999 Nucl. Phys. B **542** 413
10. Drouffe J M, Godrèche C and Camia F 1998 J. Phys. A **31** L19
11. Godrèche C and Luck J M 2001 Eur. Phys. J. B **23** 473
12. Godrèche C 2003 J. Phys. A **36** 6313
13. Cocozza-Thivent C 1985 Z. Wahr. **70** 509
14. Godrèche C and Luck J M in preparation
15. Grosskinsky S and Spohn H 2003 Bull. Braz. Math. Soc. **34** 489
16. Evans M R and Hanney T 2003 J. Phys. A **36** L44
17. Kolmogorov A N 1936 Math Ann. **112** 115
18. Kelly F 1979 *Reversibility and Stochastic Networks*, Wiley
19. Godrèche C, Levine E and Mukamel D 2005 J. Phys. A **38** L523
20. Godrèche C 2006 j. Phys. A **39**, 9055
21. Godrèche C and Luck J M 2005 J. Phys. A **38** 7215
22. Evans M R 2000 Braz. J. Phys. **30** 42
23. O'Loan O J, Evans M R and Cates M E 1998 Phys. Rev. E **58** 1404
24. Majumdar S N, Evans M R and Zia R K P 2005 Phys. Rev. Lett. **94** 180601; Evans M R, Majumdar S N and Zia R K P 2005 cond-mat/0510512
25. Grosskinsky S, Schütz G M and Spohn H 2003 J. Stat. Phys. **113** 389
26. Kafri Y, Levine E, Mukamel D, Schütz G M and Török J 2002 Phys. Rev. Lett. **89** 035702
27. Kaupuzs J, Mahnke R and Harris R J 2005 Phys. Rev. E **72** 056125
28. Feller W 1966 *An Introduction to Probability Theory and its Applications* (New-York: Wiley) vol 1
29. Cordery R, Sarker S and Tobochnik J 1981 Phys. Rev. B **24** 5402; Cornell S J, Kaski K and Stinchcombe R B 1991 Phys. Rev. B **44** 12263; Cornell S J and Bray A J 1996 Phys. Rev. E **54** 1153; Ben-Naim E and Krapivsky P L 1998 J. Stat. Phys. **93** 583; Godrèche C and Luck J M 2003 J. Phys. A **36** 9973
30. Bray A J 1994 Adv. Phys. **43** 357
31. Godrèche C and Luck J M 2002 J. Phys. Cond. Matt. **14** 1589
32. Cugliandolo L and Kurchan J 1994 J. Phys. A **27** 5749
33. Godrèche C and Luck J M 2000 J. Phys. A **33** 9141
34. Godrèche C and Luck J M 2000 J. Phys. A **33** 1151
35. Janssen H K, Schaub B and Schmittmann B 1989 Z. Phys. B **73** 539
36. Huse D A 1989 Phys. Rev. B **40** 304
37. Cugliandolo L F, Kurchan J and Peliti L 1997 Phys. Rev. **E55** 3898; Barrat A 1998 Phys. Rev. E **57** 3629; Berthier L, Barrat J L and Kurchan J 1999 Eur. Phys. J. **B11** 635

38. Kipnis C and Landim C 1999 *Scaling limits of interacting particle systems* Springer
39. Bertini L, De Sole A, Gabrielli D, Jona-Lasinio G and Landim C 2002 J. Stat. Phys. **107** 635
40. Harris R J, Rakos A and Schütz G 2005 J. Stat. Mech. P08003

Field-Theory Approaches
to Nonequilibrium Dynamics

U. C. Täuber

Department of Physics, Center for Stochastic Processes in Science and Engineering
Virginia Polytechnic Institute and State University
Blacksburg, Virginia 24061-0435, USA
tauber@vt.edu

It is explained how field-theoretic methods and the dynamic renormalisation group (RG) can be applied to study the universal scaling properties of systems that either undergo a continuous phase transition or display generic scale invariance, both near and far from thermal equilibrium. Part 1 introduces the response functional field theory representation of (nonlinear) Langevin equations. The RG is employed to compute the scaling exponents for several universality classes governing the critical dynamics near second-order phase transitions in equilibrium. The effects of reversible mode-coupling terms, quenching from random initial conditions to the critical point, and violating the detailed balance constraints are briefly discussed. It is shown how the same formalism can be applied to nonequilibrium systems such as driven diffusive lattice gases. Part 2 describes how the master equation for stochastic particle reaction processes can be mapped onto a field theory action. The RG is then used to analyse simple diffusion-limited annihilation reactions as well as generic continuous transitions from active to inactive, absorbing states, which are characterised by the power laws of (critical) directed percolation. Certain other important universality classes are mentioned, and some open issues are listed.

7.1 Critical Dynamics

Field-theoretic tools and the *renormalisation group* (RG) method have had a tremendous impact in our understanding of the *universal power laws* that emerge near equilibrium critical points (see, e.g., Refs. [1–6]), including the associated *dynamic critical phenomena* [7,8]. Our goal here is to similarly describe the scaling properties of systems driven far from thermal equilibrium, which either undergo a *continuous nonequilibrium phase transition* or display *generic scale invariance*. We are then confronted with capturing the (stochastic) dynamics of the long-wavelength modes of the "slow" degrees of freedom,

U.C. Täuber: *Field-Theory Approaches to Nonequilibrium Dynamics*, Lect. Notes Phys. **716**, 295–348 (2007)
DOI 10.1007/3-540-69684-9_7

namely the order parameter for the transition, any conserved quantities, and perhaps additional relevant variables. In these lecture notes, I aim to briefly describe how a representation in terms of a *field theory action* can be obtained for (1) general nonlinear Langevin stochastic differential equations [8, 9]; and (2) for master equations governing classical particle reaction–diffusion systems [10–12]. I will then demonstrate how the dynamic (perturbative) RG can be employed to derive the asymptotic *scaling laws* in stochastic dynamical systems; to infer the *upper critical dimension* d_c (for dimensions $d \leq d_c$, fluctuations strongly affect the universal scaling properties); and to systematically compute the *critical exponents* as well as to determine further universal properties in various intriguing dynamical model systems both near and far from equilibrium. (For considerably more details, especially on the more technical aspects, the reader is referred to Ref. [13].)

7.1.1 Continuous Phase Transitions and Critical Slowing Down

The vicinity of a *critical point* is characterised by *strong correlations* and *large fluctuations*. The system under investigation is then behaving in a highly cooperative manner, and as a consequence, the standard approximative methods of statistical mechanics, namely perturbation or cluster expansions that assume either weak interactions or short-range correlations, fail. Upon approaching an equilibrium continuous (second-order) phase transition, i.e., for $|\tau| \ll 1$, where $\tau = (T - T_c)/T_c$ measures the deviation from the critical temperature T_c, the thermal fluctuations of the *order parameter* $S(x)$ (which characterises the different thermodynamic phases, usually chosen such that the thermal average $\langle S \rangle = 0$ vanishes in the high-temperature "disordered" phase) are, in the thermodynamic limit, governed by a diverging length scale

$$\xi(\tau) \sim |\tau|^{-\nu} \ . \tag{7.1}$$

Here, we have defined the *correlation length* via the typically exponential decay of the static *cumulant* or *connected two-point correlation function* $C(x) = \langle S(x) S(0) \rangle - \langle S \rangle^2 \sim e^{-|x|/\xi}$, and ν denotes the correlation length *critical exponent*. As $T \to T_c$, $\xi \to \infty$, which entails the absence of any characteristic length scale for the order parameter fluctuations at criticality. Hence we expect the *critical correlations* to follow a power law $C(x) \sim |x|^{-(d-2+\eta)}$ in d dimensions, which defines the *Fisher exponent* η. The following *scaling ansatz* generalises this power law to $T \neq T_c$, but still in the vicinity of the critical point,

$$C(\tau, x) = |x|^{-(d-2+\eta)} \, \widetilde{C}_\pm(x/\xi) \ , \tag{7.2}$$

with two distinct regular *scaling functions* $\widetilde{C}_+(y)$ for $T > T_c$ and $\widetilde{C}_-(y)$ for $T < T_c$, respectively. For its Fourier transform $C(\tau, q) = \int d^d x \, e^{-iq \cdot x} \, C(\tau, x)$, one obtains the corresponding scaling form

$$C(\tau, q) = |q|^{-2+\eta} \, \hat{C}_\pm(q\,\xi) \ , \tag{7.3}$$

with new scaling functions $\hat{C}_\pm(p) = |p|^{2-\eta} \int d^d y \, e^{-ip \cdot y} \, |y|^{-(d-2+\eta)} \, \widetilde{C}_\pm(y)$.

As we will see in Subsect. 7.1.5, there are only *two* independent static critical exponents. Consequently, it must be possible to use the static scaling hypothesis (7.2) or (7.3), along with the definition (7.1), to express the exponents describing the thermodynamic singularities near a second-order phase transition in terms of ν and η through *scaling laws*. For example, the order parameter in the low-temperature phase ($\tau < 0$) is expected to grow as $\langle S \rangle \sim (-\tau)^{\beta}$. Let us consider Eq. (7.2) in the limit $|x| \to \infty$. In order for the $|x|$ dependence to cancel, $\widetilde{C}_{\pm}(y) \propto |y|^{d-2+\eta}$ for large $|y|$, and therefore $C(\tau, |x| \to \infty) \sim \xi^{-(d-2+\eta)} \sim |\tau|^{\nu(d-2+\eta)}$. On the other hand, $C(\tau, |x| \to \infty) \to -\langle S \rangle^2 \sim -(-\tau)^{2\beta}$ for $T < T_c$; thus we identify the order parameter critical exponent through the *hyperscaling relation*

$$\beta = \frac{\nu}{2}(d - 2 + \eta) . \tag{7.4}$$

Let us next consider the isothermal static *susceptibility* χ_{τ}, which according to the equilibrium fluctuation–response theorem is given in terms of the spatial integral of the correlation function $C(\tau, x)$: $\chi_{\tau}(\tau) = (k_B T)^{-1} \lim_{q \to 0} C(\tau, q)$. But $\hat{C}_{\pm}(p) \sim |p|^{2-\eta}$ as $p \to 0$ to ensure nonsingular behaviour, whence $\chi_{\tau}(\tau) \sim \xi^{2-\eta} \sim |\tau|^{-\nu(2-\eta)}$, and upon defining the associated thermodynamic critical exponent γ via $\chi_{\tau}(\tau) \sim |\tau|^{-\gamma}$, we obtain the scaling relation

$$\gamma = \nu(2 - \eta) . \tag{7.5}$$

The scaling laws (7.2), (7.3) as well as scaling relations such as (7.4) and (7.5) can be put on solid foundations by means of the RG procedure, based on an *effective* long-wavelength Hamiltonian $\mathcal{H}[S]$, a functional of $S(x)$, that captures the essential physics of the problem, namely the relevant symmetries in order parameter and real space, and the existence of a continuous phase transition. The probability of finding a configuration $S(x)$ at given temperature T is then given by the canonical distribution

$$\mathcal{P}_{eq}[S] \propto \exp(-\mathcal{H}[S]/k_B T) . \tag{7.6}$$

For example, the mathematical description of the critical phenomena for an $O(n)$-symmetric order parameter field $S^{\alpha}(x)$, with vector index $\alpha = 1, \ldots, n$, is based on the *Landau–Ginzburg–Wilson functional* [1–6]

$$\mathcal{H}[S] = \int d^d x \sum_{\alpha} \left[\frac{r}{2}[S^{\alpha}(x)]^2 + \frac{1}{2}[\nabla S^{\alpha}(x)]^2 \right.$$

$$\left. + \frac{u}{4!}[S^{\alpha}(x)]^2 \sum_{\beta}[S^{\beta}(x)]^2 - h^{\alpha}(x) S^{\alpha}(x) \right] , \tag{7.7}$$

where $h^{\alpha}(x)$ is the external field thermodynamically conjugate to $S^{\alpha}(x)$, $u > 0$ denotes the strength of the nonlinearity that drives the phase transformation, and r is the control parameter for the transition, i.e., $r \propto T - T_c^0$, where T_c^0 is the (mean-field) critical temperature. Spatial variations of the order parameter are energetically suppressed by the term $\sim [\nabla S^\alpha(\boldsymbol{x})]^2$, and the corresponding positive coefficient has been absorbed into the fields S^α.

We shall, however, not pursue the static theory further here, but instead proceed to a full *dynamical description* in terms of nonlinear Langevin equations [7, 8]. We will then formulate the RG within this dynamic framework, and therein demonstrate the emergence of scaling laws and the computation of critical exponents in a systematic perturbative expansion with respect to the deviation $\epsilon = d - d_c$ from the upper critical dimension.

In order to construct the desired effective stochastic dynamics near a critical point, we recall that correlated region of size ξ become quite large in the vicinity of the transition. Since the associated relaxation times for such clusters should grow with their extent, one would expect the characteristic time scale for the relaxation of the order parameter fluctuations to increase as well as $T \to T_c$, namely

$$t_c(\tau) \sim \xi(\tau)^z \sim |\tau|^{-z\nu} \, , \tag{7.8}$$

which introduces the *dynamic critical exponent* z that encodes the *critical slowing down* at the phase transition; usually $z \geq 1$. Since the typical relaxation rates therefore scale as $\omega_c(\tau) = 1/t_c(\tau) \sim |\tau|^{z\nu}$, we may utilise the static scaling variable $\boldsymbol{p} = \boldsymbol{q}\,\xi$ to generalise the crucial observation (7.8) and formulate a *dynamic scaling hypothesis* for the wavevector-dependent dispersion relation of the order parameter fluctuations [14, 15],

$$\omega_c(\tau, \boldsymbol{q}) = |\boldsymbol{q}|^z \,\hat{\omega}_\pm(\boldsymbol{q}\,\xi) \, . \tag{7.9}$$

We can then proceed to write down dynamical scaling laws by simply postulating the additional scaling variables $s = t/t_c(\tau)$ or $\omega/\omega_c(\tau, \boldsymbol{q})$. For example, as an immediate consequence we find for the time-dependent mean order parameter

$$\langle S(\tau, t) \rangle = |\tau|^\beta \, \hat{S}(t/t_c) \, , \tag{7.10}$$

with $\hat{S}(s \to \infty) = \text{const.}$, but $\hat{S}(s) \sim s^{-\beta/z\nu}$ as $s \to 0$ in order for the τ dependence to disappear. At the critical point ($\tau = 0$), this yields the power-law decay $\langle S(t) \rangle \sim t^{-\alpha}$, with

$$\alpha = \frac{\beta}{z\,\nu} = \frac{1}{2\,z}\,(d - 2 + \eta) \, . \tag{7.11}$$

Similarly, the scaling law for the *dynamic order parameter susceptibility (response function)* becomes

$$\chi(\tau, \boldsymbol{q}, \omega) = |\boldsymbol{q}|^{-2+\eta} \, \hat{\chi}_\pm(\boldsymbol{q}\,\xi, \omega\,\xi^z) \, , \tag{7.12}$$

which constitutes the dynamical generalisation of Eq. (7.3), for $\chi(\tau, \boldsymbol{q}, 0) = (k_\mathrm{B}T)^{-1} C(\tau, \boldsymbol{q})$. Upon applying the *fluctuation–dissipation theorem*, valid in thermal equilibrium, we therefrom obtain the *dynamic correlation function*

$$C(\tau, \boldsymbol{q}, \omega) = \frac{2k_{\mathrm{B}}T}{\omega} \operatorname{Im} \chi(\tau, \boldsymbol{q}, \omega) = |\boldsymbol{q}|^{-z-2+\eta} \, \hat{C}_{\pm} \left(\boldsymbol{q}\, \xi, \omega\, \xi^{z} \right) , \qquad (7.13)$$

and for its Fourier transform in real space and time,

$$C(\tau, \boldsymbol{x}, t) = \int \frac{\mathrm{d}^{d}q}{(2\pi)^{d}} \int \frac{\mathrm{d}\omega}{2\pi} \, \mathrm{e}^{\mathrm{i}(\boldsymbol{q}\cdot\boldsymbol{x}-\omega t)} \, C(\tau, \boldsymbol{q}, \omega) = |\boldsymbol{x}|^{-(d-2+\eta)} \, \tilde{C}_{\pm} \left(\boldsymbol{x}/\xi, t/\xi^{z} \right) ,$$
$$(7.14)$$

which reduces to the static limit (7.2) if we set $t = 0$.

The critical slowing down of the order parameter fluctuations near the critical point provides us with a natural *separation of time scales*. Assuming (for now) that there are no other conserved variables in the system, which would constitute additional slow modes, we may thus resort to a *coarse-grained* long-wavelength and long-time description, focusing merely on the order parameter kinetics, while subsuming all other "fast" degrees of freedom in *random "noise"* terms. This leads us to a *mesoscopic Langevin equation* for the *slow* variables $S^{\alpha}(\boldsymbol{x}, t)$ of the form

$$\frac{\partial S^{\alpha}(\boldsymbol{x}, t)}{\partial t} = F^{\alpha}[S](\boldsymbol{x}, t) + \zeta^{\alpha}(\boldsymbol{x}, t) . \qquad . \qquad (7.15)$$

In the simplest case, the *systematic* force terms here just represent purely *relaxational* dynamics towards the equilibrium configuration [16],

$$F^{\alpha}[S](\boldsymbol{x}, t) = -D \frac{\delta \mathcal{H}[S]}{\delta S^{\alpha}(x, t)} , \qquad (7.16)$$

where D represents the relaxation coefficient, and $\mathcal{H}[S]$ is again the effective Hamiltonian that governs the phase transition, e.g. given by Eq. (7.7). For the *stochastic forces* we may assume the most convenient form, and take them to simply represent Gaussian white noise with zero mean, $\langle \zeta^{\alpha}(\boldsymbol{x}, t) \rangle = 0$, but with their second moment in thermal equilibrium fixed by *Einstein's relation*

$$\langle \zeta^{\alpha}(\boldsymbol{x}, t) \, \zeta^{\beta}(\boldsymbol{x}', t') \rangle = 2k_{\mathrm{B}}T \, D \, \delta(\boldsymbol{x} - \boldsymbol{x}') \, \delta(t - t') \, \delta^{\alpha\beta} . \qquad (7.17)$$

As can be verified by means of the associated Fokker–Planck equation for the time-dependent probability distribution $\mathcal{P}[S, t]$, Eq. (7.17) guarantees that eventually $\mathcal{P}[S, t \to \infty] \to \mathcal{P}_{\mathrm{eq}}[S]$, the canonical distribution (7.6). The stochastic differential equation (7.15), with (7.16), the Hamiltonian (7.7), and the noise correlator (7.17), define the *relaxational model A* (according to the classification in Ref. [7]) for a nonconserved $O(n)$-symmetric order parameter.

If, however, the order parameter is *conserved*, we have to consider the associated continuity equation $\partial_t S^{\alpha} + \boldsymbol{\nabla} \cdot \boldsymbol{J}^{\alpha} = 0$, where typically the conserved current is given by a gradient of the field S^{α}: $\boldsymbol{J}^{\alpha} = -D \boldsymbol{\nabla} S^{\alpha} + \ldots$; as a consequence, the order parameter fluctuations will relax *diffusively* with diffusion coefficient D. The ensuing *model B* [7,16] for the relaxational critical dynamics of a conserved order parameter can be obtained by replacing $D \to -D \boldsymbol{\nabla}^{2}$ in Eqs. (7.16) and (7.17). In fact, we will henceforth treat both models A and B simultaneously by setting $D \rightarrow D(i\nabla)^a$, where $a = 0$ and $a = 2$ respectively represent the nonconserved and conserved cases. Explicitly, we thus obtain

$$
\begin{aligned}
\frac{\partial S^\alpha(\boldsymbol{x}, t)}{\partial t} &= -D(i\nabla)^a \frac{\delta \mathcal{H}[S]}{\delta S^\alpha(\boldsymbol{x}, t)} + \zeta^\alpha(\boldsymbol{x}, t) \\
&= -D(i\nabla)^a \Big[r - \nabla^2 + \frac{u}{6} \sum_\beta [S^\beta(\boldsymbol{x})]^2 \Big] S^\alpha(\boldsymbol{x}, t) \\
&\quad + D(i\nabla)^a h^\alpha(\boldsymbol{x}, t) + \zeta^\alpha(\boldsymbol{x}, t) ,
\end{aligned}
\tag{7.18}
$$

with

$$
\langle \zeta^\alpha(\boldsymbol{x}, t) \zeta^\beta(\boldsymbol{x}', t') \rangle = 2k_B T \, D(i\nabla)^a \, \delta(\boldsymbol{x} - \boldsymbol{x}') \delta(t - t') \delta^{\alpha\beta} .
\tag{7.19}
$$

Notice already that the presence or absence of a conservation law for the order parameter implies different dynamics for systems described by identical static behaviour. Before proceeding with the analysis of the relaxational models, we remark that in general there may exist additional *reversible* contributions to the systematic forces $F^\alpha[S]$, see Subsect. 7.1.6, and/or dynamical mode-couplings to additional conserved, slow fields, which effect further splitting into several distinct *dynamic universality classes* [6, 7, 13].

Let us now evaluate the dynamic response and correlation functions in the *Gaussian* (mean-field) approximation in the high-temperature phase. To this end, we set $u = 0$ and thus discard the nonlinear terms in the Hamiltonian (7.7) as well as in Eq. (7.18). The ensuing Langevin equation becomes linear in the fields S^α, and is therefore readily solved by means of Fourier transforms. Straightforward algebra and regrouping some terms yields

$$
\left[-i\omega + Dq^a \left(r + q^2 \right) \right] S^\alpha(\boldsymbol{q}, \omega) = Dq^a \, h^\alpha(\boldsymbol{q}, \omega) + \zeta^\alpha(\boldsymbol{q}, \omega) .
\tag{7.20}
$$

With $\langle \zeta^\alpha(\boldsymbol{q}, \omega) \rangle = 0$, this gives immediately

$$
\chi_0^{\alpha\beta}(\boldsymbol{q}, \omega) = \frac{\partial \langle S^\alpha(\boldsymbol{q}, \omega) \rangle}{\partial h^\beta(\boldsymbol{q}, \omega)} \Big|_{h=0} = Dq^a \, G_0(\boldsymbol{q}, \omega) \delta^{\alpha\beta} ,
\tag{7.21}
$$

with the *response propagator*

$$
G_0(\boldsymbol{q}, \omega) = \left[-i\omega + Dq^a \left(r + q^2 \right) \right]^{-1} .
\tag{7.22}
$$

As is readily established by means of the residue theorem, its Fourier back-transform in time obeys *causality*,

$$
G_0(\boldsymbol{q}, t) = \Theta(t) \, e^{-Dq^a (r+q^2) t} .
\tag{7.23}
$$

Setting $h^\alpha = 0$, and with the noise correlator (7.19) in Fourier space

$$
\langle \zeta^\alpha(\boldsymbol{q}, \omega) \zeta^\beta(\boldsymbol{q}', \omega') \rangle = 2k_B T \, Dq^a \, (2\pi)^{d+1} \delta(\boldsymbol{q} + \boldsymbol{q}') \delta(\omega + \omega') \delta^{\alpha\beta} ,
\tag{7.24}
$$

we obtain the Gaussian dynamic correlation function $\langle S^\alpha(\boldsymbol{q}, \omega) \, S^\beta(\boldsymbol{q}', \omega') \rangle_0 = C_0(\boldsymbol{q}, \omega) \, (2\pi)^{d+1} \delta(\boldsymbol{q} + \boldsymbol{q}') \, \delta(\omega + \omega')$, where

$$C_0(\boldsymbol{q}, \omega) = \frac{2k_{\rm B}TDq^a}{\omega^2 + [Dq^a(r + q^2)]^2} = 2k_{\rm B}T \, Dq^a \, |G_0(\boldsymbol{q}, \omega)|^2 \ . \tag{7.25}$$

The fluctuation–dissipation theorem (7.13) is of course satisfied; moreover, as function of wavevector and time,

$$C_0(\boldsymbol{q}, t) = \frac{k_{\rm B}T}{r + q^2} \, {\rm e}^{-Dq^a(r+q^2)\,|t|} \ . \tag{7.26}$$

In the Gaussian approximation, away from criticality ($r > 0$, $\boldsymbol{q} \neq 0$) the temporal correlations for models A and B decay exponentially, with the relaxation rate $\omega_c(r, \boldsymbol{q}) = Dq^{2+a}(1 + r/q^2)$. Upon comparison with the dynamic scaling hypothesis (7.9), we infer the mean-field scaling exponents $\nu_0 = 1/2$ and $z_0 = 2 + a$. At the critical point, a nonconserved order parameter relaxes diffusively ($z_0 = 2$) in this approximation, whereas the conserved order parameter kinetics becomes even slower, namely subdiffusive with $z_0 = 4$. Finally, invoking Eqs. (7.12), (7.13), (7.14), or simply the static limit $C_0(\boldsymbol{q}, 0) = k_{\rm B}T/(r + q^2)$, we find $\eta_0 = 0$ for the Gaussian model.

The full nonlinear Langevin equation (7.18) cannot be solved exactly. Yet a perturbation expansion with respect to the coupling u may be set up in a slightly cumbersome, but straightforward manner by direct iteration of the equations of motion [16, 17]. More elegantly, one may utilise a path-integral representation of the Langevin stochastic process [18, 19], which allows the application of all the standard tools from statistical and quantum field theory [1–6], and has the additional advantage of rendering symmetries in the problem more explicit [8, 9, 13].

7.1.2 Field Theory Representation of Langevin Equations

Our starting point is a set of coupled Langevin equations of the form (7.15) for mesoscopic, coarse-grained stochastic variables $S^\alpha(\boldsymbol{x}, t)$. For the stochastic forces, we make the simplest possible assumption of *Gaussian white noise*,

$$\langle \zeta^\alpha(\boldsymbol{x}, t) \rangle = 0 \ , \quad \langle \zeta^\alpha(\boldsymbol{x}, t) \, \zeta^\beta(\boldsymbol{x}', t') \rangle = 2L^\alpha \, \delta(\boldsymbol{x} - \boldsymbol{x}') \, \delta(t - t') \, \delta^{\alpha\beta} \ , \tag{7.27}$$

where L^α may represent a differential operator (such as the Laplacian ∇^2 for conserved fields), and even a functional of S^α. In the time interval $0 \leq t \leq t_f$, the moments (7.27) are encoded in the probability distribution

$$\mathcal{W}[\zeta] \propto \exp\left[-\frac{1}{4} \int {\rm d}^d x \int_0^{t_f} {\rm d}t \sum_\alpha \zeta^\alpha(\boldsymbol{x}, t) \left[(L^\alpha)^{-1} \zeta^\alpha(\boldsymbol{x}, t)\right]\right] \ . \tag{7.28}$$

If we now switch variables from the stochastic noise ζ^α to the fields S^α by means of the equations of motion (7.15), we obtain

$$W[\zeta]\,\mathcal{D}[\zeta] = P[S]\,\mathcal{D}[S] \propto \mathrm{e}^{-\mathcal{G}[S]}\,\mathcal{D}[S]\;, \tag{7.29}$$

with the statistical weight determined by the *Onsager–Machlup functional* [9]

$$\mathcal{G}[S] = \frac{1}{4}\int \mathrm{d}^d x \int \mathrm{d}t \sum_\alpha \left(\frac{\partial S^\alpha}{\partial t} - F^\alpha[S]\right)\left[(L^\alpha)^{-1}\left(\frac{\partial S^\alpha}{\partial t} - F^\alpha[S]\right)\right]\;. \tag{7.30}$$

Note that the Jacobian for the nonlinear variable transformation $\{\zeta^\alpha\} \to \{S^\alpha\}$ has been omitted here. In fact, the above procedure is properly defined through appropriately discretising time. If a *forward* (Itô) discretisation is applied, then indeed the associated functional determinant is a mere constant that can be absorbed in the functional measure. The functional (7.30) already represents a desired field theory action. Since the probability distribution for the stochastic forces should be normalised, $\int \mathcal{D}[\zeta]\, W[\zeta] = 1$, the associated "partition function" is unity, and carries no physical information (as opposed to static statistical field theory, where it determines the free energy and hence the entire thermodynamics). The Onsager–Machlup representation is however plagued by technical problems: Eq. (7.30) contains $(L^\alpha)^{-1}$, which for conserved variables entails the inverse Laplacian operator, i.e., a Green function in real space or the singular factor $1/q^2$ in Fourier space; moreover the nonlinearities in $F^\alpha[S]$ appear *quadratically*. Hence it is desirable to linearise the action (7.30) by means of a Hubbard–Stratonovich transformation [9].

We shall follow an alternative, more general route that completely avoids the appearance of the inverse operators $(L^\alpha)^{-1}$ in intermediate steps. Our goal is to average over noise "histories" for observables $A[S]$ that need to be expressible in terms of the stochastic fields S^α: $\langle A[S]\rangle_\zeta \propto \int \mathcal{D}[\zeta]\, A[S(\zeta)]\, W[\zeta]$. For this purpose, we employ the identity

$$1 = \int \mathcal{D}[S] \prod_\alpha \prod_{(\boldsymbol{x},t)} \delta\left(\frac{\partial S^\alpha(\boldsymbol{x},t)}{\partial t} - F^\alpha[S](\boldsymbol{x},t) - \zeta^\alpha(\boldsymbol{x},t)\right)$$

$$= \int \mathcal{D}[i\tilde{S}] \int \mathcal{D}[S] \exp\left[-\int \mathrm{d}^d x \int \mathrm{d}t \sum_\alpha \tilde{S}^\alpha \left(\frac{\partial S^\alpha}{\partial t} - F^\alpha[S] - \zeta^\alpha\right)\right]\,, \tag{7.31}$$

where the first line constitutes a rather involved representation of the unity (in a somewhat symbolic notation; again proper discretisation should be invoked here), and the second line utilises the Fourier representation of the (functional) delta distribution by means of the purely imaginary auxiliary fields $\tilde{S}$ (and factors 2π have been absorbed in its functional measure).

Inserting (7.31) and the probability distribution (7.28) into the desired stochastic noise average, we arrive at

$$\langle A[S]\rangle_\zeta \propto \int \mathcal{D}[i\tilde{S}] \int \mathcal{D}[S] \exp\left[-\int \mathrm{d}^d x \int \mathrm{d}t \sum_\alpha \tilde{S}^\alpha \left(\frac{\partial S^\alpha}{\partial t} - F^\alpha[S]\right)\right] A[S]$$

$$\times \int \mathcal{D}[\zeta] \exp\left(-\int \mathrm{d}^d x \int \mathrm{d}t \sum_\alpha \left[\frac{1}{4}\zeta^\alpha (L^\alpha)^{-1}\zeta^\alpha - \tilde{S}^\alpha\,\zeta^\alpha\right]\right)\,. \tag{7.32}$$

We may now evaluate the Gaussian integrals over the noise ζ^α, which yields

$$\langle A[S] \rangle_\zeta = \int \mathcal{D}[S]\, A[S]\, \mathcal{P}[S] \ , \quad \mathcal{P}[S] \propto \int \mathcal{D}[i\widetilde{S}]\, e^{-\mathcal{A}[\widetilde{S},S]} \ , \tag{7.33}$$

with the statistical weight now governed by the *Janssen–De Dominicis "response" functional* [9, 18, 19]

$$\mathcal{A}[\widetilde{S}, S] = \int d^d x \int_0^{t_f} dt \sum_\alpha \left[\widetilde{S}^\alpha \left(\frac{\partial S^\alpha}{\partial t} - F^\alpha[S] \right) - \widetilde{S}^\alpha L^\alpha \widetilde{S}^\alpha \right] \ . \tag{7.34}$$

Once again, we have omitted the functional determinant from the variable change $\{\zeta^\alpha\} \to \{S^\alpha\}$, and normalisation implies $\int \mathcal{D}[i\widetilde{S}] \int \mathcal{D}[S]\, e^{-\mathcal{A}[\widetilde{S},S]} = 1$. The first term in the action (7.34) encodes the temporal evolution according to the systematic terms in the Langevin equations (7.15), whereas the second term specifies the noise correlations (7.27). Since the auxiliary variables $\widetilde{S}^\alpha$, often termed Martin–Siggia–Rose response fields [20], appear only quadratically here, they may be eliminated via completing the squares and Gaussian integrations; thereby one recovers the Onsager–Machlup functional (7.30).

The Janssen–De Dominicis functional (7.34) takes the form of a $(d+1)$-dimensional statistical field theory with two independent sets of fields S^α and $\widetilde{S}^\alpha$. We may thus bring the established machinery of statistical and quantum field theory [1–6] to bear here; it should however be noted that the response functional formalism for stochastic Langevin dynamics incorporates causality in a nontrivial manner, which leads to important distinctions [8].

Let us specify the Janssen–De Dominicis functional for the purely *relaxational models* A and B [16, 17], see Eqs. (7.18) and (7.19), splitting it into the Gaussian and anharmonic parts $\mathcal{A} = \mathcal{A}_0 + \mathcal{A}_{\text{int}}$ [9], which read

$$\mathcal{A}_0[\widetilde{S}, S] = \int d^d x \int dt \sum_\alpha \left(\widetilde{S}^\alpha \left[\frac{\partial}{\partial t} + D\, (i\nabla)^a\, (r - \nabla^2) \right] S^\alpha \right.$$

$$\left. - D\, \widetilde{S}^\alpha\, (i\nabla)^a\, \widetilde{S}^\alpha - D\, \widetilde{S}^\alpha\, (i\nabla)^a\, h^\alpha \right) \ , \tag{7.35}$$

$$\mathcal{A}_{\text{int}}[\widetilde{S}, S] = D\, \frac{u}{6} \int d^d x \int dt \sum_{\alpha,\beta} \widetilde{S}^\alpha\, (i\nabla)^a\, S^\alpha\, S^\beta\, S^\beta \ . \tag{7.36}$$

Since we are interested in the vicinity of the critical point $T \approx T_c$, we have absorbed the constant $k_B T_c$ into the fields. The prescription (7.33) tells us how to compute time-dependent correlation functions $\langle S^\alpha(x, t)\, S^\beta(x', t') \rangle$. Using Eq. (7.35), the dynamic order parameter *susceptibility* follows from

$$\chi^{\alpha\beta}(x - x', t - t') = \frac{\delta \langle S^\alpha(x, t) \rangle}{\delta h^\beta(x', t')}\bigg|_{h=0} = D \left\langle S^\alpha(x, t)\, (i\nabla)^a\, \widetilde{S}^\beta(x', t') \right\rangle \ ; \tag{7.37}$$

for the simple relaxational models (only), the response function is just given by a correlator that involves an auxiliary variable, which explains why the $\widetilde{S}^\alpha$ are referred to as "response" fields. In equilibrium, one may employ the Onsager–Machlup functional (7.30) to derive the *fluctuation–dissipation theorem* [9]

$$\chi^{\alpha\beta}(\boldsymbol{x}-\boldsymbol{x}',t-t') = \Theta(t-t')\frac{\partial}{\partial t'}\left\langle S^\alpha(\boldsymbol{x},t)\,S^\beta(\boldsymbol{x}',t')\right\rangle , \qquad (7.38)$$

which is equivalent to Eq. (7.13) in Fourier space.

In order to access arbitrary correlators, we define the *generating functional*

$$\mathcal{Z}[\tilde{j},j] = \left\langle \exp\int d^d x\int dt\sum_\alpha\left(\tilde{j}^\alpha\,\tilde{S}^\alpha + j^\alpha\,S^\alpha\right)\right\rangle , \qquad (7.39)$$

wherefrom the correlation functions follow via functional derivatives,

$$\left\langle \prod_{ij} S^{\alpha_i}\,\tilde{S}^{\alpha_j}\right\rangle = \prod_{ij}\frac{\delta}{\delta j^{\alpha_i}}\,\frac{\delta}{\delta\tilde{j}^{\alpha_j}}\,\mathcal{Z}[\tilde{j},j]\bigg|_{\tilde{j}=0=j} , \qquad (7.40)$$

and the *cumulants* or *connected* correlation functions via

$$\left\langle \prod_{ij} S^{\alpha_i}\,\tilde{S}^{\alpha_j}\right\rangle_c = \prod_{ij}\frac{\delta}{\delta j^{\alpha_i}}\,\frac{\delta}{\delta\tilde{j}^{\alpha_j}}\,\ln\mathcal{Z}[\tilde{j},j]\bigg|_{\tilde{j}=0=j} . \qquad (7.41)$$

In the harmonic approximation, setting $u = 0$, $\mathcal{Z}[\tilde{j},j]$ can be evaluated explicitly (most directly in Fourier space) by means of Gaussian integration [9,13]; one thereby recovers (with $k_BT = 1$) the Gaussian response propagator (7.22) and two-point correlation function (7.25). Moreover, as a consequence of causality, $\left\langle \tilde{S}^\alpha(\boldsymbol{q},\omega)\,\tilde{S}^\beta(\boldsymbol{q}',\omega')\right\rangle_0 = 0$.

7.1.3 Outline of Dynamic Perturbation Theory

Since we cannot evaluate correlation functions with the nonlinear action (7.36) exactly, we resort to a perturbational treatment, assuming, for the time being, a small coupling strength u. The *perturbation expansion* with respect to u is constructed by rewriting the desired correlation functions in terms of averages with respect to the Gaussian action (7.35), henceforth indicated with index '0', and then expanding the exponential of $-\mathcal{A}_{\text{int}}$,

$$\left\langle \prod_{ij} S^{\alpha_i}\,\tilde{S}^{\alpha_j}\right\rangle = \frac{\left\langle \prod_{ij} S^{\alpha_i}\,\tilde{S}^{\alpha_j}\,e^{-\mathcal{A}_{\text{int}}[\tilde{S},S]}\right\rangle_0}{\left\langle e^{-\mathcal{A}_{\text{int}}[\tilde{S},S]}\right\rangle_0}$$

$$= \left\langle \prod_{ij} S^{\alpha_i}\,\tilde{S}^{\alpha_j}\sum_{l=0}^{\infty}\frac{1}{l!}\left(-\mathcal{A}_{\text{int}}[\tilde{S},S]\right)^l\right\rangle_0 . \qquad (7.42)$$

The remaining Gaussian averages, a series of polynomials in the fields S^α and $\tilde{S}^\alpha$, can be evaluated by means of *Wick's theorem*, here an immediate consequence of the Gaussian statistical weight, which states that all such averages can be written as a sum over all possible factorisations into Gaussian two-point functions $\langle S^\alpha \tilde{S}^\beta \rangle_0$, i.e., essentially the response propagator G_0, Eq. (7.22), and $\langle S^\alpha S^\beta \rangle_0$, the Gaussian correlation function C_0, Eq. (7.25). Recall that the denominator in Eq. (7.42) is exactly unity as a consequence of normalisation; alternatively, this result follows from causality in conjunction with our forward descretisation prescription, which implies that we should identify $\Theta(0) = 0$. (We remark that had we chosen another temporal discretisation rule, any apparent contributions from the denominator would be precisely cancelled by the in this case nonvanishing functional Jacobian from the variable transformation $\{\zeta^\alpha\} \to \{S^\alpha\}$.) At any rate, our stochastic field theory contains *no "vacuum" contributions*.

The many terms in the perturbation expansion (7.42) are most lucidly organised in a graphical representation, using *Feynman diagrams* with the basic elements depicted in Fig. 7.1. We represent the response propagator (7.22) by a *directed line* (here conventionally from right to left), which encodes its causal nature; the noise by a two-point *"source" vertex*, and the anharmonic term in Eq. (7.36) as a *four-point vertex*. In the diagrams representing the different terms in the perturbation series, these vertices serve as links for the propagator lines, with the fields S^α being encoded as the "incoming", and the $\tilde{S}^\alpha$ as the "outgoing" components of the lines. In Fourier space, translational invariance in space and time implies wavevector and frequency conservation at each vertex, see Fig. 7.2 below. An alternative, equivalent representation uses both the response and correlation propagators as independent elements, the latter depicted as undirected line, thereby disposing of the noise vertex, and retaining the nonlinearity in Fig. 7.1(c) as sole vertex.

Following standard field theory procedures [1–5], one establishes that the perturbation series for the *cumulants* (7.41) is given in terms of *connected* Feynman graphs only (for a detailed exposition of this and the following results, see Ref. [13]). An additional helpful reduction in the number of diagrams to be considered arises when one considers the *vertex functions*, which generalise the self-energy contributions $\Sigma(\mathbf{q}, \omega)$ in the *Dyson equation* for the response propagator, $G(\mathbf{q}, \omega)^{-1} = D q^a \chi(\mathbf{q}, \omega)^{-1} = G_0(\mathbf{q}, \omega)^{-1} - \Sigma(\mathbf{q}, \omega)$.

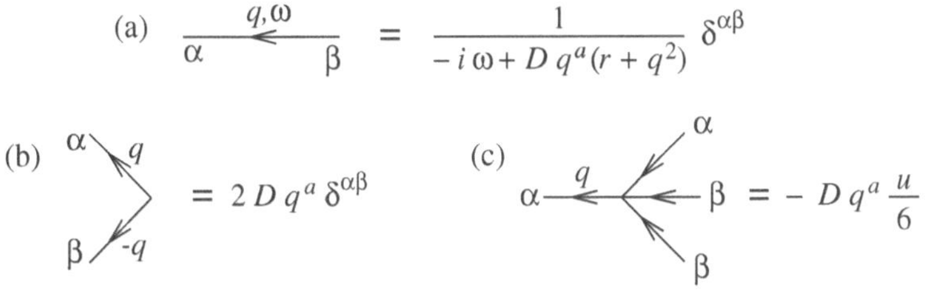

$$\text{(a)} \quad \underset{\alpha \qquad \beta}{\overset{q,\omega}{\longleftarrow}} \quad = \quad \frac{1}{-i\omega + D q^a (r + q^2)} \, \delta^{\alpha\beta}$$

$$\text{(b)} \quad \overset{\alpha \searrow q}{\underset{\beta \nearrow -q}{>}} \quad = 2 D q^a \, \delta^{\alpha\beta} \qquad \text{(c)} \quad \alpha \overset{q}{\underset{\beta}{\longleftarrow}} \overset{\alpha}{\underset{\beta}{\diagup}} \beta = -D q^a \frac{u}{6}$$

Fig. 7.1. Elements of dynamic perturbation theory for the $O(n)$-symmetric relaxational models: **(a)** response propagator; **(b)** noise vertex; **(c)** anharmonic vertex

To this end, we define the fields $\widetilde{\Phi}^\alpha = \delta \ln \mathcal{Z}/\delta \tilde{\jmath}^\alpha$ and $\Phi^\alpha = \delta \ln \mathcal{Z}/\delta \jmath^\alpha$, and introduce the new *generating functional*

$$\Gamma[\widetilde{\Phi}, \Phi] = -\ln \mathcal{Z}[\tilde{\jmath}, \jmath] + \int d^d x \int dt \sum_\alpha \left(\tilde{\jmath}^\alpha \, \widetilde{\Phi}^\alpha + \jmath^\alpha \, \Phi^\alpha \right) , \tag{7.43}$$

wherefrom the vertex functions are obtained via the functional derivatives

$$\Gamma^{(\widetilde{N},N)}_{\{\alpha_i\};\{\alpha_j\}} = \prod_i^{\widetilde{N}} \frac{\delta}{\delta \widetilde{\Phi}^{\alpha_i}} \prod_j^N \frac{\delta}{\delta \Phi^{\alpha_j}} \Gamma[\widetilde{\Phi}, \Phi]\Bigg|_{\tilde{\jmath}=0=\jmath} . \tag{7.44}$$

Diagrammatically, these quantities turn out to be represented by the possible sets of *one-particle (1PI) irreducible Feynman graphs* with N incoming and $\widetilde{N}$ outgoing "amputated" legs; i.e., these diagrams do not split into allowed subgraphs by simply cutting any single propagator line. For example, for the two-point functions a direct calculation yields the relations

$$\Gamma^{(1,1)}(\boldsymbol{q}, \omega) = D q^a \, \chi(-\boldsymbol{q}, -\omega)^{-1} = G_0(-\boldsymbol{q}, -\omega)^{-1} - \Sigma(-\boldsymbol{q}, -\omega) , \tag{7.45}$$

$$\Gamma^{(2,0)}(\boldsymbol{q}, \omega) = -\frac{C(\boldsymbol{q}, \omega)}{|G(\boldsymbol{q}, \omega)|^2} = -\frac{2D q^a}{\omega} \, \mathrm{Im} \, \Gamma^{(1,1)}(\boldsymbol{q}, \omega) , \tag{7.46}$$

where the second equation for $\Gamma^{(2,0)}$ follows from the fluctuation–dissipation theorem (7.13). Note that $\Gamma^{(0,2)}(\boldsymbol{q}, \omega) = 0$ vanishes because of causality.

The perturbation series can then be organised graphically as an expansion in successive orders with respect to the number of closed propagator *loops*. As an example, Fig. 7.2 depicts the one-loop contributions for the vertex functions $\Gamma^{(1,1)}$ and $\Gamma^{(1,3)}$ in the time domain with all required labels. One may formulate general *Feynman rules* for the construction of the diagrams and their translation into mathematical expressions for the lth order contribution to the *vertex function* $\Gamma^{(\widetilde{N},N)}$:

1. Draw all topologically different, connected *one-particle irreducible graphs* with $\widetilde{N}$ outgoing and N incoming lines connecting l relaxation vertices $\propto u$. Do *not* allow closed response loops (since in the Itô calculus $\Theta(0) = 0$).
2. Attach wavevectors $\boldsymbol{q}_i$, frequencies ω_i or times t_i, and component indices α_i to all directed lines, obeying "momentum (and energy)" conservation at each vertex.

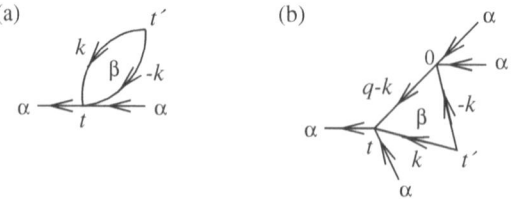

Fig. 7.2. One-loop diagrams for (a) $\Gamma^{(1,1)}$ and (b) $\Gamma^{(1,3)}$ in the time domain

3. Each directed line corresponds to a response propagator $G_0(-\boldsymbol{q}, -\omega)$ or $G_0(\boldsymbol{q}, t_i - t_j)$ in the frequency and time domain, respectively, the two-point vertex to the noise strength $2D\,q^a$, and the four-point relaxation vertex to $-D\,q^a\,u/6$. Closed loops imply integrals over the internal wavevectors and frequencies or times, subject to causality constraints, as well as sums over the internal vector indices. Apply the residue theorem to evaluate frequency integrals.

4. Multiply with -1 and the combinatorial factor counting all possible ways of connecting the propagators, l relaxation vertices, and k two-point vertices leading to topologically identical graphs, including a factor $1/l!\,k!$ originating in the expansion of $\exp(-\mathcal{A}_{\mathrm{int}}[\tilde{S}, S])$.

For later use, we provide the explicit results for the two-point vertex functions to two-loop order. After some algebra, the three diagrams in Fig. 7.3 give

$$
\begin{aligned}
\Gamma^{(1,1)}(\boldsymbol{q}, \omega) = {}& i\omega + Dq^a \left[r + q^2 + \frac{n+2}{6}\, u \int_k \frac{1}{r + k^2} \right. \\
& - \left(\frac{n+2}{6}\, u \right)^2 \int_k \frac{1}{r + k^2} \int_{k'} \frac{1}{(r + k'^2)^2} \\
& - \frac{n+2}{18}\, u^2 \int_k \frac{1}{r + k^2} \int_{k'} \frac{1}{r + k'^2} \frac{1}{r + (q - k - k')^2} \\
& \left. \times \left(1 - \frac{i\omega}{i\omega + \Delta(k) + \Delta(k') + \Delta(q - k - k')} \right) \right], \quad (7.47)
\end{aligned}
$$

where we have separated out the dynamic part in the last line, and introduced the abbreviations $\Delta(\boldsymbol{q}) = Dq^a\,(r + q^2)$ and $\int_k = \int d^d k/(2\pi)^d$ [13]. For the noise vertex, Fig. 7.4(a) yields [13]

$$
\begin{aligned}
\Gamma^{(2,0)}(\boldsymbol{q}, \omega) = {}& -2Dq^a \left[1 + Dq^a\, \frac{n+2}{18}\, u^2 \int_k \frac{1}{r + k^2} \int_{k'} \frac{1}{r + k'^2} \right. \\
& \left. \times \frac{1}{r + (q - k - k')^2}\, \mathrm{Re}\, \frac{1}{i\omega + \Delta(k) + \Delta(k') + \Delta(q - k - k')} \right] ; \quad (7.48)
\end{aligned}
$$

notice that for model B, as a consequence of the conservation law for the order parameter and ensuing wavevector dependence of the nonlinear vertex,

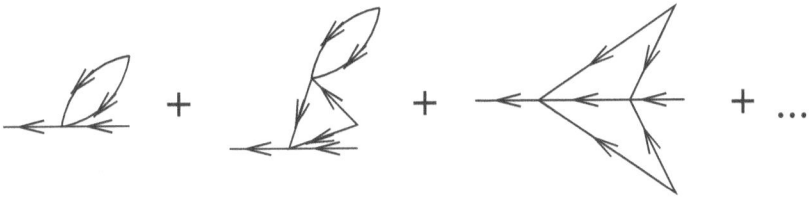

Fig. 7.3. One-particle irreducible diagrams for $\Gamma^{(1,1)}(\boldsymbol{q}, \omega)$ to second order in u

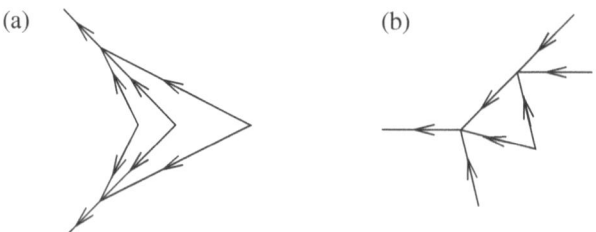

Fig. 7.4. (a) Two-loop diagram for $\Gamma^{(2,0)}(q,\omega)$; (b) one-loop graph for $\Gamma^{(1,3)}$

see Fig. 7.1(c), to *all orders* in the perturbation expansion

$$a = 2 : \ \Gamma^{(1,1)}(q = 0, \omega) = i\omega \ , \quad \left. \frac{\partial}{\partial q^2} \Gamma^{(2,0)}(q,\omega) \right|_{q=0} = -2D \ . \quad (7.49)$$

At last, with the shorthand notation $\underline{k} = (q,\omega)$, the analytical expression corresponding to the graph in Fig. 7.4(b) for the four-point vertex function at symmetrically chosen external wavevector labels reads

$$\Gamma^{(1,3)}(-3\underline{k}/2; \{\underline{k}/2\}) = D \left(\frac{3}{2} q \right)^a u \left[1 - \frac{n+8}{6} u \right.$$

$$\left. \times \int_k \frac{1}{r + k^2} \frac{1}{r + (q-k)^2} \left(1 - \frac{i\omega}{i\omega + \Delta(k) + \Delta(q-k)} \right) \right] . \ (7.50)$$

7.1.4 Renormalisation

Consider a typical loop integral, say the correction in Eq. (7.50) to the four-point vertex function $\Gamma^{(1,3)}$ at zero external frequency and momentum, whose "bare" value, without any fluctuation contributions, is u. In dimensions $d < 4$, one obtains, after introducing d-dimensional spherical coordinates and rendering the integrand dimensionless ($x = |k|/\sqrt{r}$):

$$u \int \frac{\mathrm{d}^d k}{(2\pi)^d} \frac{1}{(r + k^2)^2} = \frac{u \, r^{-2+d/2}}{2^{d-1} \pi^{d/2} \Gamma(d/2)} \int_0^\infty \frac{x^{d-1}}{(1 + x^2)^2} \, \mathrm{d}x \ , \quad (7.51)$$

where we have inserted the surface area $S_d = 2\pi^{d/2}/\Gamma(d/2)$ of the d-dimensional unit sphere, with Euler's Gamma function, $\Gamma(1 + x) = x \, \Gamma(x)$. Note that the integral on the right-hand side is finite. Thus, we see that the *effective* expansion parameter in perturbation theory is not just u, but the combinaton $u_{\mathrm{eff}} = u \, r^{(d-4)/2}$. Far away from T_c, it is small, and the perturbation expansion well-defined. However, $u_{\mathrm{eff}} \to \infty$ as $r \to 0$ for $d < 4$: we are facing *infrared (IR) divergences*, induced by the strong critical fluctuations that render the loop corrections singular. A straightforward application of perturbation theory will therefore not provide meaningful results, and we must expect the fluctuation contributions to modify the critical power laws.

Conversely, for dimensions $d \geq 4$, the integral in (7.51) develops *ultraviolet (UV) divergences* as the upper integral boundary is sent to infinity ($k = |\mathbf{k}|$),

$$\int_0^\Lambda \frac{k^{d-1}}{(r+k^2)^2}\, dk \sim \left\{ \begin{array}{ll} \ln(\Lambda^2/r) & d = 4 \\ \Lambda^{d-4} & d > 4 \end{array} \right\} \to \infty \quad \text{as } \Lambda \to \infty \, . \tag{7.52}$$

In lattice models, there is a finite wavevector cutoff, namely the Brillouin zone boundary, $\Lambda \sim (2\pi/a_0)^d$ for a hypercubic lattice with lattice constant a_0, whence physically these UV problems do not emerge. Yet we shall see that a formal treatment of these unphysical UV divergences will allow us to infer the correct power laws for the physical IR singularities associated with the critical point. The borderline dimension that separates the IR and UV singular regimes is referred to as *upper critical dimension* d_c; here $d_c = 4$. Note that at d_c, UV and IR singularities are intimately connected and appear in the form of *logarithmic divergences*, see Eq. (7.52). The situation is summarised in Table 7.1, where we have also stated that models with continuous order parameter symmetry, such as the Hamiltonian (7.7) with $n \geq 2$, do not allow long-range order in dimensions $d \leq d_{lc} = 2$ (*Mermin–Wagner–Hohenberg* theorem [21–23]). Here, d_{lc} is called the *lower critical dimension*; for the Ising model represented by Eq. (7.7) with $n = 1$, of course $d_{lc} = 1$.

Table 7.1. Mathematical and physical distinctions of the regimes $d < d_c$, $d = d_c$, and $d > d_c$, for the $O(n)$-symmetric models A and B (or static Φ^4 field theory)

Dimension Interval	Perturbation Series	Model A / B or Φ^4 Field Theory	Critical Behaviour
$d \leq d_{lc} = 2$	IR-singular UV-convergent	ill-defined u relevant	no long-range order ($n \geq 2$)
$2 < d < 4$	IR-singular UV-convergent	super-renormalisable u relevant	nonclassical exponents
$d = d_c = 4$	logarithmic IR-/ UV-divergence	renormalisable u marginal	logarithmic corrections
$d > 4$	IR-regular UV-divergent	nonrenormalisable u irrelevant	mean-field exponents

The upper critical dimension can be obtained in a more direct manner through simple *power counting*. To this end, we introduce an arbitrary momentum scale μ, i.e., define the *scaling dimensions* $[x] = \mu^{-1}$ and $[q] = \mu$. If in addition we choose $[t] = \mu^{-2-a}$, or $[\omega] = \mu^{2+a}$, then the relaxation constant becomes dimensionless, $[D] = \mu^0$. For the deviation from the critical point, we obtain $[r] = \mu^2$, and the *positive* exponent indicates that this control parameter constitutes a *relevant* coupling in the theory; as we shall see below, its renormalised counterpart grows under subsequent RG transformations. For the nonlinear coupling, one finds $[u] = \mu^{4-d}$, so it is relevant for $d < 4$: nonlinear thermal fluctuations will qualitatively affect the physical properties at the phase transition; but u becomes irrelevant for $d > 4$: one then expects mean-field (Gaussian) critical exponents. At the upper critical dimension $d_c = 4$, the nonlinear coupling u is *marginally relevant*: this will induce logarithmic corrections to the mean-field scaling laws, see Table 7.1.

It is obviously not a simple task to treat the IR-singular perturbation expansion in a meaningful, well-defined manner, and thus allow nonanalytic modifications of the critical power laws (note that mean-field scaling is completely determined by dimensional analysis or power counting). The key of the success of the RG approach is to focus on the very specific symmetry that emerges near critical points, namely *scale invariance*. There are several (largely equivalent) versions of the RG method; we shall here formulate and employ the field-theoretic variant [1–6, 8, 13]. In order to proceed, it is convenient to evaluate the loop integrals in momentum space by means of *dimensional regularisation*, whereby one assigns finite values even to UV-divergent expressions, namely the analytically continued values from the UV-finite range. For example, even for noninteger dimensions d and σ, we set

$$\int \frac{d^d k}{(2\pi)^d} \frac{k^{2\sigma}}{(\tau + k^2)^s} = \frac{\Gamma(\sigma + d/2)\,\Gamma(s - \sigma - d/2)}{2^d\,\pi^{d/2}\,\Gamma(d/2)\,\Gamma(s)}\,\tau^{\sigma - s + d/2}\,. \tag{7.53}$$

The *renormalisation* program then consists of the following steps:

1. We aim to carefully keep track of formal, unphysical UV divergences. In dimensionally regularised integrals (7.53), these appear as poles in $\epsilon = d_c - d$; their residues characterise the asymptotic UV behaviour of the field theory under consideration.

2. Therefrom we may infer the (UV) scaling properties of the control parameters of the model under a RG transformation, namely essentially a change of the momentum scale μ, while keeping the form of the action invariant. This will allow us to define suitable *running couplings*.

3. We seek *fixed points* in parameter space where certain marginal couplings (u here) do not change anymore under RG transformations. This describes a *scale-invariant* regime for the model under consideration, where the UV and IR scaling properties become intimately linked. Studying the parameter flows near a stable RG fixed point then allows us to extract the asymptotic IR power laws.

As a preliminary step, we need to take into account that the fluctuations will also shift the critical point downwards from the mean-field phase transition temperature T_c^0; i.e., we expect the transition to occur at $T_c < T_c^0$. This fluctuation-induced T_c shift can be determined by demanding that the inverse static susceptibility vanish at T_c: $\chi(\boldsymbol{q} = 0, \omega = 0)^{-1} = \tau = r - r_c$, where $\tau \sim T - T_c$ and thus $r_c = T_c - T_c^0$. Using our previous results (7.45) and (7.47), we find to first order in u (and with finite cutoff Λ),

$$r_c = -\frac{n+2}{6}\,u \int_k^\Lambda \frac{1}{r_c + k^2} + O(u^2) = -\frac{n+2}{6}\,\frac{u\,S_d\,\Lambda^{d-2}}{(2\pi)^d\,(d-2)} + O(u^2)\,. \tag{7.54}$$

Notice that this quantity depends on microscopic details (the lattice structure enters the cutoff Λ) and is thus not universal; moreover it diverges for $d \geq 2$ (quadratically near $d_c = 4$) as $\Lambda \to \infty$. We next use $r = \tau + r_c$ to write physical quantities as functions of the true distance τ from the critical point, which technically amounts to an *additive renormalisation*; e.g., the dynamic response function becomes to one-loop order

$$\chi(q,\omega)^{-1} = -\frac{i\omega}{Dq^a} + q^2 + \tau \left[1 - \frac{n+2}{6} u \int_k \frac{1}{k^2(\tau + k^2)}\right] + O(u^2) . \quad (7.55)$$

The remaining loop integral is UV-singular in dimensions $d \geq d_c = 4$.

We may now formally absorb the remaining UV divergences into *renormalised* fields and parameters, a procedure called *multiplicative renormalisation*. For the renormalised fields, we use the convention

$$S_R^\alpha = Z_S^{1/2} S^\alpha , \quad \tilde{S}_R^\alpha = Z_{\tilde{S}}^{1/2} \tilde{S}^\alpha , \quad (7.56)$$

where we have exploited the $O(n)$ rotational symmetry in using identical *renormalisation constants (Z factors)* for each component. The renormalised cumulants with N order parameter fields S^α and $\tilde{N}$ response fields $\tilde{S}^\alpha$ naturally involve the product $Z_S^{N/2} Z_{\tilde{S}}^{\tilde{N}/2}$, whence

$$\Gamma_R^{(\tilde{N},N)} = Z_{\tilde{S}}^{-\tilde{N}/2} Z_S^{-N/2} \Gamma^{(\tilde{N},N)} . \quad (7.57)$$

In a similar manner, we relate the "bare" parameters of the theory via Z factors to their renormalised counterparts, which we furthermore render dimensionless through appropriate momentum scale factors,

$$D_R = Z_D D , \quad \tau_R = Z_\tau \tau \mu^{-2} , \quad u_R = Z_u u A_d \mu^{d-4} , \quad (7.58)$$

where we have separated out the factor $A_d = \Gamma(3 - d/2)/2^{d-1} \pi^{d/2}$ for convenience. In the *minimal subtraction* scheme, the Z factors contain *only* the UV-singular terms, which in dimensional regularisation appear as poles at $\epsilon = 0$, and their residues, evaluated at $d = d_c$.

These renormalisation constants are not all independent, however; since the equilibrium fluctuation–dissipation theorem (7.38) or (7.13) must hold in the renormalised theory as well, we infer that necessarily

$$Z_D = (Z_S/Z_{\tilde{S}})^{1/2} , \quad (7.59)$$

and consequently from Eq. (7.45)

$$\chi_R = Z_S \chi . \quad (7.60)$$

Moreover, for model B with conserved order parameter Eq. (7.49) implies that to all orders in the perturbation expansion

$$a = 2 : \quad Z_{\tilde{S}} Z_S = 1 \,, \quad Z_D = Z_S \,. \tag{7.61}$$

For the following, it is crucial that the theory is *renormalisable*, i.e., a *finite* number of reparametrisations suffice to formally rid it of all UV divergences. Indeed, for the relaxational models A and B, and the static Ginzburg–Landau–Wilson Hamiltonian (7.7), all higher vertex function beyond the four-point function are UV-convergent near d_c, and there are only the *three* independent static renormalisation factors Z_S, Z_τ, and Z_u, and in addition Z_D for non-conserved order parameter dynamics. As we shall see, these directly translate into the *two* independent static critical exponents and the unrelated dynamic scaling exponent z for model A; for model B with conserved order parameter, Eq. (7.61) will yield a scaling relation between z and η.

In order to explicitly determine the renormalisation constants, we need to ensure that we stay away from the IR-singular regime. This is guaranteed by selecting as *normalisation point* either $\tau_R = 1$ (i.e., $Z_\tau \tau = \mu^2$) or $q = \mu$. Inevitably therefore, the renormalised theory depends on the corresponding arbitrary momentum scale μ. Since there are no fluctuation contributions to order u to either $\partial \Gamma^{(1,1)}(q,0)/\partial q^2$ or $\partial \Gamma^{(1,1)}(0,\omega)/\partial \omega$ (at $\tau_R = 1$), we find $Z_S = 1$ and $Z_D = 1$ within the one-loop approximation. Expressions (7.55) and (7.50) then yield with the formula (7.53)

$$Z_\tau = 1 - \frac{n+2}{6} \frac{u_R}{\epsilon} \,, \quad Z_u = 1 - \frac{n+8}{6} \frac{u_R}{\epsilon} \,. \tag{7.62}$$

To two-loop order, we may infer the field renormalisation Z_S from the static susceptibility as the singular contributions to $\partial \chi_R(q,0)/\partial q^2|_{q=0}$, and Z_D for model A, through a somewhat lengthy calculation [13], from either $\Gamma_R^{(2,0)}(0,0)$ or $\Gamma_R^{(1,1)}(0,\omega)$, with the results

$$Z_S = 1 + \frac{n+2}{144} \frac{u_R^2}{\epsilon} \,, \quad a = 0 : \ Z_D = 1 - \frac{n+2}{144} \left(6 \ln \frac{4}{3} - 1 \right) \frac{u_R^2}{\epsilon} \,. \tag{7.63}$$

7.1.5 Scaling Laws and Critical Exponents

We now wish to related the renormalised vertex functions at different inverse length scales μ. This is accomplished by simply recalling that the *unrenormalised* vertex functions obviously do *not* depend on μ,

$$0 = \mu \frac{\mathrm{d}}{\mathrm{d}\mu} \Gamma^{(\tilde{N},N)}(D, \tau, u) = \mu \frac{\mathrm{d}}{\mathrm{d}\mu} \left[Z_{\tilde{S}}^{\tilde{N}/2} Z_S^{N/2} \Gamma_R^{(\tilde{N},N)}(\mu, D_R, \tau_R, u_R) \right]. \tag{7.64}$$

In the second step, the bare quantities have been replaced with their renormalised counterparts. The innocuous statement (7.64) then implies a very nontrivial partial differential equation for the renormalised vertex functions, the desired *renormalisation group equation*,

$$\left[\mu \frac{\partial}{\partial \mu} + \frac{\tilde{N} \gamma_{\tilde{S}} + N \gamma_S}{2} + \gamma_D D_R \frac{\partial}{\partial D_R} + \gamma_\tau \tau_R \frac{\partial}{\partial \tau_R} + \beta_u \frac{\partial}{\partial u_R} \right]$$
$$\times \Gamma_R^{(\tilde{N}, N)} (\mu, D_R, \tau_R, u_R) = 0 . \tag{7.65}$$

Here we have defined *Wilson's flow functions* (the index "0" indicates that the derivatives with respect to μ are to be taken with fixed unrenormalised parameters)

$$\gamma_{\tilde{S}} = \mu \frac{\partial}{\partial \mu} \Big|_0 \ln Z_{\tilde{S}} , \quad \gamma_S = \mu \frac{\partial}{\partial \mu} \Big|_0 \ln Z_S , \tag{7.66}$$

$$\gamma_\tau = \mu \frac{\partial}{\partial \mu} \Big|_0 \ln(\tau_R/\tau) = -2 + \mu \frac{\partial}{\partial \mu} \Big|_0 \ln Z_\tau , \tag{7.67}$$

$$\gamma_D = \mu \frac{\partial}{\partial \mu} \Big|_0 \ln(D_R/D) = \frac{1}{2} (\gamma_S - \gamma_{\tilde{S}}) , \tag{7.68}$$

where we have used the relation (7.59); for model B, Eq. (7.61) gives in addition

$$\gamma_D = \gamma_S = -\gamma_{\tilde{S}} . \tag{7.69}$$

We have also introduced the *RG beta function* for the nonlinear coupling u,

$$\beta_u = \mu \frac{\partial}{\partial \mu} \Big|_0 u_R = u_R \left(d - 4 + \mu \frac{\partial}{\partial \mu} \Big|_0 \ln Z_u \right) . \tag{7.70}$$

Explicitly, Eqs. (7.63) and (7.62) yield to lowest nontrivial order, with $\epsilon = 4 - d$,

$$\gamma_S = -\frac{n+2}{72} u_R^2 + O(u_R^3) , \tag{7.71}$$

$$a = 0 : \gamma_D = \frac{n+2}{72} \left(6 \ln \frac{4}{3} - 1 \right) u_R^2 + O(u_R^3) , \tag{7.72}$$

$$\gamma_\tau = -2 + \frac{n+2}{6} u_R + O(u_R^2) , \tag{7.73}$$

$$\beta_u = u_R \left[-\epsilon + \frac{n+8}{6} u_R + O(u_R^2) \right] . \tag{7.74}$$

In the RG equation for the renormalised dynamic susceptibility, Eq. (7.60) tells us that the second term in Eq. (7.65) is to be replaced with $-\gamma_S$. Its explicit dependence on the scale μ can be factored out via $\chi_R(\mu, D_R, \tau_R, u_R, q, \omega) = \mu^{-2} \hat{\chi}_R (\tau_R, u_R, q/\mu, \omega/D_R \mu^{2+a})$, see Eq. (7.55), whence

$$\left[-2 - \gamma_S + \gamma_D D_R \frac{\partial}{\partial D_R} + \gamma_\tau \tau_R \frac{\partial}{\partial \tau_R} + \beta_u \frac{\partial}{\partial u_R} \right] \hat{\chi}_R(D_R, \tau_R, u_R) = 0 . \tag{7.75}$$

This *linear partial differential equation* is readily solved by means of the *method of characteristics*, as is Eq. (7.65) for the vertex functions. The idea is to find a curve parametrisation $\mu(\ell) = \mu\,\ell$ in the space spanned by the parameters $\tilde{D}$, $\tilde{\tau}$, and $\tilde{u}$ such that

$$\ell\frac{\mathrm{d}\tilde{D}(\ell)}{\mathrm{d}\ell} = \tilde{D}(\ell)\,\gamma_D(\ell)\ , \quad \ell\frac{\mathrm{d}\tilde{\tau}(\ell)}{\mathrm{d}\ell} = \tilde{\tau}(\ell)\,\gamma_\tau(\ell)\ , \quad \ell\frac{\mathrm{d}\tilde{u}(\ell)}{\mathrm{d}\ell} = \beta_u(\ell)\ , \quad (7.76)$$

with initial values D_R, τ_R, and u_R, respectively at $\ell = 1$. The *first-order ordinary differential equations* (7.76), with $\gamma_D(\ell) = \gamma_D(\tilde{u}(\ell))$ etc. define *running couplings* that describe how the parameters of the theory change under scale transformations $\mu \to \mu\,\ell$. The formal solutions for $\tilde{D}(\ell)$ and $\tilde{\tau}(\ell)$ read

$$\tilde{D}(\ell) = D_R \exp\left[\int_1^\ell \gamma_D(\ell')\,\frac{\mathrm{d}\ell'}{\ell'}\right]\ , \quad \tilde{\tau}(\ell) = \tau_R \exp\left[\int_1^\ell \gamma_\tau(\ell')\,\frac{\mathrm{d}\ell'}{\ell'}\right]\ . \quad (7.77)$$

For the function $\hat{\chi}(\ell) = \hat{\chi}_R(\tilde{D}(\ell), \tilde{\tau}(\ell), \tilde{u}(\ell))$, we then obtain another ordinary differential equation, namely

$$\ell\frac{\mathrm{d}\hat{\chi}(\ell)}{\mathrm{d}\ell} = [2 + \gamma_S(\ell)]\,\hat{\chi}(\ell)\ , \quad (7.78)$$

which is solved by

$$\hat{\chi}(\ell) = \hat{\chi}(1)\,\ell^2\,\exp\left[\int_1^\ell \gamma_S(\ell')\,\frac{\mathrm{d}\ell'}{\ell'}\right]\ . \quad (7.79)$$

Collecting everything, we finally arrive at

$$\chi_R(\mu, D_R, \tau_R, u_R, \boldsymbol{q}, \omega) = (\mu\,\ell)^{-2}\,\exp\left[-\int_1^\ell \gamma_S(\ell')\,\frac{\mathrm{d}\ell'}{\ell'}\right]$$
$$\times\,\hat{\chi}_R\left(\tilde{\tau}(\ell), \tilde{u}(\ell), \frac{|\boldsymbol{q}|}{\mu\,\ell}, \frac{\omega}{\tilde{D}(\ell)\,(\mu\,\ell)^{2+a}}\right)\ . \quad (7.80)$$

The solution (7.80) of the RG equation (7.75), along with the flow equations (7.76), (7.77) for the running couplings tell us how the dynamic susceptibility depends on the (momentum) scale $\mu\,\ell$ at which we consider the theory. Similar relations can be obtained for arbitrary vertex functions by solving the associated RG equations (7.65) [13]. The point here is that the right-hand side of Eq. (7.80) may be evaluated outside the IR-singular regime, by fixing one of its arguments at a finite value, say $|\boldsymbol{q}|/\mu\,\ell = 1$. The function $\hat{\chi}_R$ is regular, and can be calculated by means of perturbation theory. A *scale-invariant* regime is characterised by the renormalised nonlinear coupling u_R becoming independent of the scale $\mu\,\ell$, or $\tilde{u}(\ell) \to u^* = $ const. For an *RG fixed point* to be *infrared-stable*, we thus require

$$\beta_u(u^*) = 0\ , \quad \beta_u'(u^*) > 0\ , \quad (7.81)$$

since Eq. (7.76) then implies that $\tilde{u}(\ell \to 0) \to u^*$. Taking the limit $\ell \to 0$ thus provides the desired mapping of physical observables such as (7.80) onto the critical region. In the vicinity of an IR-stable RG fixed point, Eq. (7.77) yields the power laws $\tilde{D}(\ell) \approx D_R \ell^{\gamma_D^*}$, where $\gamma_D^* = \gamma_D(\ell \to 0) = \gamma_D(u^*)$, etc. Consequently, Eq. (7.80) reduces to

$$\chi_R(\tau_R, \mathbf{q}, \omega) \approx \mu^{-2} \ell^{-2-\gamma_S^*} \hat{\chi}_R\left(\tau_R \ell^{\gamma_\tau^*}, u^*, \frac{|\mathbf{q}|}{\mu \ell}, \frac{\omega}{D_R \mu^{2+a} \ell^{2+a+\gamma_D^*}}\right) , \quad (7.82)$$

and upon *matching* $\ell = |\mathbf{q}|/\mu$ we recover the *dynamic scaling law* (7.12) with the *critical exponents*

$$\eta = -\gamma_S^* , \quad \nu = -1/\gamma_\tau^* , \quad z = 2 + a + \gamma_D^* . \quad (7.83)$$

To one-loop order, we obtain from the RG beta function (7.74)

$$u_H^* = \frac{6\,\epsilon}{n+8} + O(\epsilon^2) . \quad (7.84)$$

Here we have indicated that our perturbative expansion for small u has effectively turned into a *dimensional expansion* in $\epsilon = d_c - d$. In dimensions $d < 4$, the *Heisenberg* fixed point u_H^* is IR-stable, since $\beta_u'(u_H^*) = \epsilon > 0$. With Eqs. (7.71) and (7.73), the identifications (7.83) then give us explicit results for the static scaling exponents, as mere functions of dimension $d = 4 - \epsilon$ and the number of order parameter components n,

$$\eta = \frac{n+2}{2\,(n+8)^2}\,\epsilon^2 + O(\epsilon^3) , \quad \frac{1}{\nu} = 2 - \frac{n+2}{n+8}\,\epsilon + O(\epsilon^2) . \quad (7.85)$$

For model A with nonconserved order parameter, the two-loop result (7.72) yields the independent dynamic critical exponent

$$a = 0 : \quad z = 2 + c\eta , \quad c = 6\ln\frac{4}{3} - 1 + O(\epsilon) ; \quad (7.86)$$

for model B with conserved order parameter, instead $\gamma_D^* = \gamma_S^* = -\eta$, whence we arrive at the *exact* scaling relation

$$a = 2 : \quad z = 4 - \eta . \quad (7.87)$$

In dimensions $d > d_c = 4$, the *Gaussian* fixed point $u_0^* = 0$ is stable ($\beta_u'(0) = -\epsilon > 0$). Therefore all anomalous dimensions disappear, i.e., $\gamma_S^* = 0 = \gamma_D^*$ and $\gamma_\tau^* = -2$, and we are left with the mean-field critical exponents $\eta_0 = 0$, $\nu_0 = 1/2$, and $z_0 = 2 + a$. Precisely at the upper critical dimension $d_c = 4$, the RG flow equation for the nonlinear coupling becomes

$$\ell\frac{d\tilde{u}(\ell)}{d\ell} = \frac{n+8}{6}\,\tilde{u}(\ell)^2 + O\!\left(\tilde{u}(\ell)^3\right) , \quad (7.88)$$

which is solved by

$$\widetilde{u}(\ell) = \frac{u_R}{1 - \frac{n+8}{6} u_R \ln \ell} \ . \tag{7.89}$$

In four dimensions, $\widetilde{u}(\ell) \to 0$, but only logarithmically slowly, which causes *logarithmic corrections* to the mean-field critical power laws. For example, upon inserting Eq. (7.89) into the flow equation (7.76), one finds $\widetilde{\tau}(\ell) \sim \tau_R \, \ell^{-2} (\ln |\ell|)^{-(n+2)/(n+8)}$; with $\widetilde{\tau}(\ell = \xi^{-1}) = O(1)$, iterative inversion yields

$$\xi(\tau_R) \sim \tau_R^{-1/2} (\ln \tau_R)^{(n+2)/2(n+8)} \ . \tag{7.90}$$

This concludes our derivation of asymptotic scaling laws for the critical dynamics of the purely relaxational models A and B, and the explicit computation of the scaling exponents in powers of $\epsilon = d_c - d$. In the following sections, I will briefly sketch how the response functional formalism and the dynamic renormalisation group can be employed to study the critical dynamics of systems with reversible mode-coupling terms, the "ageing' behaviour induced by quenching from random initial conditions to the critical point, the effects of violating the detailed balance constraints on universal dynamic critical properties, and the generically scale-invariant features of nonequilibrium systems such as driven diffusive Ising lattice gases.

7.1.6 Critical Dynamics with Reversible Mode-Couplings

In the previous chapters, we have assumed purely relaxational dynamics for the order parameter, see Eq. (7.16). In general, however, there are also *reversible* contributions to the systematic force terms F^α that enter its Langevin equation [7, 24]. Consider the Hamiltonian dynamics of *microscopic* variables, say, local spin densities, at $T = 0$: $\partial_t S_m^\alpha(x, t) = \{H[S_m], S_m^\alpha(x, t)\}$. Here, the *Poisson brackets* $\{A, B\}$ constitute the classical analog of the quantum-mechanical commutator $\frac{i}{\hbar}[A, B]$ (correspondence principle). Upon *coarse-graining*, the microscopic variables S_m^α become the *mesoscopic* hydrodynamic fields S^α. Since the set of *slow* modes should provide a complete description of the critical dynamics, we may formally expand

$$\left\{ \mathcal{H}[S], S^\alpha(\boldsymbol{x}) \right\} = \int \mathrm{d}^d x' \sum_\beta \frac{\delta \mathcal{H}[S]}{\delta S^\beta(\boldsymbol{x}')} Q^{\beta\alpha}(\boldsymbol{x}', \boldsymbol{x}) \ , \tag{7.91}$$

with the mutual Poisson brackets of the hydrodynamic variables

$$Q^{\alpha\beta}(\boldsymbol{x}, \boldsymbol{x}') = \left\{ S^\alpha(\boldsymbol{x}), S^\beta(\boldsymbol{x}') \right\} = -Q^{\beta\alpha}(\boldsymbol{x}', \boldsymbol{x}) \ . \tag{7.92}$$

By inspection of the associated Fokker–Planck equation, one may then establish an additional *equilibrium condition* in order for the time-dependent probability distribution to reach the canonical limit (7.6): $\mathcal{P}[S, t] \to \mathcal{P}_{\mathrm{eq}}[S]$ as $t \to \infty$ *provided* the probability current is *divergence-free* in the space spanned by the stochastic fields $S^\alpha(\boldsymbol{x})$:

$$\int d^d x \sum_\alpha \frac{\delta}{\delta S^\alpha(\boldsymbol{x})} \left(F^\alpha_{\text{rev}}[S] \, e^{-\mathcal{H}[S]/k_{\text{B}}T} \right) = 0 \ . \tag{7.93}$$

It turns out that this equilibrium condition is often more crucial than the Einstein relation (7.17). In order to satisfy Eq. (7.93) at $T \neq 0$, we must supplement Eq. (7.91) by a finite-temperature correction, whereupon the *reversible mode-coupling* contributions to the systematic forces become

$$F^\alpha_{\text{rev}}[S](\boldsymbol{x}) = -\int d^d x' \sum_\beta \left[Q^{\alpha\beta}(\boldsymbol{x}, \boldsymbol{x}') \frac{\delta \mathcal{H}[S]}{\delta S^\beta(\boldsymbol{x}')} - k_{\text{B}}T \frac{\delta Q^{\alpha\beta}(\boldsymbol{x}, \boldsymbol{x}')}{\delta S^\beta(\boldsymbol{x}')} \right] \ , \tag{7.94}$$

and the complete coupled set of stochastic differential equations reads

$$\frac{\partial S^\alpha(\boldsymbol{x}, t)}{\partial t} = F^\alpha_{\text{rev}}[S](\boldsymbol{x}, t) - D^\alpha (i\boldsymbol{\nabla})^{a_\alpha} \frac{\delta \mathcal{H}[S]}{\delta S^\alpha(\boldsymbol{x}, t)} + \zeta^\alpha(\boldsymbol{x}, t) \ , \tag{7.95}$$

where as before the D^α denote the relaxation coefficients, and $a^\alpha = 0$ or 2 respectively for nonconserved and conserved modes.

As an instructive example, let us consider the *Heisenberg model* for *isotropic ferromagnets*, $H[\{\boldsymbol{S}_j\}] = -\frac{1}{2} \sum_{j,k=1}^N J_{jk} \, \boldsymbol{S}_j \cdot \boldsymbol{S}_k$, where the spin operators satisfy the usual commutation relations $[S_j^\alpha, S_k^\beta] = i\hbar \sum_\gamma \epsilon^{\alpha\beta\gamma} S_j^\gamma \, \delta_{jk}$. The corresponding Poisson brackets for the magnetisation density read

$$Q^{\alpha\beta}(\boldsymbol{x}, \boldsymbol{x}') = -g \sum_\gamma \epsilon^{\alpha\beta\gamma} S^\gamma(\boldsymbol{x}) \, \delta(\boldsymbol{x} - \boldsymbol{x}') \ , \tag{7.96}$$

where the purely *dynamical coupling* g incorporates various factors that emerge upon coarse-graining and taking the continuum limit. The second contribution in Eq. (7.94) vanishes, since it reduces to a contraction of the antisymmetric tensor $\epsilon^{\alpha\beta\gamma}$ with the Kronecker symbol $\delta^{\beta\gamma}$, whence we arrive at the Langevin equations governing the critical dynamics of the three order parameter components for isotropic ferromagnets [25]

$$\frac{\partial \boldsymbol{S}(\boldsymbol{x}, t)}{\partial t} = -g \, \boldsymbol{S}(\boldsymbol{x}, t) \times \frac{\delta \mathcal{H}[S]}{\delta \boldsymbol{S}(\boldsymbol{x}, t)} + D \boldsymbol{\nabla}^2 \frac{\delta \mathcal{H}[S]}{\delta \boldsymbol{S}(\boldsymbol{x}, t)} + \boldsymbol{\zeta}(\boldsymbol{x}, t) \ , \tag{7.97}$$

with $\langle \boldsymbol{\zeta}(\boldsymbol{x}, t) \rangle = 0$. Since $[H[\{\boldsymbol{S}_j\}], \sum_k S_k^\alpha] = 0$, the total magnetisation is conserved, whence the noise correlators should be taken as

$$\langle \zeta^\alpha(\boldsymbol{x}, t) \, \zeta^\beta(\boldsymbol{x}', t') \rangle = -2D \, k_{\text{B}}T \, \boldsymbol{\nabla}^2 \delta(\boldsymbol{x} - \boldsymbol{x}') \, \delta(t - t') \, \delta^{\alpha\beta} \ . \tag{7.98}$$

The vector product term in Eq. (7.97) describes the spin precession in the local effective magnetic field $\delta \mathcal{H}[S]/\delta \boldsymbol{S}$, which includes a contribution induced by the exchange interaction.

The Langevin equation (7.97) and (7.98) with the Hamiltonian (7.7) for $n = 3$ define the so-called *model J* [7]. In addition to the model B response functional (7.35) and (7.36) with $a = 2$ (setting $k_B T = 1$ again), the reversible force in Eq. (7.97) leads to an additional contribution to the action

$$\mathcal{A}_{\rm mc}[\widetilde{S}, S] = -g \int d^d x \int dt \sum_{\alpha,\beta,\gamma} \epsilon^{\alpha\beta\gamma} \widetilde{S}^\alpha S^\beta \left(\nabla^2 S^\gamma + h^\gamma\right) , \qquad (7.99)$$

which gives rise to an additional *mode-coupling vertex*, as depicted in Fig. 7.5(a). Power counting yields the scaling dimension $[g] = \mu^{3-d/2}$ for the associated coupling strength, whence we expect a *dynamical* upper critical dimension $d'_c = 6$. However, since we are investigating a system in thermal equilibrium, we can treat its thermodynamics and static properties separately from its dynamics. Obviously therefore, the static critical exponents must still be given (to lowest nontrivial order and for $d < d_c = 4$) by Eq. (7.85) for the three-component Heisenberg model with $O(3)$ rotational symmetry. Therefore our sole task is to find the dynamic critical exponent z.

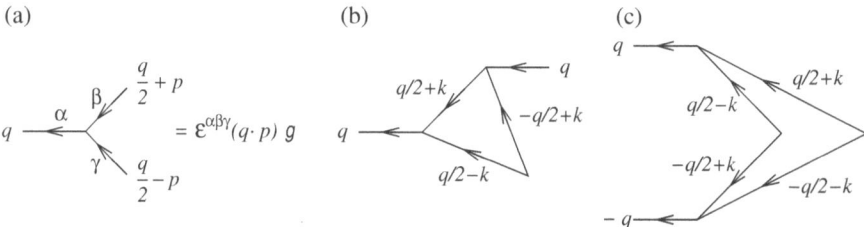

Fig. 7.5. (**a**) Mode-coupling three-point vertex for model J. One-loop Feynman diagrams for the propagator (**b**) and noise vertex (**c**) renormalisations in model J. The same graphs (b), (c) apply for driven diffusive systems (Sect. 7.1.9)

Remarkably, z is entirely fixed by the symmetries of the problem and can be determined exactly. To this end, we exploit the fact that the S^α are the generators of the rotation group; indeed, it follows from Eq. (7.99) that applying a time-dependent external field $h^\gamma(t)$ induces a contribution

$$\left\langle S^\alpha(\boldsymbol{x}, t)\right\rangle_h = g \int_0^t dt' \sum_\beta \epsilon^{\alpha\beta\gamma} \left\langle S^\beta(\boldsymbol{x}, t')\right\rangle_h h^\gamma(t) \qquad (7.100)$$

to the average magnetisation. As a consequence, we obtain for the *nonlinear susceptibility* $R^{\alpha;\beta\gamma} = \delta^2 \langle S^\alpha\rangle / \delta h^\beta \, \delta h^\gamma |_{h=0}$,

$$\int d^d x' \, R^{\alpha;\beta\gamma}(\boldsymbol{x}, t; \boldsymbol{x} - \boldsymbol{x}', t - t') = g \, \epsilon^{\alpha\beta\gamma} \, \chi^{\beta\beta}(\boldsymbol{x}, t) \, \Theta(t) \, \Theta(t - t') . \qquad (7.101)$$

An analogous expression must hold after renormalisation as well. If we define the dimensionless renormalised mode-coupling according to

$$g_R^2 = Z_g \, g^2 \, B_d \, \mu^{d-6} , \qquad f_R = g_R^2 / D_R^2 , \qquad (7.102)$$

where $B_d = \Gamma(4 - d/2)/2^d \, d \, \pi^{d/2}$, Eq. (7.101) implies the identity [9]

$$Z_g = Z_S \ . \tag{7.103}$$

For the RG beta function associated with the effective coupling entering the loop corrections, we thus infer

$$\beta_f = \mu \frac{\partial}{\partial \mu}\bigg|_0 f_R = f_R \left(d - 6 + \gamma_S - 2\gamma_D \right) \ . \tag{7.104}$$

Consequently, at any *nontrivial* IR-stable RG fixed point $0 < f^* < \infty$, we have the *exact* scaling relation, valid to *all* orders in perturbation theory,

$$d < 6 : \ z = 4 + \gamma_D^* = 4 + \frac{d - 6 + \gamma_S^*}{2} = \frac{d + 2 - \eta}{2} \ . \tag{7.105}$$

Since the resulting value for the dynamic exponent, $z \approx 5/2$ in three dimensions, is markedly smaller than the model B mean-field $z_0 = 4$, we conclude that the reversible spin precession kinetics speeds up the order parameter dynamics considerably [7,9,25].

An explicit one-loop calculation, either for the propagator self-energy $\Gamma^{(1,1)}(\boldsymbol{q}, \omega)$, depicted in Fig. 7.5(b), or the noise vertex $\Gamma^{(2,0)}(\boldsymbol{q}, \omega)$, shown in Fig. 7.5(c), yields [9,13]

$$\gamma_D = -f_R + O(u_R^2, f_R^2) \ , \tag{7.106}$$

which along with $\gamma_S = 0 + O(u_R^2, f_R^2)$ confirms that there exists a nontrivial mode-coupling RG fixed point

$$f_J^* = \frac{\varepsilon}{2} + O(\varepsilon^2) \ , \tag{7.107}$$

where $\varepsilon = 6 - d$, which is IR-stable for $d < 6$. As $\eta = 0$ for $d > 4$, we indeed recover the mean-field dynamic exponent $z_0 = 4$ in $d \geq 6$ dimensions. With the leading singularity thus isolated, the regular scaling functions can be computed numerically to high accuracy within a self-consistent one-loop approximation that also goes under the name *mode-coupling theory*. Details of this procedure, an alternative derivation, and many results of mode-coupling theory as applied to the critical dynamics of magnets and comparisons with experimental data can be found in Ref. [26].

Typically, reversible force terms of the form (7.94) involve dynamical couplings of the order parameter to *other* conserved, slow variables. In addition, there may also be static couplings to conserved fields in the Hamiltonian. These various possibilities give rise to a range of different *dynamic universality classes* for near-equilibrium critical dynamics [7]. We shall not pursue these further here (for a partial account within the field-theoretic RG approach, see Ref. [13]), but instead proceed and now consider nonequilibrium effects.

7.1.7 Critical Relaxation, Initial Slip, and Ageing

We begin with a brief discussion of the *coarsening* dynamics of systems described by model A/B kinetics that are rapidly *quenched* from a disordered state at $T \gg T_c$ to the critical point $T \approx T_c$ [27, 28]. The situation may be modeled as a relaxation from *Gaussian random initial conditions*, i.e., the probability distribution for the order parameter at $t = 0$ can be taken as

$$\mathcal{P}[S, t = 0] \propto e^{-\mathcal{H}_0[S]} = \exp\left(-\frac{\Delta}{2} \int d^d x \sum_\alpha [S^\alpha(\boldsymbol{x}, 0) - a^\alpha(\boldsymbol{x})]^2\right) , \quad (7.108)$$

where the functions $a^\alpha(\boldsymbol{x})$ specify the most likely initial configurations. Power counting for the parameter Δ gives $[\Delta] = \mu^2$, whence it is a relevant perturbation that will flow to $\Delta \to \infty$ under the RG. Asymptotically, therefore, the system will be governed by sharp *Dirichlet boundary conditions*. Whereas the response propagators remains a causal function of the time difference between applied perturbation and effect, $G_0(\boldsymbol{q}, t - t') = \Theta(t - t') e^{-Dq^a (r + q^2)(t - t')}$, see Eq. (7.23), time translation invariance is broken by the initial state in the *Dirichlet correlator* of the Gaussian model,

$$C_D(\boldsymbol{q}; t, t') = \frac{1}{r + q^2} \left(e^{-Dq^a (r + q^2)|t - t'|} - e^{-Dq^a (r + q^2)(t + t')}\right) . \quad (7.109)$$

Away from criticality, i.e., for $r > 0$ and $\boldsymbol{q} \neq 0$, temporal correlations decay exponentially fast, and the system quickly approaches the stationary equilibrium state. However, as $T \to T_c$, the equilibration time diverges according to $t_c \sim |\tau|^{-z\nu} \to \infty$, and the system never reaches thermal equilibrium. Two-time correlation functions will then depend on both times separately, in a specific manner to be addressed below, a phenomenon termed critical *"ageing"* (for more details, see Refs. [29, 30]).

The field-theoretic treatment of the model A/B dynamical action (7.35), (7.36) with the initial term (7.108) follows the theory of *boundary critical phenomena* [31]. However, it turns out that additional singularities on the temporal "surface" at $t + t' = 0$ appear only for model A, and can be incorporated into a single new renormalisation factor; to one-loop order, one finds [27, 28]

$$a = 0 : \quad \widetilde{S}_R^\alpha(\boldsymbol{x}, 0) = (Z_0 Z_{\widetilde{S}})^{1/2} \widetilde{S}^\alpha(\boldsymbol{x}, 0) , \quad Z_0 = 1 - \frac{n + 2}{6} \frac{u_R}{\epsilon} . \quad (7.110)$$

This in turn leads to a *single* independent critical exponent associated with the initial time relaxation, the *initial slip exponent*, which becomes for the purely relaxational models A and B with nonconserved and conserved order parameter:

$$a = 0 : \quad \theta = \frac{\gamma_0^*}{2z} = \frac{n + 2}{4(n + 8)} \epsilon + O(\epsilon^2) , \quad a = 2 : \quad \theta = 0 . \quad (7.111)$$

In order to obtain the *short-time* scaling laws for the dynamic response and correlation functions in the *ageing limit* $t'/t \to 0$, one requires additional information that can be garnered from the short-distance *operator product expansion* for the fields,

$$t \to 0 : \quad \tilde{S}(\boldsymbol{x}, t) = \tilde{\sigma}(t)\,\tilde{S}_0(\boldsymbol{x}) , \quad S(\boldsymbol{x}, t) = \sigma(t)\,\tilde{S}_0(\boldsymbol{x}) . \tag{7.112}$$

Subsequent analysis then yields eventually [27, 28]

$$\chi(\boldsymbol{q}; t, t' \to 0) = |\boldsymbol{q}|^{z-2+\eta} \left(\frac{t}{t'}\right)^{\theta} \hat{\chi}_0(\boldsymbol{q}\,\xi, |\boldsymbol{q}|^z\, Dt) , \tag{7.113}$$

$$C(\boldsymbol{q}; t, t' \to 0) = |\boldsymbol{q}|^{-2+\eta} \left(\frac{t}{t'}\right)^{\theta-1} \hat{C}_0(\boldsymbol{q}\,\xi, |\boldsymbol{q}|^z\, Dt) , \tag{7.114}$$

and for the time dependence of the mean order parameter

$$\langle S(t) \rangle = S_0\, t^{\theta'}\, \hat{S}\left(S_0\, t^{\theta'+\beta/z\nu}\right) , \tag{7.115}$$

$$a = 0 : \quad \theta' = \theta - \frac{z - 2 + \eta}{z} , \quad a = 2 : \quad \theta' = \theta = 0 . \tag{7.116}$$

One may also compute the universal *fluctuation–dissipation ratios* in this nonequilibrium ageing regime [29, 30]. It emerges, though, that these depend on the quantity under investigation, which prohibits a unique definition of an effective nonequilibrium temperature for critical ageing. The method sketched above can be extended to models with reversible mode-couplings [32]. For model J capturing the critical dynamics of isotropic ferromagnets, one finds

$$\theta = \frac{z - 4 + \eta}{z} = -\frac{6 - d - \eta}{d + 2 - \eta} ; \tag{7.117}$$

in systems where a *nonconserved* order parameter is dynamically coupled to other conserved modes, the initial slip exponent θ is actually *not* a universal number, but depends on the width of the initial distribution [32].

7.1.8 Nonequilibrium Relaxational Critical Dynamics

Next we address the question [33], What happens if the detailed balance conditions (7.17) and (7.93) are violated? To start, we change the noise strength $D \to \tilde{D}$ in the purely relaxational models A and B, which (in our units) violates the Einstein relation (7.17). However, this modification can obviously be absorbed into a rescaled *effective temperature*, $k_B T \to k_B T' = \tilde{D}/D$. Formally this is established by means of the dynamical action (7.34), which now reads

$$\mathcal{A}[\tilde{S}, S] = \int \mathrm{d}^d x \int \mathrm{d}t \sum_\alpha \tilde{S}^\alpha \left[\partial_t\, S^\alpha + D\, (i\boldsymbol{\nabla})^a \left(r - \boldsymbol{\nabla}^2\right) S^\alpha \right.$$

$$\left. - \tilde{D}\, (i\boldsymbol{\nabla})^a\, \tilde{S}^\alpha + D\, \frac{u}{6}\, (i\boldsymbol{\nabla})^a\, S^\alpha \sum_\beta S^\beta S^\beta \right] . \tag{7.118}$$

Upon simple rescaling $\widetilde{S}^\alpha \to \widetilde{S}'^\alpha = \widetilde{S}^\alpha \sqrt{\widetilde{D}/D}$, $S^\alpha \to S'^\alpha = S^\alpha \sqrt{D/\widetilde{D}}$, the response functional (7.118) recovers its equilibrium form, albeit with modified nonlinear coupling $u \to \widetilde{u} = u\,\widetilde{D}/D$. However, the *universal asymptotic* properties of these models are governed by the Heisenberg fixed point (7.84), and the specific value of the (renormalised) coupling, which only serves as the initial condition for the RG flow, does not matter. In fact, the relaxational dynamics of the kinetic Ising model with Glauber dynamics (model A with $n = 1$) is known to be quite stable against nonequilibrium perturbations [34, 35], even if these break the Ising Z_2 symmetry [36]. For model J the above rescaling modifies in a similar manner merely the mode-coupling strength in Eq. (7.99), namely $g \to \widetilde{g} = g\sqrt{\widetilde{D}/D}$ [37]. Again, since the dynamic critical behaviour is governed by the universal fixed point (7.107), thermal equilibrium becomes effectively *restored* at criticality. More generally, it has been established that *isotropic* detailed balance violations do not affect the universal properties in other models for critical dynamics that contain additional conserved variables either: the equilibrium RG fixed points tend to be asymptotically stable [33].

In systems with *conserved* order parameter, however, we may in addition introduce spatially *anisotropic* violations of Einstein's relation; for example, in model B one can allow for anisotropic relaxation $-D\,\nabla^2 \to -D_\perp\,\nabla_\perp^2 - D_\parallel\,\nabla_\parallel^2$, with different rates in two spatial subsectors and concomitantly anisotropic noise correlations $-\widetilde{D}\,\nabla^2 \to -\widetilde{D}_\perp\,\nabla_\perp^2 - \widetilde{D}_\parallel\,\nabla_\parallel^2$. We have thus produced a truly nonequilibrium situation provided $\widetilde{D}_\perp/D_\perp \neq \widetilde{D}_\parallel/D_\parallel$, which we may interpret as having effectively coupled the longitudinal and transverse spatial sectors to heat baths with different temperatures $T_\perp < T_\parallel$, say [38].

Evaluating the fluctuation-induced shift of the transition temperature, see Eq. (7.54) one finds not surprisingly that the *transverse* sector softens first, while the longitudinal sector remains noncritical. This suggests that we can neglect the nonlinear longitudinal fluctuations as well as the $\nabla_\parallel^4$ term in the propagator. These features are indeed encoded in the corresponding *anisotropic* scaling: $[q_\perp] = \mu$, $[q_\parallel] = \mu^2$, $[\omega] = \mu^4$, whence $[\widetilde{D}_\perp] = [D_\perp] = \mu^0$, and $[\widetilde{D}_\parallel] = [D_\parallel] = \mu^{-2}$ become irrelevant. Upon renaming $D = D_\perp$ and $c = r_\parallel D_\parallel/D_\perp$, this ultimately leads to the *randomly driven* or *two-temperature model B* [39, 40] as the effective theory describing the phase transition:

$$\frac{\partial S^\alpha(\boldsymbol{x},t)}{\partial t} = D\left[\nabla_\perp^2\left(r - \nabla_\perp^2\right) + c\,\nabla_\parallel^2\right] S^\alpha(\boldsymbol{x},t)$$
$$+\frac{D\,\widetilde{u}}{6}\,\nabla_\perp^2\,S^\alpha(\boldsymbol{x},t)\sum_\beta [S^\beta(\boldsymbol{x},t)]^2 + \zeta^\alpha(\boldsymbol{x},t)\,, \quad (7.119)$$

with the noise correlations

$$\langle \zeta^\alpha(\boldsymbol{x},t)\,\zeta^\beta(\boldsymbol{x}',t')\rangle = -2D\,\nabla_\perp^2\,\delta(\boldsymbol{x} - \boldsymbol{x}')\,\delta(t - t')\,\delta^{\alpha\beta}\,. \quad (7.120)$$

Quite remarkably, the Langevin equation (7.119) can be derived as an *equilibrium* diffusive relaxational kinetics

$$\frac{\partial S^\alpha(\boldsymbol{x}, t)}{\partial t} = D \, \boldsymbol{\nabla}_\perp^2 \, \frac{\delta \mathcal{H}_{\text{eff}}[S]}{\delta S^\alpha(\boldsymbol{x}, t)} + \zeta^\alpha(\boldsymbol{x}, t) \tag{7.121}$$

from an effective *long-range* Hamiltonian

$$\mathcal{H}_{\text{eff}}[S] = \int \frac{\mathrm{d}^d q}{(2\pi)^d} \frac{q_\perp^2 (r + q_\perp^2) + c \, q_\parallel^2}{2 \, q_\perp^2} \sum_\alpha |S^\alpha(\boldsymbol{q})|^2$$

$$+ \frac{\widetilde{u}}{4!} \int \mathrm{d}^d x \sum_{\alpha, \beta} [S^\alpha(\boldsymbol{x})]^2 \, [S^\beta(\boldsymbol{x})]^2 \ . \tag{7.122}$$

Power counting gives $[\widetilde{u}] = \mu^{4 - d_\parallel - d}$: the spatial anisotropy suppresses longitudinal fluctuations and lower the upper critical dimension to $d_c = 4 - d_\parallel$. The anisotropic correlations encoded in Eq. (7.122) also reduce the lower critical dimension and affect the nature of the ordered phase [39, 41].

The scaling law for, e.g., the dynamic response function takes the form

$$\chi(\tau_\perp, \boldsymbol{q}_\perp, \boldsymbol{q}_\parallel, \omega) = |\boldsymbol{q}_\perp|^{-2 + \eta} \, \hat{\chi}\left(\frac{\tau}{|\boldsymbol{q}_\perp|^{1/\nu}}, \frac{\sqrt{c} \, |\boldsymbol{q}_\parallel|}{|\boldsymbol{q}_\perp|^{1 + \Delta}}, \frac{\omega}{D \, |\boldsymbol{q}_\perp|^z}\right) \ , \tag{7.123}$$

where we have introduced a new *anisotropy exponent* Δ. Since the nonlinear coupling $\widetilde{u}$ only affects the transverse sector, we find to *all* orders in the perturbation expansion:

$$\Gamma^{(1,1)}(\boldsymbol{q}_\perp = 0, \boldsymbol{q}_\parallel, \omega) = \mathrm{i}\omega + D \, c \, q_\parallel^2 \ , \tag{7.124}$$

and consequently obtain the Z factor identity

$$Z_c = Z_D^{-1} = Z_S^{-1} \ , \tag{7.125}$$

which at any IR-stable RG fixed point implies the *exact* scaling relations

$$z = 4 - \eta \ , \quad \Delta = 1 - \frac{\gamma_c^*}{2} = 1 - \frac{\eta}{2} = \frac{z}{2} - 1 \ , \tag{7.126}$$

whereas the scaling exponents for the longitudinal sector read

$$z_\parallel = \frac{z}{1 + \Delta} = 2 \ , \quad \nu_\parallel = \nu \, (1 + \Delta) = \frac{\nu}{2} \, (4 - \eta) \ . \tag{7.127}$$

As for the equilibrium model B, the only independent critical exponents to be determined are η and ν. To one-loop order, only the combinatorics of the Feynman diagrams (see Fig. 7.2) enters their explicit values, whence one finds for $d < d_c = 4 - d_\parallel$ formally identical results as for the usual Ginzburg–Landau–Wilson Hamiltonian (7.7),

$$\eta = 0 + O(\epsilon^2) , \quad \frac{1}{\nu} = 2 - \frac{n+2}{n+8}\epsilon + O(\epsilon^2) , \qquad (7.128)$$

albeit with *different* $\epsilon = 4 - d - d_\parallel$. To two-loop order, however, the anisotropy manifestly affects the evaluation of the loop contributions, and the value for η deviates from the expression in Eq. (7.85) [40].

Interestingly, an analogously constructed nonequilibrium *two-temperature model J* with reversible mode-coupling vertex *cannot* be cast into a form that is equivalent to an equilibrium system, for owing to the emerging anisotropy, the condition (7.93) cannot be satisfied. A one-loop RG analysis yields a run-away flow, and no stable RG fixed point is found [38]. Similar behaviour ensues in other anisotropic nonequilibrium variants of critical dynamics models with conserved order parameter; the precise interpretation of the apparent instability is as yet unclear [33].

7.1.9 Driven Diffusive Systems

Finally, we wish to consider Langevin representations of genuinely nonequilibrium systems, namely driven diffusive lattice gases (for a comprehensive overview, see Ref. [42]). First we address the coarse-grained continuum version of the asymmetric exclusion process, i.e., hard-core repulsive particles that hop preferentially in one direction. We describe this system in terms of a *conserved* particle density, whose fluctuations we denote with $S(\boldsymbol{x},t)$, such that $\langle S \rangle = 0$, obeying a *continuity equation* $\partial_t S(\boldsymbol{x},t) + \boldsymbol{\nabla} \cdot \boldsymbol{J}(\boldsymbol{x},t) = 0$. We assume the system to be driven along the "$\parallel$' direction; in the transverse sector (of dimension $d_\perp = d - 1$) we thus just have a noisy *diffusion current* $\boldsymbol{J}_\perp = -D\boldsymbol{\nabla}_\perp S + \boldsymbol{\eta}$, whereas there is a nonlinear term, stemming from the hard-core interactions, in the current along the direction of the external drive, with $J_{0\parallel} = \text{const.}$: $J_\parallel = J_{0\parallel} - Dc\nabla_\parallel S - \frac{1}{2}Dg S^2 + \eta_\parallel$. For the stochastic currents, we assume Gaussian white noise $\langle \eta_i \rangle = 0 = \langle \eta_\parallel \rangle$ and $\langle \eta_i(\boldsymbol{x},t)\,\eta_j(\boldsymbol{x}',t') \rangle = 2D\,\delta(\boldsymbol{x}-\boldsymbol{x}')\,\delta(t-t')\,\delta_{ij}$, $\langle \eta_\parallel(\boldsymbol{x},t)\,\eta_\parallel(\boldsymbol{x}',t') \rangle = 2D\,\tilde{c}\,\delta(\boldsymbol{x}-\boldsymbol{x}')\,\delta(t-t')$. Notice that since we are not in thermal equilibrium, Einstein's relation need not be fulfilled. We can however always rescale the field to satisfy it in the transverse sector; the ratio $w = \tilde{c}/c$ then measures the deviation from equilibrium. These considerations yield the generic Langevin equation for the density fluctuations in *driven diffusive systems (DDS)* [43, 44]

$$\frac{\partial S(\boldsymbol{x},t)}{\partial t} = D\left(\boldsymbol{\nabla}_\perp^2 + c\,\nabla_\parallel^2\right) S(\boldsymbol{x},t) + \frac{Dg}{2}\,\nabla_\parallel S(\boldsymbol{x},t)^2 + \zeta(\boldsymbol{x},t) , \qquad (7.129)$$

with conserved noise $\zeta = -\boldsymbol{\nabla}_\perp \cdot \boldsymbol{\eta} - \nabla_\parallel \eta_\parallel$, where $\langle \zeta \rangle = 0$ and

$$\langle \zeta(\boldsymbol{x},t)\,\zeta(\boldsymbol{x}',t') \rangle = -2D\left(\boldsymbol{\nabla}_\perp^2 + \tilde{c}\,\nabla_\parallel^2\right)\delta(\boldsymbol{x}-\boldsymbol{x}')\,\delta(t-t') . \qquad (7.130)$$

Notice that the drive term $\propto g$ breaks both the system's spatial reflection symmetry and the Ising Z_2 symmetry $S \to -S$.

The corresponding Janssen–De Dominicis response functional (7.34) reads

$$\mathcal{A}[\tilde{S}, S] = \int d^d x \int dt\, \tilde{S} \left[\frac{\partial S}{\partial t} - D \left(\nabla_\perp^2 + c \nabla_\parallel^2 \right) S \right.$$
$$\left. + D \left(\nabla_\perp^2 + \tilde{c} \nabla_\parallel^2 \right) \tilde{S} - \frac{D g}{2} \nabla_\parallel S^2 \right] . \tag{7.131}$$

It describes a *"massless"* theory, hence we expect the system to be *generically scale-invariant*, without the need to tune it to a special point in parameter space. The nonlinear drive term will induce anomalous scaling in the drive direction, different from ordinary diffusive behaviour. In the transverse sector, however, we have to *all* orders in the perturbation expansion simply

$$\Gamma^{(1,1)}(q_\perp, q_\parallel = 0, \omega) = i\omega + D q_\perp^2 , \quad \Gamma^{(2,0)}(q_\perp, q_\parallel = 0, \omega) = -2D q_\perp^2 , \tag{7.132}$$

since the nonlinear three-point vertex, which is of the form depicted in Fig. 7.5(a), is proportional to $iq_\parallel$. Consequently,

$$Z_{\tilde{S}} = Z_S = Z_D = 1 , \tag{7.133}$$

which immediately implies

$$\eta = 0 , \quad z = 2 . \tag{7.134}$$

Moreover, the nonlinear coupling g itself does not renormalise either as a consequence of *Galilean invariance*. Namely, the Langevin equation (7.129) and the action (7.131) are left invariant under Galilean transformations

$$S'(x'_\perp, x'_\parallel, t') = S(x_\perp, x_\parallel - Dgv\, t, t) - v ; \tag{7.135}$$

thus, the boost velocity v must scale as the field S under renormalisation, and since the product Dgv must be invariant under the RG, this leaves us with

$$Z_g = Z_D^{-1} Z_S^{-1} = 1 . \tag{7.136}$$

The effective nonlinear coupling governing the perturbation expansion in terms of loop diagrams turns out to be $g^2/c^{3/2}$; if we define its renormalised counterpart as

$$v_R = Z_c^{3/2} v\, C_d \mu^{d-2} , \tag{7.137}$$

with the convenient choice $C_d = \Gamma(2 - d/2)/2^{d-1}\pi^{d/2}$, we see that the associated RG beta function becomes

$$\beta_v = v_R \left(d - 2 - \frac{3}{2} \gamma_c \right) . \tag{7.138}$$

At *any* nontrivial RG fixed point $0 < v^* < \infty$, therefore $\gamma_c^* = \frac{2}{3}(d - 2)$. We thus infer that below the upper critical dimension $d_c = 2$ for DDS, the longitudinal scaling exponents are fixed by the system's symmetry [43, 44],

$$\Delta = -\frac{\gamma_c^*}{2} = \frac{2-d}{3} \ , \quad z_\| = \frac{2}{1+\Delta} = \frac{6}{5-d} \ . \tag{7.139}$$

An explicit one-loop calculation for the two-point vertex functions, see Fig. 7.5(b) and (c), yields

$$\gamma_c = -\frac{v_R}{16}\left(3+w_R\right) \ , \quad \gamma_{\tilde{c}} = -\frac{v_R}{32}\left(3w_R^{-1}+2+3w_R\right) \ , \tag{7.140}$$

$$\beta_w = w_R\left(\gamma_{\tilde{c}} - \gamma_c\right) = -\frac{v_R}{32}\left(w_R-1\right)\left(w_R-3\right) \ . \tag{7.141}$$

This establishes that in fact the fixed point $w^* = 1$ is IR-stable (provided $0 < v^* < \infty$), which means that asymptotically the Einstein relation is satisfied in the longitudinal sector as well [43].

In this context, it is instructive to make an intriguing connection with the *noisy Burgers equation* [45], describing simplified fluid dynamics in terms of a velocity field $u(x,t)$:

$$\frac{\partial u(x,t)}{\partial t} + \frac{Dg}{2}\,\nabla\left[u(x,t)^2\right] = D\,\nabla^2 u(x,t) + \zeta(x,t) \ , \tag{7.142}$$

$$\langle\zeta_i\rangle = 0 \ , \ \langle\zeta_i(x,t)\,\zeta_j(x',t')\rangle = -2D\,\nabla_i\nabla_j\,\delta(x-x')\,\delta(t-t') \ . \tag{7.143}$$

For $Dg = 1$, the nonlinearity is just the usual fluid advection term. In one dimension, the Burgers equation (7.142) becomes *identical* with the DDS Langevin equation (7.129), so we immediately infer its anomalous dynamic critical exponent $z_\| = 3/2$. At least in one dimension therefore, it should represent an *equilibrium* system which asymptotically approaches the canonical distribution (7.6), where the Hamiltonian is simply the fluid's kinetic energy (and we have set $k_BT = 1$). So let us check the equilibrium condition (7.93) with $\mathcal{P}_{\mathrm{eq}}[u] \propto \exp\left[-\frac{1}{2}\int u(x)^2\,\mathrm{d}^d x\right]$:

$$\int \mathrm{d}^d x\,\frac{\delta}{\delta u(x,t)}\cdot\left[\nabla u(x,t)^2\right]\mathrm{e}^{-\frac{1}{2}\int u(x',t)^2\,\mathrm{d}^d x'}$$

$$= \int\left[2\nabla\cdot u(x,t) - u(x,t)\cdot\nabla u(x,t)^2\right]\mathrm{d}^d x\,\mathrm{e}^{-\frac{1}{2}\int u(x',t)^2\,\mathrm{d}^d x'} \ .$$

With appropriate boundary conditions, the first term here vanishes, but the second one does so *only* in $d = 1$: $-\int u\,(\mathrm{d}u^2/\mathrm{d}x)\,\mathrm{d}x = \int u^2\,(\mathrm{d}u/\mathrm{d}x)\,\mathrm{d}x = \frac{1}{3}\int(\mathrm{d}u^3/\mathrm{d}x)\,\mathrm{d}x = 0$. Driven diffusive systems in one dimension are therefore subject to a *"hidden" fluctuation–dissipation theorem*.

To conclude this part on Langevin dynamics, let us briefly consider the *driven model B* or *critical DDS* [42], which corresponds to a driven Ising lattice gas near its critical point. Here, a conserved scalar field S undergoes a second-order phase transition, but similar to the randomly driven case, again only the *transverse* sector is critical. Upon adding the DDS drive term from Eq. (7.129) to the Langevin equation (7.119), we obtain

$$\frac{\partial S(x,t)}{\partial t} = D \left[\nabla_\perp^2 \left(r - \nabla_\perp^2 \right) + c \nabla_\parallel^2 \right] S(x,t) + \frac{D\tilde{u}}{6} \nabla_\perp^2 S(x,t)^3$$

$$+ \frac{Dg}{2} \nabla_\parallel S(x,t)^2 + \zeta(x,t) , \tag{7.144}$$

with the (scalar) noise specified in Eq. (7.120). The response functional thus becomes

$$\mathcal{A}[\tilde{S}, S] = \int d^d x \int dt \, \tilde{S} \left[\frac{\partial S}{\partial t} - D \left[\nabla_\perp^2 \left(r - \nabla_\perp^2 \right) + c \nabla_\parallel^2 \right] S \right.$$

$$\left. + D \left(\nabla_\perp^2 \tilde{S} - \frac{\tilde{u}}{6} \nabla_\perp^2 S^3 - \frac{g}{2} \nabla_\parallel S^2 \right) \right] . \tag{7.145}$$

Power counting gives $[g^2] = \mu^{5-d}$, so the upper critical dimension here is $d_c = 5$, and $[\tilde{u}] = \mu^{3-d}$. The nonlinearity $\propto \tilde{u}$ is thus *irrelevant* and can be omitted if we wish to determine the asymptotic universal scaling laws; but recall that it is responsible for the phase transition in the system. The remaining vertex is then proportional to $iq_\parallel$, whence Eqs. (7.132) and (7.133) hold for critical DDS as well, and the transverse critical exponents are just those of the Gaussian model B,

$$\eta = 0 , \quad \nu = 1/2 , \quad z = 4 . \tag{7.146}$$

In addition, Galilean invariance with respect to Eq. (7.135) and therefore Eq. (7.136) hold as before. With the renormalised nonlinear drive strength defined similarly to Eq. (7.137), but a different geometric constant and the scale factor μ^{d-5}, the associated RG beta function reads

$$\beta_v = v_R \left(d - 5 - \frac{3}{2} \gamma_c \right) , \tag{7.147}$$

which again allows us to determine the longitudinal scaling exponents to *all* orders in perturbation theory, for $d < d_c = 5$,

$$\Delta = 1 - \frac{\gamma_c^*}{2} = \frac{8-d}{3} , \quad z_\parallel = \frac{4}{1+\Delta} = \frac{12}{11-d} . \tag{7.148}$$

It is worthwhile mentioning a few marked differences to the two-temperature model B discussed in Subsect. 7.1.8: In DDS, there are obviously nonzero *three-point* correlations, and in the driven critical model B the upper critical dimension is $d_c = 5$ as opposed to $d_c = 4 - d_\parallel$ for the randomly driven version. Notice also that the latter is characterised by nontrivial static critical exponents, but the kinetics is purely diffusive along the drive direction, $z_\parallel = 2$. Conversely for the driven model B, only the longitudinal scaling exponents are non-Gaussian.

7.2 Reaction–Diffusion Systems

We now turn our attention to stochastic interacting particle systems, whose microscopic dynamics is defined through a (classical) master equation. Below, we shall see how the latter can be mapped onto a stochastic quasi-Hamiltonian in a second-quantised bosonic operator representation [10–12]. Taking the continuum limit on the basis of coherent-state path integrals then yields a field theory action that may be analysed by the very same RG methods as described before in Subsects. 7.1.3–7.1.5 (for more details, see the recent overview [12]).

7.2.1 Chemical Reactions and Population Dynamics

Our goal is to study systems of "particles" $A, B, \dots$ that propagate through hopping to nearest neighbors on a d-dimensional lattice, or via diffusion in the continuum. Upon encounter, or spontaneously, with given stochastic rates, these particles may undergo species changes, annihilate, or produce offspring. At large densities, the characteristic time scales of the kinetics will be governed by the reaction rates, and the system is said to be *reaction-limited*. In contrast, at low densities, any reactions that require at least two particles to be in proximity will be *diffusion-limited*: the basic time scale will be set by the hopping rate or diffusion coefficient.

As a first approximation to the dynamics of such "chemical" reactions, let us assume homogeneous mixing of each species. We may then hope to be able to capture the kinetics in terms of *rate equations* for each particle concentration or mean density. Note that such a description neglects any spatial fluctuations and correlations in the system, and is therefore in character a mean-field approximation. As a first illustration consider the *annihilation* of $k - l > 0$ particles of species A in the *irreversible* kth-order reaction $k\,A \to l\,A$, with rate λ. The corresponding rate equation employs a factorisation of the probability of encountering k particles at the same point to simply the kth power of the concentration $a(t)$,

$$\dot{a}(t) = -(k - l)\,\lambda\,a(t)^k \ . \tag{7.149}$$

This ordinary differential equation is readily solved, with the result

$$k = 1 \ : \ a(t) = a(0)\,e^{-\lambda t} \ , \tag{7.150}$$

$$k \geq 2 \ : \ a(t) = \left[a(0)^{1-k} + (k - l)(k - 1)\,\lambda\,t\right]^{-1/(k-1)} \ . \tag{7.151}$$

For simple "radioactive" decay ($k = 1$), we of course obtain an exponential time dependence, as appropriate for statistically independent events. For pair ($k = 2$) and higher-order ($k \geq 3$) processes, however, we find algebraic long-time behaviour, $a(t) \to (\lambda t)^{-1/(k-1)}$, with an amplitude that becomes independent of the initial density $a(0)$. The absence of a characteristic time scale hints at cooperative effects, and we have to ask if and under which circumstances correlations might qualitatively affect the asymptotic long-time power laws. For according to Smoluchowski theory [12], we would expect the annihilation reactions to produce *depletion zones* in sufficiently low dimensions $d \leq d_c$, which would in turn induce a considerable *slowing down* of the density decay, see Subsect. 7.2.3. For *two-species* pair annihilation $A + B \to \emptyset$ (without mixing), another complication emerges, namely particle species *segregation* in dimensions for $d \leq d_s$; the regions dominated by either species become largely inert, and the annihilation reactions are confined to rather sharp *fronts* [12].

Competition between particle decay and production processes, e.g., in the reactions $A \to \emptyset$ (with rate κ), $A \rightleftharpoons A + A$ (with forward and back rates σ and λ, respectively), leads to even richer scenarios, as can already be inferred from the associated rate equation

$$\dot{a}(t) = (\sigma - \kappa)\,a(t) - \lambda\,a(t)^2 \ . \tag{7.152}$$

For $\sigma < \kappa$, clearly $a(t) \sim e^{-(\kappa - \sigma)\,t} \to 0$ as $t \to \infty$. The system eventually enters an *inactive* state, which even in the fully stochastic model is *absorbing*, since once there is no particle left, no process whatsoever can drive the system out of the empty state again. On the other hand, for $\sigma > \kappa$, we encounter an *active* state with $a(t) \to a_\infty = (\sigma - \kappa)/\lambda$ exponentially, with rate $\sim \sigma - \kappa$. We have thus identified a nonequilibrium *continuous* phase transition at $\sigma_c = \kappa$. Indeed, as in equilibrium critical phenomena, the critical point is governed by characteristic power laws; for example, the asymptotic particle density $a_\infty \sim (\sigma - \sigma_c)^\beta$, and the critical density decay $a(t) \sim (\lambda t)^{-\alpha}$ with $\beta_0 = 1 = \alpha_0$ in the mean-field approximation. The following natural questions then arise: What are the *critical exponents* once statistical fluctuations are properly included in the analysis? Can we, as in equilibrium systems, identify and characterise certain *universality classes*, and which microscopic or overall, global features determine them and their critical dimension?

Already the previous set of reactions may also be viewed as a (crude) model for the *population dynamics* of a single species. In the same language, we may also formulate a stochastic version of the classic *Lotka–Volterra predator–prey competition model* [46]: if by themselves, the "predators" A die out according to $A \to \emptyset$, with rate κ, whereas the prey reproduce $B \to B + B$ with rate σ, and thus proliferate with a Malthusian population explosion. The predators are kept alive and the prey under control through *predation*, here modeled as the reaction $A + B \to A + A$: with rate λ, a prey is "eaten" by a predator, who simultaneously produces an offspring. The coupled kinetic rate equations for this system read

$$\dot{a}(t) = \lambda\,a(t)\,b(t) - \kappa\,a(t)\ , \quad \dot{b}(t) = \sigma\,b(t) - \lambda\,a(t)\,b(t)\ . \tag{7.153}$$

It is straightforward to show that the quantity $K(t) = \lambda[a(t)+b(t)]-\sigma \ln a(t) - \kappa \ln b(t)$ is a constant of motion for this coupled system of differential equations, i.e., $\dot{K}(t) = 0$. As a consequence, the system is governed by regular population *oscillations*, whose frequency and amplitude are fully determined by the *initial* conditions. Clearly, this is not a very realistic feature (albeit mathematically appealing), and moreover Eqs. (7.153) are known to be quite unstable with respect to model modifications [46]. Indeed, if one includes spatial degrees of freedom and takes account of the full stochasticity of the processes involved, the system's behaviour turns out to be much richer [47]: In the species coexistence phase, one encounters for sufficiently large values of the predation rate an incessant sequence of *"pursuit and evasion"* waves that form quite complex dynamical patterns. In finite systems, these induce *erratic* population oscillations whose features are however independent of the initial configuration, but whose amplitude vanishes in the thermodynamic limit. Moreover, if locally the prey "carrying capacity" is limited (corresponding to restricting the maximum site occupation number per site on a lattice), there appears an *extinction threshold* for the predator population that separates the absorbing state of a system filled with prey from the active coexistence regime through a continuous phase transition [47].

These examples all call for a systematic approach to include stochastic fluctuations in the mathematical description of interacting reaction–diffusion systems that would be conducive to the application of field-theoretic tools, and thus allow us to bring the powerful machinery of the dynamic renormalisation group to bear on these problems. In the following, we shall describe such a general method [48–50] which allows a representation of the classical master equation in terms of a coherent-state path integral and its subsequent analysis by means of the RG (for overviews, see Refs. [10–12]).

7.2.2 Field Theory Representation of Master Equations

The above interacting particle systems, when defined on a d-dimensional lattice with sites i, are fully characterised by the set of occupation integer numbers $n_i = 0, 1, 2, \ldots$ for each particle species. The *master equation* then describes the temporal evolution of the configurational probability distribution $P(\{n_i\}; t)$ through a *balance* of gain and loss terms. For example, for the *binary annihilation* and *coagulation reactions* $A + A \to \emptyset$ with rate λ and $A + A \to A$ with rate λ', the master equation on a specific site i reads

$$\frac{\partial P(n_i; t)}{\partial t} = \lambda (n_i + 2)(n_i + 1) P(\ldots, n_i + 2, \ldots; t)$$
$$+\lambda' (n_i + 1) n_i P(\ldots, n_i + 1, \ldots; t)$$
$$-(\lambda + \lambda') n_i (n_i - 1) P(\ldots, n_i, \ldots; t) , \qquad (7.154)$$

with initially $P(\{n_i\}, 0) = \prod_i P(n_i)$, e.g., a *Poisson* distribution $P(n_i) = \bar{n}_0^{n_i} e^{-\bar{n}_0}/n_i!$. Since the reactions all change the site occupation numbers by integer values, a *second-quantised Fock space representation* is particularly useful [48–50]. To this end, we introduce the *bosonic operator algebra*

$$\left[a_i, a_j\right] = 0 = \left[a_i^\dagger, a_j^\dagger\right] , \quad \left[a_i, a_j^\dagger\right] = \delta_{ij} . \qquad (7.155)$$

From these commutation relations one establishes in the standard manner that a_i and $a_i^\dagger$ constitute lowering and raising ladder operators, from which we may construct the particle number eigenstates $|n_i\rangle$,

$$a_i\,|n_i\rangle = n_i\,|n_i - 1\rangle\ , \quad a_i^\dagger\,|n_i\rangle = |n_i + 1\rangle\ , \quad a_i^\dagger\,a_i\,|n_i\rangle = n_i\,|n_i\rangle\ . \quad (7.156)$$

(Notice that we have chosen a different normalisation than in ordinary quantum mechanics.) A state with n_i particles on sites i is then obtained from the empty vaccum state $|0\rangle$, defined through $a_i\,|0\rangle = 0$, as the product state

$$|\{n_i\}\rangle = \prod_i \left(a_i^\dagger\right)^{n_i}|0\rangle\ . \quad (7.157)$$

To make contact with the time-dependent configuration probability, we introduce the formal *state vector*

$$|\Phi(t)\rangle = \sum_{\{n_i\}} P(\{n_i\}; t)\,|\{n_i\}\rangle\ , \quad (7.158)$$

whereupon the linear time evolution according to the master equation is translated into an *"imaginary-time" Schrödinger equation*

$$\frac{\partial|\Phi(t)\rangle}{\partial t} = -H\,|\Phi(t)\rangle\ , \quad |\Phi(t)\rangle = \mathrm{e}^{-H\,t}\,|\Phi(0)\rangle\ . \quad (7.159)$$

The *stochastic quasi-Hamiltonian* (rather, the time evolution or Liouville operator) for the on-site reaction processes is a sum of local terms, $H_{\mathrm{reac}} = \sum_i H_i(a_i^\dagger, a_i)$; e.g., for the binary annihilation and coagulation reactions,

$$H_i(a_i^\dagger, a_i) = -\lambda\left(1 - a_i^{\dagger 2}\right)a_i^2 - \lambda'\left(1 - a_i^\dagger\right)a_i^\dagger\,a_i^2\ . \quad (7.160)$$

The two contributions for each process may be physically interpreted as follows: The first term corresponds to the actual *process* under consideration, and describes how many particles are annihilated and (re-)created in each reaction. The second term gives the *"order"* of each reaction, i.e., the number operator $a_i^\dagger\,a_i$ appears to the kth power, but in normal-ordered form as $a_i^{\dagger\,k}\,a_i^k$, for a kth-order process. Note that the reaction Hamiltonians such as (7.160) are *non-Hermitean*, reflecting the particle creations and destructions. In a similar manner, hopping between neighbouring sites $\langle ij\rangle$ is represented in this formalism through

$$H_{\mathrm{diff}} = D\sum_{\langle ij\rangle} \left(a_i^\dagger - a_j^\dagger\right)\left(a_i - a_j\right)\ . \quad (7.161)$$

Our goal is of course to compute averages with respect to the configurational probability distribution $P(\{n_i\}; t)$; this is achieved by means of the *projection state* $\langle\mathcal{P}| = \langle 0|\prod_i \mathrm{e}^{a_i}$, which satisfies $\langle\mathcal{P}|0\rangle = 1$ and $\langle\mathcal{P}|a_i^\dagger = \langle\mathcal{P}|$, since $[e^{a_i}, a_j^\dagger] = e^{a_i} \delta_{ij}$. For the desired *statistical averages* of observables that must be expressible in terms of the occupation numbers $\{n_i\}$, we then obtain

$$\langle F(t) \rangle = \sum_{\{n_i\}} F(\{n_i\}) P(\{n_i\}; t) = \langle \mathcal{P}| F(\{a_i^\dagger a_i\}) |\Phi(t)\rangle . \tag{7.162}$$

Let first us explore the consequences of *probability conservation*, i.e., $1 = \langle \mathcal{P}|\Phi(t)\rangle = \langle \mathcal{P}|e^{-Ht}|\Phi(0)\rangle$. This requires $\langle \mathcal{P}|H = 0$; upon commuting $e^{\sum_i a_i}$ with H, effectively the creation operators become shifted $a_i^\dagger \to 1 + a_i^\dagger$, whence this condition is fulfilled provided $H_i(a_i^\dagger \to 1, a_i) = 0$, which is indeed satisfied by our explicit expressions (7.160) and (7.161). By this prescription, we may also in averages replace $a_i^\dagger a_i \to a_i$, i.e., the *particle density* becomes $a(t) = \langle a_i \rangle$, and the two-point operator $a_i^\dagger a_i a_j^\dagger a_j \to a_i \delta_{ij} + a_i a_j$.

In the bosonic operator representation above, we have assumed that there exist no restrictions on the particle occupation numbers n_i on each site. If, however, there is a maximum $n_i \leq 2s+1$, one may instead employ a representation in terms of spin s operators. For example, particle exclusion systems with $n_i = 0$ or 1 can thus be mapped onto non-Hermitean spin $1/2$ "quantum" systems. Specifically in one dimension, such representations in terms of integrable spin chains have proved a fruitful tool; for overviews, see Refs. [51–54]. An alternative approach uses the bosonic theory, but encodes the site occupation restrictions through appropriate exponentials in the number operators $e^{-a_i^\dagger a_i}$ [55].

We may now follow an established route in quantum many-particle theory [56] and proceed towards a field theory representation through constructing the *path integral* equivalent to the "Schrödinger" dynamics (7.159) based on *coherent states*, which are right eigenstates of the annihilation operator, $a_i|\phi_i\rangle = \phi_i|\phi_i\rangle$, with complex eigenvalues ϕ_i. Explicitly, one finds

$$|\phi_i\rangle = \exp\left(-\frac{1}{2}|\phi_i|^2 + \phi_i a_i^\dagger\right) |0\rangle , \tag{7.163}$$

satisfying the *overlap* and (over-)*completeness relations*

$$\langle \phi_j|\phi_i\rangle = \exp\left(-\frac{1}{2}|\phi_i|^2 - \frac{1}{2}|\phi_j|^2 + \phi_j^* \phi_i\right) , \quad \int \prod_i \frac{d^2\phi_i}{\pi} |\{\phi_i\}\rangle \langle\{\phi_i\}| = 1 . \tag{7.164}$$

Upon splitting the temporal evolution (7.159) into infinitesimal steps, and inserting Eq. (7.164) at each time step, standard procedures (elaborated in detail in Ref. [12]) yield eventually

$$\langle F(t) \rangle \propto \int \prod_i \mathcal{D}[\phi_i] \mathcal{D}[\phi_i^*] F(\{\phi_i\}) e^{-A[\phi_i^*, \phi_i]} , \tag{7.165}$$

with the *action*

$$\mathcal{A}[\phi_i^*, \phi_i] = \sum_i \left(-\phi_i(t_f) + \int_0^{t_f} dt \left[\phi_i^* \frac{\partial \phi_i}{\partial t} + H_i(\phi_i^*, \phi_i) \right] - \bar{n}_0 \, \phi_i^*(0) \right) , \quad (7.166)$$

where the first term originates from the projection state, and the last one from the initial Poisson distribution. Notice that in the Hamiltonian, the creation and annihilation operators $a_i^\dagger$ and a_i are simply replaced with the complex numbers ϕ_i^* and ϕ_i, respectively.

Taking the *continuum limit*, $\phi_i(t) \to \psi(\boldsymbol{x}, t)$, $\phi_i^*(t) \to \hat{\psi}(\boldsymbol{x}, t)$, the "bulk" part of the action becomes

$$\mathcal{A}[\hat{\psi}, \psi] = \int d^d x \int dt \left[\hat{\psi} \left(\frac{\partial}{\partial t} - D \boldsymbol{\nabla}^2 \right) \psi + \mathcal{H}_{\text{reac}}(\hat{\psi}, \psi) \right] , \quad (7.167)$$

where the hopping term (7.161) has naturally turned into a diffusion propagator. We have thus arrived at a *microscopic* stochastic field theory for reaction–diffusion processes, with *no* assumptions whatsoever on the form of the (internal) noise. This is a crucial ingredient for nonequilibrium dynamics, and we may now use Eq. (7.167) as a basis for systematic coarse-graining and the renormalisation group analysis. Returning to our example of pair annihilation and coagulation, the reaction part of the action (7.167) reads

$$\mathcal{H}_{\text{reac}}(\hat{\psi}, \psi) = -\lambda \left(1 - \hat{\psi}^2 \right) \psi^2 - \lambda' \left(1 - \hat{\psi} \right) \hat{\psi} \, \psi^2 , \quad (7.168)$$

see Eq. (7.160). Let us have a look at the *classical field equations*, namely $\delta \mathcal{A}/\delta \hat{\psi} = 0$, which is always solved by $\hat{\psi} = 1$, reflecting probability conservation, and $\delta \mathcal{A}/\delta \hat{\psi} = 0$, which, upon inserting $\hat{\psi} = 1$ gives here

$$\frac{\partial \psi(\boldsymbol{x}, t)}{\partial t} = D \boldsymbol{\nabla}^2 \psi(\boldsymbol{x}, t) - (2\lambda + \lambda') \psi(\boldsymbol{x}, t)^2 , \quad (7.169)$$

i.e., essentially the mean-field rate equation for the local particle density $\psi(\boldsymbol{x}, t)$, see Eq. (7.149), supplemented with diffusion. The field theory action (7.167), derived from the master equation (7.154), then provides a means of including fluctuations in our analysis.

Before we proceed with this program, it is instructive to perform a shift in the field $\hat{\psi}$ about the mean-field solution, $\hat{\psi}(x, t) = 1 + \tilde{\psi}(x, t)$, whereupon the reaction Hamiltonian density (7.168) becomes

$$\mathcal{H}_{\text{reac}}(\tilde{\psi}, \psi) = (2\lambda + \lambda') \tilde{\psi} \, \psi^2 + (\lambda + \lambda') \tilde{\psi}^2 \, \psi^2 . \quad (7.170)$$

In addition to the diffusion propagator, the annihilation and coagulation processes thus give *identical* three- and four-point vertices; aside from non-universal amplitudes, one should therefore obtain *identical* scaling behaviour for both binary reactions in the asymptotic long-time limit [57]. Lastly, we remark that if we interpret the action $\mathcal{A}[\tilde{\psi}, \psi]$ as a response functional (7.34), despite the fields $\tilde{\psi}$ not being purely imaginary, our field theory becomes formally equivalent to a "Langevin" equation, wherein additive noise is added to

Eq. (7.169), albeit with *negative* correlator $L[\psi] = -(\lambda + \lambda')\,\psi^2$, which represents *"imaginary" multiplicative noise*. This Langevin description is thus not well-defined; however, one may render the noise correlator positive through a nonlinear *Cole–Hopf transformation* $\widetilde{\psi} = e^{\tilde\rho}$, $\psi = \rho\,e^{-\tilde\rho}$ such that $\widetilde{\psi}\,\psi = \rho$, with Jacobian 1, but at the expense of "diffusion noise" $\propto D\,\rho\,(\boldsymbol{\nabla}\tilde\rho)^2$ in the action [58]. In summary, binary (and higher-order) annihilation and coagulation processes cannot be cast into a Langevin framework in any simple manner.

7.2.3 Diffusion-Limited Single-Species Annihilation Processes

We begin by analysing diffusion-limited single-species annihilation $k\,A \to \emptyset$ [57,59]. The corresponding field theory action (7.167) reads

$$A[\hat\psi, \psi] = \int \mathrm{d}^d x \int \mathrm{d}t \left[\hat\psi \left(\frac{\partial}{\partial t} - D\boldsymbol{\nabla}^2 \right) \psi - \lambda \left(1 - \hat\psi^k \right) \psi^k \right] , \qquad (7.171)$$

which for $k \geq 3$ allows no (obvious) equivalent Langevin description. Straightforward power counting gives the scaling dimension for the annihilation rate, $[\lambda] = \mu^{2-(k-1)d}$, which suggests the upper critical dimension $d_c(k) = 2/(k-1)$. Thus we expect mean-field behaviour $\sim (\lambda\,t)^{-1/(k-1)}$, see Eq. (7.151), in any physical dimension for $k > 3$, logarithmic corrections at $d_c = 1$ for $k = 3$ and at $d_c = 2$ for $k = 2$, and nonclassical power laws for pair annihilation only in one dimension. The field theory defined by the action (7.171) has two vertices, the "annihilation" sink with k incoming lines only, and the "scattering" vertex with k incoming and k outgoing lines. Neither allows for propagator renormalisation, hence the model remains massless with *exact* scaling exponents $\eta = 0$ and $z = 2$, i.e., diffusive dynamics.

In addition, the entire perturbation expansion for the renormalisation for the annihilation vertices is merely a geometric series of the one-loop diagram, see Fig. 7.6 for the pair annihilation case $(k = 2)$. If we define the renormalised effective coupling according to

$$g_R = Z_g \frac{\lambda}{D} B_{kd}\,\mu^{-2(1-d/d_c)} , \qquad (7.172)$$

where $B_{kd} = k!\,\Gamma(2-d/d_c)\,d_c/k^{d/2}\,(4\pi)^{d/d_c}$, we obtain for the single nontrivial renormalisation constant

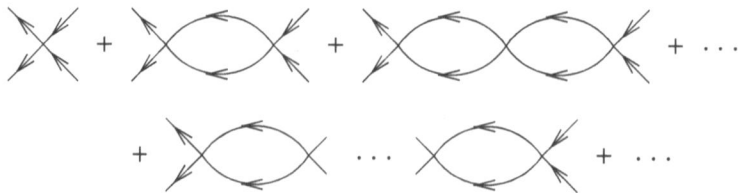

Fig. 7.6. Vertex renormalisation for diffusion-limited binary annihilation $A+A \to \emptyset$

$$Z_g^{-1} = 1 + \frac{\lambda B_{kd} \mu^{-2(1-d/d_c)}}{D(d_c - d)} \tag{7.173}$$

to *all* orders. Consequently, the associated RG beta function becomes

$$\beta_g = \mu \frac{\partial}{\partial \mu}\bigg|_0 g_R = -\frac{2 g_R}{d_c}(d - d_c + g_R) , \tag{7.174}$$

with the Gaussian fixed point $g_0^* = 0$ stable for $d > d_c(k) = 2/(k-1)$, leading to the mean-field power laws (7.151), whereas for $d < d_c(k)$ the flow approaches

$$g^* = d_c(k) - d . \tag{7.175}$$

Since the particle density has scaling dimension $[a] = \mu^d$, we may write $a_R(\mu, D_R, n_0, g_R) = \mu^d \hat{a}_R(D_R, n_0 \mu^{-d}, g_R)$, where we have retained the dependence on the initial density n_0. Since the fields and the diffusion constant do not renormalise ($\gamma_D = 0$ and $\gamma_{n_0} = -d$), the RG equation for the density takes the form

$$\left[d - d n_0 \frac{\partial}{\partial n_0} + \beta_g \frac{\partial}{\partial g_R} \right] \hat{a}_R \left(n_0 \mu^{-d}, g_R \right) = 0 , \tag{7.176}$$

see Eq. (7.75). With the characteristics set equal to $\mu \ell = (Dt)^{-1/2}$, the solution of the RG equation (7.176) near the IR-stable RG fixed point g^* becomes

$$a_R(n_0, t) \sim (D\mu^2 t)^{-d/2} \hat{a}_R \left(n_0 (D\mu^2 t)^{d/2}, g^* \right) . \tag{7.177}$$

Under the RG, the first argument in Eq. (7.177) flows to infinity. One therefore needs to establish that the result for the scaling function $\hat{a}$ is finite to *all* orders in the initial density [12,59]. One then finds the following asymptotic long-time behaviour for pair annihilation below the critical dimension [57,59],

$$k = 2, \ d < 2 : \ a(t) \sim (Dt)^{-d/2} . \tag{7.178}$$

At the critical dimension, $\tilde{g}(\ell) \to 0$ logarithmically slowly, and the process is still diffusion-limited; this gives

$$k = 2, \ d = 2 : \ a(t) \sim (Dt)^{-1} \ln(Dt) , \tag{7.179}$$

$$k = 3, \ d = 1 : \ a(t) \sim \left[(Dt)^{-1} \ln(Dt) \right]^{1/2} . \tag{7.180}$$

7.2.4 Segregation for Multi-Species Pair Annihilation

In pair annihilation reactions of two *distinct* species $A + B \to \emptyset$, where *no* reactions between the same species are allowed, a novel phenomenon emerges in sufficiently low dimensions $d \leq d_s$, namely particle segregation in separate spatial domains, with the decay processes restricted to sharp reaction fronts on their boundaries [60]. Note that the reaction $A + B \to \emptyset$ preserves the difference of particle numbers (even locally), i.e., there is a *conservation law* for $c(t) = a(t) - b(t) = c(0)$ [61]. The rate equations for the concentrations

$$\dot{a}(t) = -\lambda\, a(t)\, b(t) = \dot{b}(t) \tag{7.181}$$

are for *equal* initial densities $a(0) = b(0)$ solved by the single-species pair annihilation mean-field power law

$$a(t) = b(t) \sim (\lambda t)^{-1} , \tag{7.182}$$

whereas for *unequal* initial densities $c(0) = a(0) - b(0) > 0$, say, the majority species $a(t) \to a_\infty = c(0) > 0$ as $t \to \infty$, and the minority population disappears, $b(t) \to 0$. From Eq. (7.181) we obtain for $d > d_c = 2$ the exponential approach

$$a(t) - a_\infty \sim b(t) \sim e^{-c(0)\,\lambda t} . \tag{7.183}$$

Mapping the associated master equation onto a continuum field theory (7.167), the reaction term now reads (with the fields ψ and φ representing the A and B particles, respectively) [62]

$$\mathcal{H}_{\text{reac}}(\hat{\psi}, \psi, \hat{\varphi}, \varphi) = -\lambda \left(1 - \hat{\psi}\, \hat{\varphi} \right) \psi\, \varphi . \tag{7.184}$$

As in the single-species case, there is no propagator renormalisation, and moreover the Feynman diagrams for the renormalised reaction vertex are of precisely the same form as for $A + A \to \emptyset$, see Fig. 7.6. Thus, for unequal initial densities, $c(0) > 0$, the mean-field power law $\sim \lambda t$ in the exponent of Eq. (7.183) becomes again replaced with $(Dt)^{d/2}$ in dimensions $d \le d_c = 2$, leading to *stretched exponential* time dependence,

$$d < 2 : \ln b(t) \sim -t^{d/2}, \quad d = 2 : \ln b(t) \sim -t/\ln(Dt) . \tag{7.185}$$

However, species segregation for *equal* initial densities, $a(0) = b(0)$, even supersedes the slowing down due to the reaction rate renormalisation. As confirmed by a thorough RG analysis, this effect can be captured within the classical field equations [62]. To this end, we add *diffusion* terms (with equal diffusivities) to the rate equations (7.181) for the now *local* particle densities,

$$\left(\frac{\partial}{\partial t} - D\,\nabla^2 \right) a(\boldsymbol{x}, t) = -\lambda\, a(\boldsymbol{x}, t)\, b(\boldsymbol{x}, t) = \left(\frac{\partial}{\partial t} - D\,\nabla^2 \right) b(\boldsymbol{x}, t) . \tag{7.186}$$

The local concentration difference $c(\boldsymbol{x}, t)$ thus becomes a purely *diffusive mode*, $\partial_t\, c(\boldsymbol{x}, t) = D\,\nabla^2 c(\boldsymbol{x}, t)$, and we employ the *diffusion Green function*

$$G_0(\boldsymbol{q}, t) = \Theta(t)\, e^{-D\,q^2 t} , \quad G_0(\boldsymbol{x}, t) = \frac{\Theta(t)}{(4\pi\, Dt)^{d/2}}\, e^{-\boldsymbol{x}^2/4Dt} , \tag{7.187}$$

compare Eq. (7.23), to solve the initial value problem,

$$c(\boldsymbol{x}, t) = \int d^d x' \, G_0(\boldsymbol{x} - \boldsymbol{x}', t) \, c(\boldsymbol{x}', 0) \ . \tag{7.188}$$

Let us furthermore assume a *Poisson distribution* for the *initial* density correlations (indicated by an overbar), $\overline{a(\boldsymbol{x}, 0) \, a(\boldsymbol{x}', 0)} = a(0)^2 + a(0) \, \delta(\boldsymbol{x} - \boldsymbol{x}') = \overline{b(\boldsymbol{x}, 0) \, b(\boldsymbol{x}', 0)}$ and $\overline{a(\boldsymbol{x}, 0) \, b(\boldsymbol{x}'(0)} = a(0)^2$, which implies $\overline{c(\boldsymbol{x}, 0) \, c(\boldsymbol{x}', 0)} = 2 \, a(0) \delta(\boldsymbol{x} - \boldsymbol{x}')$. Averaging over the initial conditions then yields with Eq. (7.188)

$$\overline{c(\boldsymbol{x}, t)^2} = 2 \, a(0) \int d^d x' \, G_0(\boldsymbol{x} - \boldsymbol{x}', t)^2 = 2 \, a(0) \, (8\pi \, Dt)^{-d/2} \ ; \tag{7.189}$$

since the distribution for c will be a Gaussian, we thus obtain for the *local density excess* originating in a random initial fluctuation,

$$\overline{|c(\boldsymbol{x}, t)|} = \sqrt{\frac{2}{\pi} \, \overline{c(\boldsymbol{x}, t)^2}} = 2 \sqrt{\frac{a(0)}{\pi}} \, (8\pi \, Dt)^{-d/4} \ . \tag{7.190}$$

In dimensions $d < d_s = 4$ these density fluctuations decay *slower* than the overall particle number $\sim t^{-1}$ for $d > 2$ and $\sim t^{-d/2}$ for $d < 2$ in a homogeneous system. Species segregation into A- and B-rich domains renders the particle distribution *nonuniform*, and the density decay is governed by the slow power law (7.190), $a(t) \sim b(t) \sim (Dt)^{-d/4}$.

For very *special* initial states, however, the situation can be different. For example, consider hard-core particles (or $\lambda \to \infty$) regularly arranged in an alternating manner $\dots ABABABABAB \dots$ on a one-dimensional chain. The reactions $A + B \to \emptyset$ preserve this arrangement, whence the distinction between A and B particles becomes meaningless, and one indeed recovers the $t^{-1/2}$ power law from the single-species pair annihilation reaction.

Let us at last generalise to q-*species* annihilation $A_i + A_j \to \emptyset$, with $1 \le i < j \le q$, with equal initial densities $a_i(0)$ as well as uniform diffusion and reaction rates. For $q > 2$, there exists *no* conservation law in the stochastic system, and one may argue, based on the study of fluctuations in the associated Fokker–Planck equation, that segregation happens only for $d < d_s(q) = 4/(q-1)$ [63]. In any physical dimension $d \ge 2$, one should therefore see the same behaviour as for the single-species reaction $A + A \to \emptyset$; this is actually obvious for $q = \infty$, since in this case the probability for particles of the same species to ever meet is zero, whence the species labeling becomes irrelevant. In one dimension, with its special topology, segregation does occur, and for generic initial conditions one finds the decay law [63]

$$a_i(t) \sim t^{-\alpha(q)} + C \, t^{-1/2} \ , \qquad \alpha(q) = \frac{q-1}{2q} \ , \tag{7.191}$$

which recovers $\alpha(2) = 1/4$ and $\alpha(\infty) = 1/2$. Once again, in special situations, e.g., the alignment $\dots ABCDABCDABCD \dots$ for $q = 4$, the single-species scaling ensues. There are also curious *cyclic* variants, for example if for four species we only allow the reactions $A + B \to \emptyset$, $B + C \to \emptyset$, $C + D \to \emptyset$, and $D + A \to \emptyset$. We may then obviously identify $A = C$ and $B = D$, which leads back to the case of two-species pair annihilation. Generally, within essentially mean-field theory one finds for cyclic multi-species annihilation processes $a_i(t) \sim t^{-\alpha(q,d)}$, where for

$$2 < d_s(q) = \begin{cases} 4 & q = 2, 4, 6, \dots \\ 4\cos(\pi/q) & q = 3, 5, 7, \dots \end{cases} : \alpha(q, d) = d/d_s(q) . \quad (7.192)$$

Remarkably, for five species this yields the borderline dimension $d_s(5) = 1 + \sqrt{5}$ for segregation to occur, hence nontrivial decay exponents $\alpha(5, 2) = \frac{1}{2}(\sqrt{5} - 1)$ in $d = 2$ and $\alpha(5, 3) = \frac{3}{4}(\sqrt{5} - 1)$ in $d = 3$ that involve the golden ratio [64].

7.2.5 Active to Absorbing State Transitions and Directed Percolation

Let us now return to the *competing* single-species reactions $A \to \emptyset$ (rate κ), $A \to A + A$ (rate σ), and, in order to limit the particle density in the active phase, $A + A \to A$ (rate λ). Adding diffusion to the rate equation (7.152), we arrive at the *Fisher–Kolmogorov equation* of biology and ecology [46],

$$\frac{\partial a(\boldsymbol{x}, t)}{\partial t} = -D\left(r - \boldsymbol{\nabla}^2\right) a(\boldsymbol{x}, t) - \lambda\, a(\boldsymbol{x}, t)^2 , \quad (7.193)$$

where $r = (\kappa - \sigma)/D$. As discussed in Subsect. 7.2.1, it predicts a *continuous* transition from an active to an inactive, *absorbing* state to occur at $r = 0$. If we define the associated *critical exponents* in close analogy to equilibrium critical phenomena, see Subsect. (7.1.1), the partial differential equation (7.193) yields the Gaussian exponent values $\eta_0 = 0$, $\nu_0 = 1/2$, $z_0 = 2$, and $\alpha_0 = 1 = \beta_0$.

By the methods outlined in Subsect. 7.2.2, we may construct the coherent-state path integral (7.167) for the associated master equation,

$$A[\hat{\psi}, \psi] = \int d^d x \int dt \left[\hat{\psi} \left(\frac{\partial}{\partial t} - D\boldsymbol{\nabla}^2 \right) \psi - \kappa \left(1 - \hat{\psi}\right) \psi \right.$$
$$\left. + \sigma \left(1 - \hat{\psi}\right) \hat{\psi} \psi - \lambda \left(1 - \hat{\psi}\right) \hat{\psi} \psi^2 \right] . \quad (7.194)$$

Upon shifting the field $\hat{\psi}$ about its stationary value 1 and rescaling according to $\hat{\psi}(\boldsymbol{x}, t) = 1 + \sqrt{\sigma/\lambda}\, \tilde{S}(\boldsymbol{x}, t)$ and $\psi(\boldsymbol{x}, t) = \sqrt{\lambda/\sigma}\, S(\boldsymbol{x}, t)$, the action becomes

$$A[\tilde{S}, S] = \int d^d x \int dt \left[\tilde{S} \left(\frac{\partial}{\partial t} + D\left(r - \boldsymbol{\nabla}^2\right) \right) S - u \left(\tilde{S} - S \right) \tilde{S}\, S + \lambda \tilde{S}^2\, S^2 \right] . \quad (7.195)$$

Thus, the three-point vertices have been scaled to identical coupling strengths $u = \sqrt{\sigma \lambda}$, which represents the effective coupling of the perturbation expansion, see Fig. 7.8 below. Its scaling dimension is $[u] = \mu^{2-d/2}$, whence we infer the upper critical dimension $d_c = 4$. The four-point vertex $\propto \lambda$, with

$[\lambda] = \mu^{2-d}$, is thus *irrelevant* in the RG sense, and can be dropped for the computation of universal, asymptotic scaling properties.

The action (7.195) with $\lambda = 0$ is known as *Reggeon field theory* [65], and its basic characteristic is its invariance under *rapidity inversion* $S(\boldsymbol{x},t) \leftrightarrow -\widetilde{S}(\boldsymbol{x},-t)$. If we interpret Eq. (7.195) as a response functional, we see that it becomes formally equivalent to a stochastic process with multiplicative noise ($\langle\zeta(\boldsymbol{x},t)\rangle = 0$) captured by the Langevin equation [66,67]

$$\frac{\partial S(\boldsymbol{x},t)}{\partial t} = -D\left(r - \boldsymbol{\nabla}^2\right) S(\boldsymbol{x},t) - u\, S(\boldsymbol{x},t)^2 + \zeta(\boldsymbol{x},t) , \qquad (7.196)$$

$$\langle\zeta(\boldsymbol{x},t)\,\zeta(\boldsymbol{x}',t')\rangle = 2u\, S(\boldsymbol{x},t)\,\delta(\boldsymbol{x}-\boldsymbol{x}')\,\delta(t-t') \qquad (7.197)$$

(for a more accurate mapping procedure, see Ref. [68]), which is essentially a noisy Fisher–Kolmogorov equation (7.193), with the noise correlator (7.197) ensuring that the fluctuations indeed cease in the absorbing state where $\langle S\rangle = 0$. It has moreover been established [69–71] that the action (7.195) describes the scaling properties of critical *directed percolation (DP)* clusters [72], illustrated in Fig. 7.7. Indeed, if the DP growth direction is labeled as "time" t, we see that the structure of the DP clusters emerges from the basic decay, branching, and coagulation reactions encoded in Eq. (7.194).

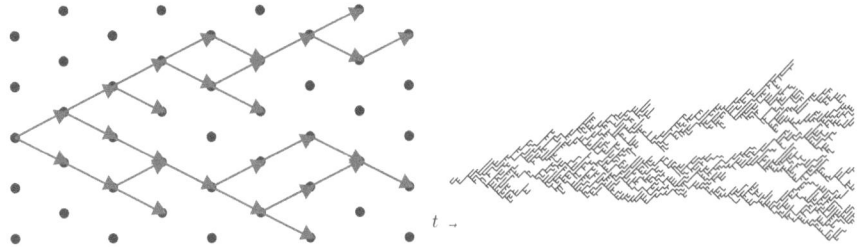

Fig. 7.7. Directed percolation process (*left*) and critical DP cluster (*right*)

In fact, the field theory action should govern the scaling properties of *generic* continuous nonequilibrium phase transitions from active to inactive, absorbing states, namely for an order parameter with Markovian stochastic dynamics that is decoupled from any other slow variable, and in the absence of quenched randomness [71,73]. This *DP conjecture* follows from the following *phenomenological* approach [68] to *simple epidemic processes (SEP)*, or epidemics with recovery [46]:

1. A *"susceptible" medium* becomes locally *"infected"*, depending on the density n of neighboring "sick" individuals. The infected regions recover after a brief time interval.
2. The state $n = 0$ is *absorbing*. It describes the *extinction* of the "disease'.
3. The disease spreads out *diffusively* via the short-range infection 1. of neighboring susceptible regions.

4. Microscopic fast degrees of freedom are incorporated as *local noise* or stochastic forces that respect statement 2., i.e., the noise alone cannot regenerate the disease.

These ingredients are captured by the coarse-grained mesoscopic Langevin equation $\partial_t n = D\left(\nabla^2 - R[n]\right) n + \zeta$ with a reaction functional $R[n]$, and s stochastic noise correlator of the form $L[n] = n\,N[n]$. Near the extinction threshold, we may expand $R[n] = r + u\,n + \ldots$, $N[n] = v + \ldots$, and higher-order terms turn out to be *irrelevant* in the RG sense. Upon rescaling, we recover the Reggeon field theory action (7.195) for DP as the corresponding response functional (7.34).

We now proceed to an explicit evaluation of the DP critical exponents to one-loop order, closely following the recipes given in Subsects. 7.1.3–7.1.5. The lowest-order fluctuation contribution to the two-point vertex function $\Gamma^{(1,1)}(q,\omega)$ (propagator self-energy) is depicted in Fig. 7.8(a). The Feynman rules of Subsect. 7.1.3 yield the corresponding analytic expression

$$\Gamma^{(1,1)}(q,\omega) = i\omega + D\left(r + q^2\right) + \frac{u^2}{D}\int_k \frac{1}{i\omega/2D + r + q^2/4 + k^2}\,. \quad (7.198)$$

The *criticality condition* $\Gamma^{(1,1)}(0,0) = 0$ at $r = r_c$ provides us with the fluctuation-induced shift of the percolation threshold

$$r_c = -\frac{u^2}{D^2}\int_k^\Lambda \frac{1}{r_c + k^2} + O(u^4)\,. \quad (7.199)$$

Inserting $\tau = r - r_c$ into Eq. (7.198), we then find to this order

$$\Gamma^{(1,1)}(q,\omega) = i\omega + D\left(\tau + q^2\right) - \frac{u^2}{D}\int_k \frac{i\omega/2D + \tau + q^2/4}{k^2\left(i\omega/2D + \tau + q^2/4 + k^2\right)}\,, \quad (7.200)$$

and the diagram in Fig. 7.8(b) for the three-point vertex functions, evaluated at zero external wavevectors and frequencies, gives

$$\Gamma^{(1,2)}(\{\underline{0}\}) = -\Gamma^{(2,1)}(\{\underline{0}\}) = -2u\left(1 - \frac{2u^2}{D^2}\int_k \frac{1}{(\tau + k^2)^2}\right)\,. \quad (7.201)$$

For the renormalisation factors, we use again the conventions (7.56) and (7.58), but with

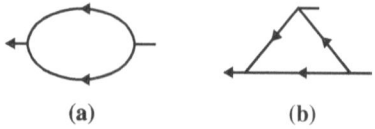

(a)   **(b)**

Fig. 7.8. DP renormalisation: one-loop diagrams for the vertex functions (a) $\Gamma^{(1,1)}$ (propagator self-energy), and (b) $\Gamma^{(1,2)} = -\Gamma^{(2,1)}$ (nonlinear vertices)

$$u_R = Z_u \, u \, A_d^{1/2} \, \mu^{(d-4)/2} \; . \tag{7.202}$$

Because of rapidity inversion invariance, $Z_{\tilde{S}} = Z_S$. With Eq. (7.53) the derivatives of $\Gamma^{(1,1)}$ with respect to ω, $\boldsymbol{q}^2$, and τ, as well as the one-loop result for $\Gamma^{(1,2)}$ in Eq. (7.201), all evaluated at the normalisation point $\tau_R = 1$, provide us with the Z factors

$$Z_S = 1 - \frac{u^2}{2D^2} \frac{A_d \, \mu^{-\epsilon}}{\epsilon} \; , \quad Z_D = 1 + \frac{u^2}{4D^2} \frac{A_d \, \mu^{-\epsilon}}{\epsilon} \; ,$$

$$Z_\tau = 1 - \frac{3u^2}{4D^2} \frac{A_d \, \mu^{-\epsilon}}{\epsilon} \; , \quad Z_u = 1 - \frac{5u^2}{4D^2} \frac{A_d \, \mu^{-\epsilon}}{\epsilon} \; . \tag{7.203}$$

From these we infer the RG flow functions

$$\gamma_S = v_R/2 \; , \quad \gamma_D = -v_R/4 \; , \quad \gamma_\tau = -2 + 3v_R/4 \; , \tag{7.204}$$

with the renormalised effective coupling

$$v_R = \frac{Z_u^2}{Z_D^2} \frac{u^2}{D^2} A_d \, \mu^{d-4} \; , \tag{7.205}$$

whose RG beta function is to this order

$$\beta_v = v_R \left[-\epsilon + 3v_R + O(v_R^2) \right] \; . \tag{7.206}$$

For $d > d_c = 4$, the Gaussian fixed point $v_0^* = 0$ is stable, and we recover the mean-field critical exponents. For $\epsilon = 4 - d > 0$, we find the nontrivial IR-stable RG fixed point

$$v^* = \epsilon/3 + O(\epsilon^2) \; . \tag{7.207}$$

Setting up and solving the RG equation (7.65) for the vertex function proceeds just as in Subsect. 7.1.5. With the identifications (7.83) (with $a = 0$) we thus obtain the DP *critical exponents* to first order in ϵ,

$$\eta = -\frac{\epsilon}{6} + O(\epsilon^2) \; , \quad \frac{1}{\nu} = 2 - \frac{\epsilon}{4} + O(\epsilon^2) \; , \quad z = 2 - \frac{\epsilon}{12} + O(\epsilon^2) \; . \tag{7.208}$$

In the vicinity of v^*, the solution of the RG equation for the *order parameter* reads, recalling that $[S] = \mu^{d/2}$,

$$\langle S_R(\tau_R, t) \rangle \approx \mu^{d/2} \, \ell^{(d-\gamma_S^*)/2} \, \hat{S} \left(\tau_R \, \ell^{\gamma_\tau^*}, v_R^*, D_R \, \mu^2 \, \ell^{2+\gamma_D^*} t \right) \; , \tag{7.209}$$

which leads to the following scaling relations and explicit exponent values,

$$\beta = \frac{\nu(d+\eta)}{2} = 1 - \frac{\epsilon}{6} + O(\epsilon^2) \; , \quad \alpha = \frac{\beta}{z\nu} = 1 - \frac{\epsilon}{4} + O(\epsilon^2) \; . \tag{7.210}$$

The scaling exponents for critical directed percolation are known analytically for a plethora of physical quantities (but the reader should beware that various different conventions are used in the literature); for the two-loop results to order ϵ^2 in the perturbative dimensional expansion, see Ref. [68]. In Table 7.2, we compare the $O(\epsilon)$ values with the results from Monte Carlo computer simulations, which allow the DP critical exponents to be measured to high precision (for recent overviews on simulation results for DP and other absorbing state phase transitions, see Refs. [74, 75]). Yet unfortunately, there are to date hardly any real experiments that would confirm the DP conjecture [71, 73] and actually measure the scaling exponents for this prominent nonequilibrium universality class.

Table 7.2. Comparison of the DP critical exponent values from Monte Carlo simulations with the results from the ϵ expansion

Scaling exponent	$d = 1$	$d = 2$	$d = 4 - \epsilon$
$\xi \sim \lvert\tau\rvert^{-\nu}$	$\nu \approx 1.100$	$\nu \approx 0.735$	$\nu = 1/2 + \epsilon/16 + O(\epsilon^2)$
$t_c \sim \xi^z \sim \lvert\tau\rvert^{-z\nu}$	$z \approx 1.576$	$z \approx 1.73$	$z = 2 - \epsilon/12 + O(\epsilon^2)$
$a_\infty \sim \lvert\tau\rvert^{\beta}$	$\beta \approx 0.2765$	$\beta \approx 0.584$	$\beta = 1 - \epsilon/6 + O(\epsilon^2)$
$a_c(t) \sim t^{-\alpha}$	$\alpha \approx 0.160$	$\alpha \approx 0.46$	$\alpha = 1 - \epsilon/4 + O(\epsilon^2)$

7.2.6 Dynamic Isotropic Percolation and Multi-Species Variants

An interesting variant of active to absorbing state phase transitions emerges when we modify the SEP rules (1) and (2) in Subsect. 7.2.5 to

1. The susceptible medium becomes infected, depending on the densities n and m of sick individuals and the "debris", respectively. After a brief time interval, the sick individuals decay into immune debris, which ultimately stops the disease locally by exhausting the supply of susceptible regions.
2. The states with $n = 0$ and any spatial distribution of m are absorbing, and describe the extinction of the disease.

Here, the debris is given by the accumulated decay products,

$$m(\boldsymbol{x}, t) = \kappa \int_{-\infty}^{t} n(\boldsymbol{x}, t')\, dt' \ . \tag{7.211}$$

After rescaling, this general epidemic process (GEP) or epidemic with removal [46] is described in terms of the mesoscopic Langevin equation [76]

$$\frac{\partial S(\boldsymbol{x}, t)}{\partial t} = -D\left(r - \boldsymbol{\nabla}^2\right) S(\boldsymbol{x}, t) - D u\, S(\boldsymbol{x}, t) \int_{-\infty}^{t} S(\boldsymbol{x}, t')\, Dt' + \zeta(\boldsymbol{x}, t) \ , \tag{7.212}$$

with noise correlator (7.197). The associated response functional reads [77, 78]

$$A[\tilde{S}, S] = \int d^d x \int dt \left[\tilde{S} \left(\frac{\partial}{\partial t} + D \left(r - \nabla^2 \right) \right) S - u \, \tilde{S}^2 \, S + D \, u \, S \int^t S(t') \right].$$
(7.213)

For the field theory thus defined, one may take the *quasistatic limit* by introducing the fields

$$\tilde{\varphi}(\boldsymbol{x}) = \tilde{S}(\boldsymbol{x}, t \to \infty) \ , \quad \varphi(\boldsymbol{x}) = D \int_{-\infty}^{\infty} S(\boldsymbol{x}, t') \, dt' \ .$$
(7.214)

For $t \to \infty$, the action (7.213) thus becomes

$$A_{\text{qst}}[\tilde{\varphi}, \varphi] = \int d^d x \, \tilde{\varphi} \left[r - \nabla^2 - u \left(\tilde{\varphi} - \varphi \right) \right] \varphi \ ,$$
(7.215)

which is known to describe the critical exponents of *isotropic percolation* [79]. An isotropic percolation cluster is shown in Fig. 7.9, to be contrasted with the anisotropic scaling evident in Fig. 7.7(b). The upper critical dimension of isotropic percolation is $d_c = 6$, and an explicit calculation, with the diagrams of Fig. 7.8, but involving the static propagators $G_0(\boldsymbol{q}) = 1/(r + \boldsymbol{q}^2)$, yields the following critical exponents for isotropic percolation, to first order in $\epsilon = 6 - d$,

$$\eta = -\frac{\epsilon}{21} + O(\epsilon^2) \ , \quad \frac{1}{\nu} = 2 - \frac{5\epsilon}{21} + O(\epsilon^2) \ , \quad \beta = 1 - \frac{\epsilon}{7} + O(\epsilon^2) \ . \quad (7.216)$$

In order to calculate the dynamic critical exponent for this *dynamic isotropic percolation (dIP)* universality class, we must return to the full action (7.213). Once again, with the diagrams of Fig. 7.8, but now involving a temporally nonlocal three-point vertex, one then arrives at

$$z = 2 - \frac{\epsilon}{6} + O(\epsilon^2) \ .$$
(7.217)

For a variety of two-loop results, the reader is referred to Ref. [68]. It is also possible to describe the *crossover* from isotropic to directed percolation within this field-theoretic framework [80, 81].

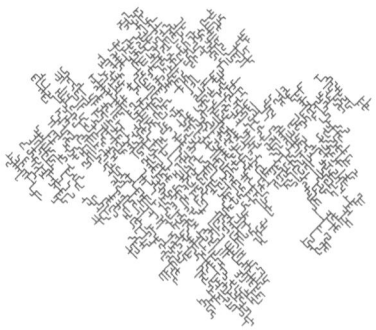

Fig. 7.9. Isotropic percolation cluster

Let us next consider *multi-species* variants of directed percolation processes, which can be obtained in the particle language by coupling the DP reactions $A_i \to \emptyset$, $A_i \rightleftharpoons A_i + A_i$ via processes of the form $A_i \rightleftharpoons A_j + A_j$ (with $j \neq i$); or directly by the corresponding generalisation within the Langevin representation with $\langle \zeta_i(\boldsymbol{x}, t) \rangle = 0$,

$$\frac{\partial S_i}{\partial t} = D_i \left(\boldsymbol{\nabla}^2 - R_i[S_i] \right) S_i + \zeta_i \ , \quad R_i[S_i] = r_i + \sum_j g_{ij} S_j + \dots \ , \quad (7.218)$$

$$\langle \zeta_i(\boldsymbol{x}, t) \zeta_j(\boldsymbol{x}', t') \rangle = 2 S_i N_i[S_i] \, \delta(\boldsymbol{x} - \boldsymbol{x}') \, \delta(t - t') \, \delta_{ij} \ , \, N_i[S_i] = u_i + \dots \ (7.219)$$

The ensuing renormalisation factors turn out to be precisely as for single-species DP, and consequently the *generical* critical behaviour even in such multi-species systems is governed by the DP universality class [58]. For example, the predator extinction threshold for the stochastic Lotka–Volterra system mentioned in Subsect. 7.2.1 is characterised by the DP exponents as well [47]. But these reactions also *generate* $A_i \to A_j$, causing additional terms $\sum_{j \neq i} g_j S_j$ in Eq. (7.218). Asymptotically, the inter-species couplings become *unidirectional*, which allows for the appearance of special *multicritical points* when several $r_i = 0$ simultaneously [82]. This leads to a *hierarchy* of order parameter exponents β_k on the kth level of a unidirectional cascade, with

$$\beta_1 = 1 - \frac{\epsilon}{6} + O(\epsilon^2) \ , \quad \beta_2 = \frac{1}{2} - \frac{13\,\epsilon}{96} + O(\epsilon^2) \ , \dots , \quad \beta_k = \frac{1}{2^k} - O(\epsilon) \ ; \quad (7.220)$$

for the associated *crossover exponent*, one can show $\Phi = 1$ to *all* orders [58]. Quite analogous features emerge for *multi-species dIP processes* [58, 68].

7.2.7 Concluding Remarks

In these lecture notes, I have described how stochastic processes can be mapped onto field theory representations, starting either from a mesoscopic Langevin equation for the coarse-grained densities of the relevant order parameter fields and conserved quantities, or from a more microsopic master equation for interacting particle systems. The dynamic renormalisation group method can then be employed to study and characterise the universal scaling behaviour near continuous phase transitions both in and far from thermal equilibrium, and for systems that generically display scale-invariant features. While the critical dynamics near equilibrium phase transitions has been thoroughly investigated experimentally in the past three decades, regrettably such direct experimental verification of the by now considerable amount of theoretical work on *nonequilibrium* systems is largely amiss. In this respect, applications of the expertise gained in the nonequilibrium statistical mechanics of complex cooperative behaviour to biological systems might prove fruitful and constitutes a promising venture. One must bear in mind, however, that nonuniversal features are often crucial for the relevant questions in biology.

In part, the lack of clearcut experimental evidence may be due to the fact that asymptotic universal properties are perhaps less prominent in accessible nonequilibrium systems, owing to long crossover times. Yet fluctuations do tend to play a more important role in systems that are driven away from thermal equilibrium, and the concept of universality classes, despite the undoubtedly much increased richness in dynamical systems, should still be useful. For example, we have seen that the directed percolation universality class quite generically describes the critical properties of phase transitions from active to inactive, absorbing states, which abound in nature. The few exceptions to this rule either require the coupling to another conserved mode [83, 84]; the presence, on a mesoscopic level, of additional symmetries that preclude the spontaneous decay $A \to \emptyset$ as in the so-called *parity-conserving (PC)* universality class, represented by branching and annihilating random walks $A \to (n + 1) A$ with n *even*, and $A + A \to \emptyset$ [85] (for recent developments based on nonperturbative RG approaches, see Ref. [86]); or the absence of any first-order reactions, as in the (by now rather notorious) *pair contact process with diffusion (PCPD)* [87], which has so far eluded a successful field-theoretic treatment [88]. A possible explanation for the fact that DP exponents have not been measured ubiquitously (yet) could be the instability towards *quenched disorder* in the reaction rates [89].

In reaction–diffusion systems, a complete classification of the scaling properties in *multi-species* systems remains incomplete, aside from pair annihilation and DP-like processes, and still constitutes a quite formidable program (for a recent overview over the present situation from a field-theoretic viewpoint, see Ref. [12]). This is even more evident for nonequilibrium systems in general, even when maintained in driven *steady states*. Field-theoretic methods and the dynamic renormalisation group represent powerful tools that I believe will continue to crucially complement exact solutions (usually of one-dimensional models), other approximative approaches, and computer simulations, in our quest to further elucidate the intriguing cooperative behaviour of strongly interacting and fluctuating many-particle systems.

Acknowledgements

This work has been supported in part by the U.S. National Science Foundation through grant NSF DMR-0308548. I am indebted to many colleagues, students, and friends, from and with whom I had the pleasure to learn and research the material presented here; specifically I would like to mention Vamsi Akkineni, Michael Bulenda, John Cardy, Olivier Deloubrière, Daniel Fisher, Reinhard Folk, Erwin Frey, Ivan Georgiev, Yadin Goldschmidt, Peter Grassberger, Henk Hilhorst, Haye Hinrichsen, Martin Howard, Terry Hwa, Hannes Janssen, Bernhard Kaufmann, Mauro Mobilia, David Nelson, Beth Reid, Zoltán Rácz, Jaime Santos, Beate Schmittmann, Franz Schwabl, Steffen Trimper, Ben Vollmayr-Lee, Mark Washenberger, Fred van Wijland, and Royce Zia. Lastly, I would like to thank the organisers of the very enjoyable

Luxembourg Summer School on *Ageing and the Glass Transition* for their kind invitation, and my colleagues at the Laboratoire de Physique Théorique, Université de Paris-Sud Orsay, France and at the Rudolf Peierls Centre for Theoretical Physics, University of Oxford, U.K., where these lecture notes were conceived and written, for their warm hospitality.

References

1. P. Ramond: *Field theory – a modern primer*, (Benjamin/Cummings, Reading 1981)
2. D.J. Amit: *Field theory, the renormalization group, and critical phenomena* (World Scientific, Singapore 1984)
3. C. Itzykson and J.M. Drouffe: *Statistical field theory* (Cambridge University Press, Cambridge 1989)
4. M. Le Bellac: *Quantum and statistical field theory*, (Oxford University Press, Oxford 1991)
5. J. Zinn-Justin: *Quantum field theory and critical phenomena* (Clarendon Press, Oxford 1993)
6. J. Cardy: *Scaling and renormalization in statistical physics* (Cambridge University Press, Cambridge 1996)
7. P.C. Hohenberg and B.I. Halperin: Rev. Mod. Phys. **49**, 435 (1977)
8. H.K. Janssen: Field-theoretic methods applied to critical dynamics. In: *Dynamical critical phenomena and related topics, Lecture Notes in Physics*, vol. 104, ed by C.P. Enz (Springer, Heidelberg 1979), pp. 26–47
9. R. Bausch, H.K. Janssen, and H. Wagner: Z. Phys. **B24**, 113 (1976)
10. J.L. Cardy: Renormalisation group approach to reaction-diffusion problems. In: *Proceedings of Mathematical Beauty of Physics*, ed by J.-B. Zuber, Adv. Ser. in Math. Phys. **24**, 113 (1997)
11. D.C. Mattis and M.L. Glasser: Rev. Mod. Phys. **70**, 979 (1998)
12. U.C. Täuber, M.J. Howard, and B.P. Vollmayr-Lee: J. Phys. A: Math. Gen. **38**, R79 (2005)
13. U.C. Täuber: *Critical dynamics: a field theory approach to equilibrium and non-equilibrium scaling behavior*, in preparation (to be published at Cambridge University Press, Cambridge); for completed chapters, see: http://www.phys.vt.edu/~tauber/utaeuber.html
14. R.A. Ferrell, N. Menyhàrd, H. Schmidt, F. Schwabl, and P. Szépfalusy: Phys. Rev. Lett. **18**, 891 (1967); Ann. of Phys. **47**, 565 (1968)
15. B.I. Halperin and P.C. Hohenberg: Phys. Rev. **177**, 952 (1969)
16. B.I. Halperin, P.C. Hohenberg, and S.-k. Ma: Phys. Rev. Lett. **29**, 1548 (1972)
17. C. De Dominicis, E. Brézin, and J. Zinn-Justin: Phys. Rev. **B12**, 4945 (1975)
18. H.K. Janssen: Z. Phys. **B23**, 377 (1976)
19. C. De Dominicis: J. Physique Colloque **37**, C2247 (1976)
20. P.C. Martin, E.D. Siggia, and H.A. Rose: Phys. Rev. **A8**, 423 (1973)
21. H. Wagner: Z. Phys. **195**, 273 (1966)
22. N.D. Mermin and H. Wagner: Phys. Rev. Lett. **17**, 1133 (1966)
23. P.C. Hohenberg: Phys. Rev. **158**, 383 (1967)
24. P.M. Chaikin and T.C. Lubensky: *Principles of condensed matter physics*, (Cambridge University Press, Cambridge 1995)

25. S.-k. Ma and G.F. Mazenko: Phys. Rev. Lett. **33**, 1383 (1974); Phys. Rev. **B11**, 4077 (1975)
26. E. Frey and F. Schwabl: Adv. Phys. **43**, 577 (1994)
27. H.K. Janssen, B. Schaub, and B. Schmittmann: Z. Phys. **B73**, 539 (1989)
28. H.K. Janssen: On the renormalized field theory of nonlinear critical relaxation. In: *From phase transitions to chaos*, ed by G. Györgyi, I. Kondor, L. Sasvári, and T. Tél (World Scientific, Singapore 1992), pp. 68–91.
29. P. Calabrese and A. Gambassi: Phys. Rev. **E66**, 066101 (2002); J. Phys. A: Math. Gen. **38**, R133 (2005)
30. A. Gambassi: In: *Proceedings of the International Summer School "Ageing and the Glass Transition"*, to appear in J. Phys. Conf. Proc. (July 2006)
31. H.W. Diehl: In: *Phase Transitions and Critical Phenomena*, vol. 10, ed by C. Domb and J.L. Lebowitz (Academic Press, London 1986)
32. K. Oerding and H.K. Janssen: J. Phys. A: Math. Gen. **26**, 5295 (1993)
33. U.C. Täuber, V.K. Akkineni, and J.E. Santos: Phys. Rev. Lett. **88**, 045702 (2002)
34. F. Haake, M. Lewenstein, and M. Wilkens: Z. Phys. **B55**, 211 (1984)
35. G. Grinstein, C. Jayaprakash, and Y. He: Phys. Rev. Lett. **55**, 2527 (1985)
36. K.E. Bassler and B. Schmittmann: Phys. Rev. Lett. **73**, 3343 (1994)
37. U.C. Täuber and Z. Rácz: Phys. Rev. **E55**, 4120 (1997)
38. U.C. Täuber, J.E. Santos, and Z. Rácz: Eur. Phys. J. **B7**, 309 (1999)
39. B. Schmittmann and R.K.P. Zia: Phys. Rev. Lett. **66**, 357 (1991)
40. B. Schmittmann: Europhys. Lett. **24**, 109 (1993)
41. K.E. Bassler and Z. Rácz: Phys. Rev. Lett. **73**, 1320 (1994); Phys. Rev. **E52**, R9 (1995)
42. B. Schmittmann and R.K.P. Zia: Statistical mechanics of driven diffusive systems. In: *Phase Transitions and Critical Phenomena*, vol. 17, ed by C. Domb and J.L. Lebowitz (Academic Press, London 1995)
43. H.K. Janssen and B. Schmittmann: Z. Phys. **B63**, 517 (1986)
44. K.-t. Leung and J.L. Cardy: J. Stat. Phys. **44**, 567 (1986)
45. D. Forster, D.R. Nelson, and M.J. Stephen: Phys. Rev. **A16**, 732 (1977)
46. J.D. Murray: *Mathematical Biology*, vols. I/II (Springer, New York, 3rd ed 2002)
47. M. Mobilia, I.T. Georgiev, and U.C. Täuber: e-print **q-bio.PE/0508043** (2005)
48. M. Doi: J. Phys. A: Math. Gen. **9**, 1465 & 1479 (1976)
49. P. Grassberger and M. Scheunert: Fortschr. Physik **28**, 547 (1980)
50. L. Peliti: J. Physique **46**, 1469 (1985)
51. F.C. Alcaraz, M. Droz, M. Henkel, and V. Rittenberg: Ann. of Phys. **230**, 250 (1994)
52. M. Henkel, E. Orlandini, and J. Santos: Ann. of Phys. **259**, 163 (1997)
53. G.M. Schütz: In: *Phase Transitions and Critical Phenomena*, vol. 19, ed by C. Domb and J.L. Lebowitz (Academic Press, London 2001)
54. R. Stinchcombe: Adv. Phys. **50**, 431 (2001)
55. F. van Wijland: Phys. Rev. **E63**, 022101 (2001)
56. J.W. Negele and H. Orland: *Quantum many-particle systems* (Addison-Wesley, Redwood City 1988)
57. L. Peliti: J. Phys. A: Math. Gen. **19**, L365 (1986)
58. H.K. Janssen: J. Stat. Phys. **103**, 801 (2001)
59. B.P. Lee: J. Phys. A: Math. Gen. **27**, 2633 (1994)
60. B.P. Lee and J. Cardy: Phys. Rev. **E50**, R3287 (1994)

61. D. Toussaint and F. Wilczek: J. Chem. Phys. **78**, 2642 (1983)
62. B.P. Lee and J. Cardy: J. Stat. Phys. **80**, 971 (1995)
63. O. Deloubrière, H.J. Hilhorst, and U.C. Täuber: Phys. Rev. Lett. **89**, 250601 (2002); H.J. Hilhorst, O. Deloubrière, M.J. Washenberger, and U.C. Täuber: J. Phys. A: Math. Gen. **37**, 7063 (2004)
64. H.J. Hilhorst, M.J. Washenberger, and U.C. Täuber: J. Stat. Mech. P10002 (2004)
65. M. Moshe: Phys. Rep. **37C**, 255 (1978)
66. P. Grassberger and K. Sundermeyer: Phys. Lett. **B77**, 220 (1978)
67. P. Grassberger and A. De La Torre: Ann. of Phys. **122**, 373 (1979)
68. H.K. Janssen and U.C. Täuber: Ann. of Phys. **315**, 147 (2005)
69. S.P. Obukhov: Physica **A101**, 145 (1980)
70. J.L. Cardy and R.L. Sugar: J. Phys. A: Math. Gen. **13**, L423 (1980)
71. H.K. Janssen: Z. Phys. **B42**, 151 (1981)
72. W. Kinzel: In: *Percolation structures and processes*, ed by G. Deutsch, R. Zallen, and J. Adler (Hilger, Bristol 1983)
73. P. Grassberger, Z. Phys. **B47**, 365 (1982)
74. H. Hinrichsen: Adv. Phys. **49**, 815 (2001)
75. G. Ódor: Rev. Mod. Phys. **76**, 663 (2004)
76. P. Grassberger: Math. Biosc. **63**, 157 (1983)
77. H.K. Janssen: Z. Phys. **B58**, 311 (1985)
78. J.L. Cardy and P. Grassberger: J. Phys. A: Math. Gen. **18**, L267 (1985)
79. J. Benzoni and J.L. Cardy: J. Phys. A: Math. Gen. **17**, 179 (1984)
80. E. Frey, U.C. Täuber, and F. Schwabl: Europhys. Lett. **26**, 413 (1994); Phys. Rev. **E49**, 5058 (1994)
81. H.K. Janssen and O. Stenull: Phys. Rev. **E62**, 3173 (2000)
82. U.C. Täuber, M.J. Howard, and H. Hinrichsen: Phys. Rev. Lett. **80**, 2165 (1998); Y.Y. Goldschmidt, H. Hinrichsen, M.J. Howard, and U.C. Täuber: Phys. Rev. **E59**, 6381 (1999)
83. R. Kree, B. Schaub, and B. Schmittmann: Phys. Rev. **A39**, 2214 (1989)
84. F. van Wijland, K. Oerding, and H. Hilhorst: Physica **A251**, 179 (1998); K. Oerding, F. van Wijland, J.P. Leroy, and H. Hilhorst: J. Stat. Phys. **99**, 1365 (2000)
85. J. Cardy and U.C. Täuber: Phys. Rev. Lett. **77**, 4780 (1996); J. Stat. Phys. **90**, 1 (1998)
86. L. Canet, H. Chaté, and B. Delamotte: Phys. Rev. Lett. **92**, 255703 (2004); L. Canet, H. Chaté, B. Delamotte, I. Dornic, and M.A. Muñoz: e-print cond-mat/0505170 (2005)
87. M. Henkel and H. Hinrichsen: J. Phys. A: Math. Gen. **37**, R117 (2004)
88. H.K. Janssen, F. van Wijland, O. Deloubrière, and U.C. Täuber: Phys. Rev. **E70**, 056114 (2004)
89. H.K. Janssen: Phys. Rev. **E55**, 6253 (1997)

Lecture Notes in Physics

For information about earlier volumes
please contact your bookseller or Springer
LNP Online archive: springerlink.com